Mountain Tourism and Ecological Impacts:

Himalayan Region and Beyond

Soumendra Nath Biswas
Assam University, Silchar, India

Pradip Kumar
Assam University, Silchar, India

IGI Global
PUBLISHER of TIMELY KNOWLEDGE

A volume in the Advances in Hospitality, Tourism, and the Services Industry (AHTSI) Book Series

Published in the United States of America by
IGI Global
Business Science Reference (an imprint of IGI Global)
701 E. Chocolate Avenue
Hershey PA, USA 17033
Tel: 717-533-8845
Fax: 717-533-8661
E-mail: cust@igi-global.com
Web site: http://www.igi-global.com

Library of Congress Cataloging-in-Publication Data

Names: Biswas, Soumendra Nath, editor. | Kumar, Pradip, 1982- editor.
Title: Mountain tourism and ecological impacts : Himalayan Region and beyond / edited by Soumendra Biswas, Pradip Kumar.
Description: Hershey, PA : Business Science Reference, [2024] | Includes bibliographical references and index. | Summary: "A sustainable approach to mountain tourism especially in the Himalayan region is essential to promote long-term growth, optimum use of resources, and maintain carrying capacity. This book will focus on challenges and prospects of tourism development in the great Himalayan region"-- Provided by publisher.
Identifiers: LCCN 2023048647 (print) | LCCN 2023048648 (ebook) | ISBN 9798369308233 (h/c) | ISBN 9798369308240 (eISBN)
Subjects: LCSH: Mountain tourism--Himalaya Mountains Region. | Ecotourism--Himalaya Mountains Region. | Nature--Effect of human beings on--Himalaya Mountains Region. | Himalaya Mountains Region--Environmental conditions. | Sustainable development--Himalaya Mountains Region. | Natural resources--Himalaya Mountains Region--Management.
Classification: LCC G156.5.M68 M684 2024 (print) | LCC G156.5.M68 (ebook) | DDC 338.4/791549604--dc23/eng/DLC20231130
LC record available at https://lccn.loc.gov/2023048647
LC ebook record available at https://lccn.loc.gov/2023048648

This book is published in the IGI Global book series Advances in Hospitality, Tourism, and the Services Industry (AHTSI) (ISSN: 2475-6547; eISSN: 2475-6555)

British Cataloguing in Publication Data
A Cataloguing in Publication record for this book is available from the British Library.

Advances in Hospitality, Tourism, and the Services Industry (AHTSI) Book Series

Maximiliano Korstanje
University of Palermo, Argentina

ISSN:2475-6547
EISSN:2475-6555

Mission

Globally, the hospitality, travel, tourism, and services industries generate a significant percentage of revenue and represent a large portion of the business world. Even in tough economic times, these industries thrive as individuals continue to spend on leisure and recreation activities as well as services.

The Advances in Hospitality, Tourism, and the Services Industry (AHTSI) book series offers diverse publications relating to the management, promotion, and profitability of the leisure, recreation, and services industries. Highlighting current research pertaining to various topics within the realm of hospitality, travel, tourism, and services management, the titles found within the AHTSI book series are pertinent to the research and professional needs of managers, business practitioners, researchers, and upper-level students studying in the field.

Coverage

- Service Training
- Food and Beverage Management
- Travel Agency Management
- Hotel Management
- Health and Wellness Tourism
- Casino Management
- Service Management
- Sustainable Tourism
- Cruise Marketing and Sales
- Leisure & Business Travel

IGI Global is currently accepting manuscripts for publication within this series. To submit a proposal for a volume in this series, please contact our Acquisition Editors at Acquisitions@igi-global.com or visit: http://www.igi-global.com/publish/.

The Advances in Hospitality, Tourism, and the Services Industry (AHTSI) Book Series (ISSN 2475-6547) is published by IGI Global, 701 E. Chocolate Avenue, Hershey, PA 17033-1240, USA, www.igi-global.com. This series is composed of titles available for purchase individually; each title is edited to be contextually exclusive from any other title within the series. For pricing and ordering information please visit http://www.igi-global.com/book-series/advances-hospitality-tourism-services-industry/121014. Postmaster: Send all address changes to above address.

Titles in this Series

For a list of additional titles in this series, please visit: http://www.igi-global.com/book-series/advances-hospitality-tourism-services-industry/121014

Service Innovations in Tourism Metaverse, Immersive Technologies, and Digital Twin
Surabhi Singh (G.L. Bajaj Institute of Management and Research, India)
Business Science Reference • copyright 2024 • 295pp • H/C (ISBN: 9798369311035) • US $275.00 (our price)

Utilizing Smart Technology and AI in Hybrid Tourism and Hospitality
Sanjeev Kumar (Lovely Professional University, India) Mohammad Badruddoza Talukder (Daffodil Institute of IT, Bangladesh) and Ana Pego (Nova University of Lisbon, Portugal)
Business Science Reference • copyright 2024 • 367pp • H/C (ISBN: 9798369319789) • US $290.00 (our price)

Innovative Technologies for Increasing Service Productivity
Sridhar Manohar (Chitkara Business School, Chitkara University, India) Amit Mittal (Chitkara Business School, Chitkara University, India) Sekar Raju (Ivy College of Business, Iowa State University, USA) and Arjun J. Nair (Chitkara University, India)
Business Science Reference • copyright 2024 • 404pp • H/C (ISBN: 9798369320198) • US $290.00 (our price)

Supporting Environmental Stability Through Ecotourism
Muhammad Abrar (Lyallpur Business School, Government College University, Faisalabad, Pakistan) Muhammad Asim Nawaz (Lyallpur Business School, Government College University, Faisalabad, Pakistan) Faiqa Kiran (Lyallpur Business School, Government College University, Faisalabad, Pakistan) and Fahad Asmi (School of Business, Charles Sturt University, Bathurst, Australia)
Business Science Reference • copyright 2024 • 296pp • H/C (ISBN: 9798369310304) • US $265.00 (our price)

Career Pathways and Professional Identities for Front-Line Workers in the Service Industries
Mark Peter Vitale (University of Phoenix, USA)
Business Science Reference • copyright 2024 • 370pp • H/C (ISBN: 9781668425282) • US $255.00 (our price)

Experiences, Advantages, and Economic Dimensions of Pilgrimage Routes
Vítor João Pereira Domingues Martinho (Agricultural School, CERNAS-IPV Research Centre, Polytechnic Institute of Viseu, Portugal) João Augusto Guerra da Rocha Nunes (School of Education, Polytechnic Institute of Viseu, Portugal) Maria Lúcia de Jesus Pato (Agricultural School, CERNAS-IPV Research Centre, Polytechnic Institute of Viseu, Portugal) and Liliana Andrade de Matos Castilho (School of Education, Polytechnic Institute of Viseu, Portugal)
Business Science Reference • copyright 2023 • 367pp • H/C (ISBN: 9781668499238) • US $240.00 (our price)

IGI Global
PUBLISHER of TIMELY KNOWLEDGE

701 East Chocolate Avenue, Hershey, PA 17033, USA
Tel: 717-533-8845 x100 • Fax: 717-533-8661
E-Mail: cust@igi-global.com • www.igi-global.com

Table of Contents

Detailed Table of Contents

This study was conducted with the goals of determining the current state of the adventure tourism-related market and organizations already in operation, as well as attempting to zero in on the factors that will be most important in the development of adventure tourism as a commercial enterprise in Ladakh, Himachal Pradesh, and Arunachal Pradesh in the future. The study includes a discussion on the prospects of the adventure tourism industry as well as its probable potential to change the country in such a way that it moves closer to the adventure tourism business. In the theoretical framework, the positive aspects of adventure tourism were discussed; in the study effort, evidence that support the theoretical notion was brought forth. The theory section of the study discusses a variety of subjects pertaining to adventure tourism, including human psychology in relation to it, the history of adventure tourism, and the current market scenario of adventure tourism-related commercial organizations in Ladakh, Himachal Pradesh, and Arunachal Pradesh as chosen study areas.

This study focuses on adventure-based tourism development in popular chapter station of Sikkim that affects tourist movement and socio-economic life of the local communities. This chapter begins with the meaning and importance of adventure tourism and its impact on the tourism business, local societies, and environment of the tourism destination of Sikkim. Qualitative interviews were conducted with tourists, tourism operators and local communities. Findings of this study reveal that regulation of protected areas, legislation, and government programs play an important part in the development of sustainable adventure-based tourism in the study area. Nature education, development of business concepts, and safety of the tourists are the important points of findings. The role of the government and local communities are also highlighted here. These findings are used to suggest implications for policymakers and practitioners.

Religious tourism activities are now the emerging trends in the Himalayan region along with adventure tourism. The Himalayan region is the abode of many Goddesses. Factors responsible towards exploiting and promoting religious tourism in Himalayan region have been carefully analyzed with reference to Dev Bhoomi, Uttarakhand, but many issues still need more attention. This research chapter uses surveying of 150 stakeholders in relation towards assessment of factors responsible towards development of religious tourism in Himalayan region with special reference to Uttarakhand. Research results are analyzed on the basis of six factors influencing the growth and development of religious tourism in Himalayan region: transportation, accommodation, safety and security, environment, economy, and attitude of the local people. The result analysis depicts factors responsible for promoting religious tourism in Uttarakhand as only quite average. This chapter has proposed some important solutions towards promoting religious tourism in the Himalayan region with special reference to Uttarakhand.

This chapter delves into the realm of pilgrimage tourism in Himalayan region of India, aiming to shed light on the multifaceted aspects of this religious and cultural phenomenon. Through a comprehensive investigation, the study strives to achieve three primary research objectives. Firstly, it seeks to discern the demographic profiles and travel patterns of pilgrims traversing the sacred landscapes of India. Secondly, it endeavours to uncover the myriad challenges and hardships encountered by tourists embarking on their spiritual quests, encompassing issues ranging from logistical hindrances to cultural disparities. Lastly, drawing upon the findings from the preceding objectives, the chapter aspires to present a set of well-informed recommendations geared towards the enhancement of the pilgrimage tourism experience in Himalayan regions of India. This chapter presents valuable insights to propel the evolution and enrichment of pilgrimage tourism in culturally diverse and spiritually vibrant nations.

The establishment of protected areas, the promotion of responsible tourist behavior, and the adoption of sustainable tourism practices: the effectiveness of these measures will be assessed based on their contribution to biodiversity conservation, local community empowerment, and overall environmental sustainability. The chapter will provide recommendations for the future management of mountain tourism in the Himalayan region and beyond. These recommendations will include the need for stronger regulations and enforcement mechanisms to ensure that tourism activities do not harm delicate ecosystems. Additionally, it will emphasize the importance of community involvement and education in promoting sustainable tourism practices and raising awareness about the value of biodiversity. In conclusion, mountain tourism in the Himalayan region and beyond has the potential to promote economic growth and cultural exchange. However, it is crucial to acknowledge and address the ecological impacts associated with this form of tourism.

The tourism sector has seen great development in recent years and expanding its forms to target different segments of tourists, and Ethnic tourism is one such form, which has been a common strategy used by many nations to promote economic and cultural growth and aid in the preservation of ethnic heritage. The eastern Himalayan region includes Bhutan, southern, central, and eastern Nepal along with Northeast India. Each region has different characteristics and features, which have contributed to the region's rich ecological and cultural diversity. The Northeastern region of the Eastern Himalayas is dominated by many indigenous communities which can be distinguished on traditional values, lifestyle, and customs. Ethnic tourism in the northeastern region has a considerable number of positive impacts which can enhance the value of the present tourism scenario of the region. Thus, this study shall focus on the prospects of this special interest tourism form, while addressing the need for improvement of planning strategies for the growth of ethnic tourism.

Adventure tourism in the Indian Himalayan Region is a dynamic and rapidly growing sector that offers travellers unique opportunities to explore the breathtaking landscapes and cultural diversity of this region. This chapter provides a comprehensive overview of adventure tourism in the Indian Himalayan. The chapter begins by discussing the regions of Indian Himalayas and its drainage systems. Furthermore, the chapter delves into the Indian Himalayan region and adventure activities. The chapter also discuss about public and private agencies for promotion of adventure tourism in Indian Himalayan Region, which play a pivotal role in ensuring the safety and sustainability of adventure tourism activities. Later, the chapter will discuss about various adventure training institute in Indian Himalayan Region. Lastly, the chapter explores future prospects and emerging trends in adventure tourism within the Indian Himalayan Region. It discusses the role of changing traveler preferences, and the importance of community engagement in shaping the industry's trajectory.

In the world of tourism, accommodation plays a very crucial role. Accommodation is the key to the

Meghalaya, Jaintia Hills, is one of the States 12 Districts. East and West Jaintia are the two districts that make up Jaintia Hills. Jaintia hills is a popular tourist destination with a lot of natural, peaceful, and picturesque views. It is situated in a mountainous area. Compared to East Jaintia Hills, West Jaintia Hills District has a higher level of tourism activity. This chapter is designed based on secondary data and observation method which attempts to investigate the tourism development in West Jaintia Hills District, Meghalaya, in the foothills of Eastern Himalayas.

Preface

Welcome to *Mountain Tourism and Ecological Impacts: Himalayan Region and Beyond.* As editors, Soumendra Nath Biswas and Pradip Kumar, we are thrilled to present this comprehensive reference book that delves into the intricate tapestry of tourism development in one of the world's most captivating regions—the Himalayas.

The Himalayan region, with its awe-inspiring landscape, diverse topography, unique climate, rich biodiversity, and vibrant ethnic communities, sets the stage for a dynamic interplay between tourism, ecology, and local livelihoods. This book emerges as a crucial contribution, unraveling the multifaceted dimensions of mountain tourism, its potential for local economic growth, and the imperative for sustainable practices.

The United Nations recognizes the pivotal role of sustainable tourism in fostering livelihoods, alleviating poverty, and conserving the environment in mountainous areas. This acknowledgment resonates deeply with the ethos of our exploration. Our focus extends beyond the surface beauty of the Himalayas to encompass the profound impact of tourism on the lives of those who call this majestic region home.

In these pages, we aim to address the dual nature of mountain tourism—the promise it holds for positive change and the potential negative consequences it may bring. Climate change, ever-looming, adds another layer of complexity to the challenges faced by Himalayan destinations. Hence, a sustainable approach becomes paramount, ensuring long-term growth, responsible resource utilization, and the preservation of the delicate balance within the ecosystem.

This book is a comprehensive resource for a diverse audience. From academics and researchers to industry practitioners and policymakers, it offers a nuanced understanding of the importance, scope, challenges, and prospects of tourism development in the Himalayan region. The thematic chapters cover a spectrum of topics, ranging from nature-based tourism and niche tourism through naturopathy, Ayurveda, and spa, to opportunities for ethnic community development.

Chapter 1 critically examines the adventure tourism market in Ladakh, Himachal Pradesh, and Arunachal Pradesh. The study explores current market conditions, organizational landscapes, and factors shaping the future of adventure tourism. It discusses theoretical frameworks, including the positive aspects of adventure tourism, human psychology, historical context, and the existing market scenario in the chosen study areas.

Focusing on Sikkim, chapter 2 analyzes the impact of adventure-based tourism on local communities and tourist movement. Qualitative interviews with tourists, operators, and locals provide insights into the regulatory role of protected areas, legislation, and government programs. The study emphasizes nature education, business development, and tourist safety while highlighting the crucial roles of government and local communities.

Chapter 3 explores the emerging trend of religious tourism in the Himalayan region, with a focus on Uttarakhand. Utilizing a survey of 150 stakeholders, it assesses factors influencing religious tourism development. The analysis encompasses transportation, accommodation, safety, security, environment, economy, and local attitudes. The study identifies areas needing attention and proposes solutions to promote religious tourism in Uttarakhand.

Delving into pilgrimage tourism, chapter 4 aims to uncover the demographic profiles, travel patterns, challenges, and hardships faced by pilgrims in the Himalayan regions of India. It presents recommendations based on findings to enhance the pilgrimage tourism experience, emphasizing cultural and spiritual enrichment.

Focusing on protected areas and sustainable tourism practices, chapter 5 assesses their effectiveness in contributing to biodiversity conservation, local community empowerment, and environmental sustainability. It proposes recommendations for stronger regulations, community involvement, and education to address ecological impacts associated with mountain tourism.

Chapter 6 explores ethnic tourism in the Eastern Himalayan region, covering Bhutan, southern, central, and eastern Nepal, and Northeast India. It emphasizes the diversity of indigenous communities, traditional values, and the positive impacts of ethnic tourism on the region's cultural and ecological richness.

Providing a comprehensive overview, chapter 7 delves into adventure tourism in the Indian Himalayan region. It discusses the geography, drainage systems, adventure activities, public and private agencies, adventure training institutes, and future prospects. The chapter emphasizes community engagement and changing traveler preferences in shaping the industry.

Chapter 8 examines factors influencing tourists to choose government-owned accommodation in the Himalayan region, specifically in West Bengal. It analyzes the psychological attraction to such accommodations and aims to understand why tourists are drawn to them.

Focusing on rural tourism in Sikkim, chapter 9 explores its potential as a driver of sustainable development. It analyzes factors contributing to sustainable rural growth, including economic benefits, employment, and community development. The study emphasizes pro-poor tourism, environmental conservation, and cultural preservation.

Investigating the impacts of climate change on Himalayan tourism destinations, chapter 10 explores altered weather patterns, consequences for adventure tourism activities, and socio-economic repercussions for local communities. It emphasizes the need for sustainable practices, collaboration, and effective policies at national and international levels.

Focusing on Darjeeling and surrounding areas, chapter 11 explores the effects of nature-based tourism on the local economy and society. It discusses the dual nature of nature-based tourism, contributing to conservation and economic benefits while highlighting the importance of proper planning and management.

Examining the impacts of over-tourism in the Himalayan region, chapter 12 reviews existing literature to understand its pros and cons. It suggests sustainable visitor management practices to mitigate the ecological, socio-cultural, and economic impacts, offering valuable insights for policymakers and local communities.

Chapter 13 focuses on the potential of homestays in the Northeastern Himalayan region, considering their contribution to community-based tourism. It identifies key areas for promotion, emphasizing sustainable community-based tourism and the benefits it offers to the region.

Highlighting the Himalayas' potential as a global wellness destination, chapter 14 explores opportunities for nature-based therapy and its impact on physical, psychological, social, and spiritual well-being. It underscores the unique relationship between humans and nature in the Himalayan region.

Examining Lahaul-Pangi's environmental challenges, chapter 15 focuses on handloom and handicrafts as potential revenue sources. It presents findings on regional artisans, including income, education levels, and design proficiency. The chapter discusses training workshops designed to empower artisans and contribute to regional development.

Chapter 16 aims to re-brand Munnar as a "Mountain Destination" by examining current resources and surveying millennial visitors' expectations. It develops a new product line based on global trends and seeks input from policymakers, government representatives, and community leaders for the re-branding initiative.

Examining the pros and cons of protecting the Himalayan ecology amidst tourism growth, chapter 17 reviews historical and cultural treasures. It addresses concerns such as resource consumption, environmental damage, waste management, and loss of indigenous knowledge. The research advocates for sustainable practices, community involvement, and effective policy frameworks.

Chapter 18 investigates tourism development in West Jaintia Hills District, Meghalaya, situated in the foothills of the Eastern Himalayas. It explores prospects and challenges, considering the district's natural, peaceful, and picturesque views. The study relies on secondary data and observational methods to provide insights into tourism development in the region.

This book is aimed for students at both undergraduate and postgraduate levels in various disciplines such as Hospitality and Tourism Management, Business Management, Cultural Studies, History, Geography, Anthropology, and more, this book serves as an invaluable reference. Additionally, professionals in public and private organizations working in the realm of tourism and hospitality will find insights and perspectives to inform their practices.

The chapters span a wide array of topics, each shedding light on crucial aspects of mountain tourism, including eco-tourism, wildlife tourism, adventure tourism, religious tourism, and community development. Delving into the specifics, the book explores topics like disaster management, environmental catastrophes, topography, and ecology of the Himalayan region, climate change impacts, safety and security measures, and transportation in Himalayan tourism destinations.

We hope that this compilation fosters a deeper appreciation for the delicate balance required in mountain tourism, inspiring responsible practices for the benefit of both visitors and the cherished Himalayan environment.

Happy reading!

Editors,

Soumendra Nath Biswas
Assam University, India

Pradip Kumar
Assam University, India

Chapter 1

A Study on Adventure Tourism and Mountaineering Practice as a Business Perspective Among the Northern Region of India

Lokesh Kumar
https://orcid.org/0000-0002-0757-466X
Assam University, India

Jaya Bharath Reddy
Acharya Nagarjuna University, India

ABSTRACT

This study was conducted with the goals of determining the current state of the adventure tourism-related market and organizations already in operation, as well as attempting to zero in on the factors that will be most important in the development of adventure tourism as a commercial enterprise in Ladakh, Himachal Pradesh, and Arunachal Pradesh in the future. The study includes a discussion on the prospects of the adventure tourism industry as well as its probable potential to change the country in such a way that it moves closer to the adventure tourism business. In the theoretical framework, the positive aspects of adventure tourism were discussed; in the study effort, evidence that support the theoretical notion was brought forth. The theory section of the study discusses a variety of subjects pertaining to adventure tourism, including human psychology in relation to it, the history of adventure tourism, and the current market scenario of adventure tourism-related commercial organizations in Ladakh, Himachal Pradesh, and Arunachal Pradesh as chosen study areas.

1. INTRODUCTION

The etymological origins of tourism can be traced back thousands of years. Tourism and travel are directly related, and historically, travel was mostly done for business, religious, and military purposes.

DOI: 10.4018/979-8-3693-0823-3.ch001

Today, the tourist industry has grown to include a wide range of actors, as well as economic, sociological, cultural, political, environmental, and spatial factors.The tourist industry as a whole is growing at a rapid rate, and one sector that is seeing this growth is adventure tourism and mountaineering practice. It is a particular sort of niche tourism that entails investigation or travel to isolated locations where visitors should be prepared for the unexpected. An adventure vacation is a beach trip. Adventure travel often entails visiting unfriendly, difficult-to-reach locations. It could also entail doing things that are risky in some way and call for a lot of grit and effort (Anandan, 2004).

The term "adventure tourism" refers to an outdoor travel activity that can be classified as a leisure travel activity with a requirement for physical risk and hazard. Adventure tourism is a thrill-seeking pastime where the adventurer is willing to take on both difficult and easy difficulties. Adventure tourism is a distinct type of tourism where visitors are willing to accept areas with little or no infrastructure development in exchange for an authentic experience. Adventurers are welcome and take part in the adventure activities gladly and without any negative emotions, despite the risks of death and injury. Mountaineering involves coming face to face with a challenging natural environment and oneself. This sport's content involves climbing, time outdoors, and taking in stunning scenery, as well as struggle, difficulties, and risksbecause mountaineering takes place in a setting where human needs are not taken into account. Mountaineering began as attempts to reach the highest peak of vast unclimbed mountains. It has different sub-branches depending on whether the path is on rock, snow, ice, or it is flat andthese sub-branches include classic rock, ice, alpine, and sport climbing in addition to summer mountaineering, winter mountaineering, and other sub-branches (Bashir and Muthueleckuvan, 2018). Because mountaineering is evolving, it might be difficult to distinguish between mountaineering tourism and mountaineering enjoyment. Mountaineering is fragmenting and blending with tourism as mountain adventure tourism expands its range of traditional activities (walking and climbing). In actuality, climbing is being absorbed by tourism and becoming less distinct inside a more broadly defined consumer culture.

2. BACKGROUND OF THE STUDY

The world's most dynamic sector right now is the travel and tourism industry. A new era of improved cultural longevity and economic importance for all nations may be ushered in by rising international tourism (Ajitha, 2004). This sector of the economy is well-known throughout the world, in both developed and developing nations, and it aids in the good social development of both. The national economies, labor markets, land-use regulations, and tax receipts of numerous countries have all been impacted by tourism (Clawson and Johnson, 2004). Adventure tourism is the act of visiting a location specifically with the intention of partaking in one or more adventurous activities. Any time of year, anywhere in the world, adventure tourism is possible (Sung, Morrison and O'Leary, 1996). It always calls for a minimum of one night of hotel, if not many more. In recent years, the number of tourists looking for thrilling new experiences has soared. Outside of the bustling summer months, travellers seeking a genuine experience frequently look for festivals off the beaten route. For travellers looking for distinctive tourist experiences, adventure tourism has emerged as a pillar. As consumers look for authentic experiences in today's globalized economy, adventure tourism is on the rise. The desire to discover new and interesting vacation spots allows these uncharted locales to look for uncharted areas where they might have amazing experiences (Kang, Hsu and Wolfe, 2003). Therefore, the researcher makes an effort to investigate the practice of adventure tourism and mountaineering from an economic viewpoint in northern India, particularly in Ladakh, Himachal Pradesh, and Arunachal Pradesh.

3. TOURISM IN INDIA

Tourism is the largest service industry in India and its GDP is 6.23 percent. It employs 8.78% of the labor force in the nation. India sees a 5 million rise in tourists per year. 527 million people are native, including immigrants. The Indian tourist industry is estimated to be worth $100 billion in the US, with a $275 billion increase expected during the following ten years. The most well-known PR agency in India is the Ministry of Tourism. The government has started the "Incredible India" campaign to draw tourists. The campaign's slogan is "Atithi Devo Bhava", which means "Our guest is our God." Not everything is as pleasant as it seems. India struggles greatly and lags behind many other nations due to a lack of hotel rooms. Only 25,000 visitor hotels were available in India one year ago. The difficulty of obtaining a visa to enter the nation contributes in part to the decline in tourism. Indian airports are crowded, and visibility for visitors is very poor. However, in the upcoming years, tourism increase may be anticipated (Arunmozhi and Panneerselvam, 2013).

3.1 Indian Himalayan Range

The Indian Himalayan Range (IHR) is a Himalayan watercourse that cuts across the hilly regions of Assam and West Bengal, as well as Jammu and Kashmir, Ladakh, Uttarakhand, Sikkim, and Arunachal Pradesh. A sizable portion of the Indian subcontinent receives water from the region, which is rich in biodiversity. The Siwalik foothills are where the IHR physiographic mountain range starts in the south and continues north (trans-Himalaya) to the Tibetan plateau. The Himadri (Greater Himalayas), Himachal (Lesser Himalayas), and Siwalik (Outer Himalayas) are geographically divided along important geological water lines. Since they all begin in hilly areas and cross flat plains, the Indus, the Sutlej, the Kali, and the Brahmaputra Rivers all have ancient origins (Mangoch, S., & Jain, 2022).

India's major geographic advantage in adventure tourism in India is that it benefits greatly from its favorable geographic location due to its abundant natural and ecotourism resources:

- 70% of the Himalayas;
- 7,000 kilometers of coastline;
- One of the three nations with both hot and cold deserts in the world;
- Ranks sixth in terms of the number of UNESCO Natural Heritage sites;
- Ranks tenth overall in terms of the area covered by forests.

3.2 India's potentiality for Adventuretourism

India ranks relatively low in the world of adventure tourism, despite its natural resources and potential. India has a lot of potential to develop into a major hub for adventure tourism and sports worldwide. India has a great deal of potential to draw adventure travellers from around the world. The geographical features of the nation, from North to South and East to West, give the growth of adventure tourism options a competitive edge.

3.3 Adventure Tourism

Adventure tourism is a specific sort of tourism that involves some level of risk (actual or perceived) and may call for specialized knowledge and physical exertion. Any travel activity that involves physical exertion, cultural interaction, and a connection with the natural world is considered adventure tourism. Hard adventure and soft adventure are the two main categories of adventure tourism experiences.

The Hard Adventure is a travel exercise that includes more risky components and difficult physical demands. To engage in Hard Adventure activities, you must have a higher level of confidence, enthusiasm, knowledge, and specialized abilities. Adventurers are pushed by this type of tourism out of their comfort zones and into perilous terrain. Out of all adventurers, the Hard Adventurists make up a smaller percentage. Climbing (mountain, rock, and ice), caving, rafting, mountain biking, aero sports, trekking, and other extreme activities are considered hard adventures. A travel exercise called a "soft adventure" refers to a comfortable adventure. There is no need for training or expert supervision in this type of adventure, making it a very safe sort of adventure tourism. This kind of adventure tourism is most popular among adventurers. The activities that fall under the category of soft adventure include cultural art tours, wildlife viewing, camping,fishing, hiking andsnowshoeing on gradually varying terrain, sunbathing, relaxed bicycle tours, volunteer opportunities,skiing, or just horseback riding (Ajay Singh Katoch, 2021).

3.4 Adventure Tourism Promotes Fundamental Principles of Sustainable Tourism

The fundamental principles of sustainable tourism are incorporated and promoted by adventure tourism. It draws tourists outside of the busiest times, promotes the natural and cultural treasures of a place, encourages the preservation of those assets, and develops resilient and devoted travellers. Adventure tourism is tenacious, draws valuable clients, and helps local economies. Practitioners of adventure tourism promote sustainable travel because adventure tourism is impossible without pristine natural environments and meaningful cultural encounters.

3.5 Mountaineering: Concept

Mountaineering is the term used to describe human activities that include climbing high mountains. This includes everything from high-altitude hiking to full-on mountaineering. Over the course of its history, mountaineering has, in a manner not dissimilar to that which has occurred with regard to other types of qualified tourism, developed extensive specialization. In this way, it encompasses a wide range of activities, from straightforward ascents on terrain that is quite easy and located a few hours away from civilization to strenuous ascents that take multiple days and are located on mountains that are difficult to reach. A hiker or climber is put in harm's way in any scenario, regardless of the difficulty of the route technically or the length of the ascent, because the high-mountain environment presents a multitude of risks.

3.6 Indian Mountaineering Foundation

The Indian Mountaineering Foundation (IMF) is the premier mountaineering and allied sports governed by the national body. Climbing and mountaineering on both the national and international levels are supported, promoted, and regulated in India by the IMF. IMF is also involved in the preservation of mountains, the accumulation of knowledge, and the dissemination of information regarding the Indian Himalayas. The International Mountaineering Foundation plays an important part in the promotion of adventure tourism, in particular mountaineering and trekking in the Himalayas (tourism.gov, 2022).

3.7 Commercial Side of Adventure Tourism

Agriculture and tourism throughout India are the primary sources of revenue for the region's inhabitants, allowing them to maintain their standard of living. Recently, in the year 2015, hotels and organizations connected to the tourism industry celebrated their golden jubilees, an event that has been highly significant for the tourism industry. The modern tourist industry in north India has been the envy of the rest of the globe for the past fifty years. This is because it has given visitors to Ladakh, Himachal Pradesh, and Arunachal Pradeshover that time period some of the best experiences and memories of their lives. Despite this, the country has not been able to fully utilize all of its resources and has not been able to derive the most benefit from its tourism industry. Both of these things should have been possible. The moment has come for the nations of the Himalayas to investigate all of the newly opened doors in the field of adventure tourism in order to derive the greatest possible financial advantages from these endeavors.

3.8 Current Situation of Hotel, Travel, and Tour Business

The year 2015 is a landmark year for the tourist industry because it was the year they celebrated their half-century mark of uninterrupted service in the field. The fact that tourism-related groups are holding celebrations to mark this occasion is a sign that the tourist industry is developing with the passage of time, growing more robust, and managing to thrive despite facing severe challenges for a number of years. Over the course of the last half-century, the business of adventure tourism has undergone significant development that is predominantly beneficial. Along with the rise of hospitality businesses and the incorporation of new and developing ideas, the sector has moved a step forward towards optimism over the course of the past half-century. This can be attributed to the fact that the number of tourists who arrive each year is expanding.

3.9 Initiatives Taken by the Government

There is a lot of potential for adventure tourism in India, and the country has a lot of different resources for adventure tourism. The growth of adventure tourism in India is still in the beginning stages at this point. The best opportunities for adventure tourism can be found in India's beaches, deserts, mountains, hill stations, forests, rivers, and wildlife, among other things. The Ministry of Tourism under the Government of India has taken a number of different actions to facilitate the growth of the country's adventure tourism industry. The Indian Ministry of Tourism, the Adventure Tour Operators Association of India (ATOAI), and the Indian Mountaineering Federation (IMF) are all working together to put India on the international map of adventure tourism. In addition, the Ministry of Tourism developed a set of standards for safety and quality requirements for adventure tourism. These rules encompass water-based, air-based, and land-based activities that fall under the umbrella of adventure tourism. In addition, the government of India has developed outdoor and adventure institutes in the country. These institutes offer a variety of courses related to adventure sports, and they are equipped with all of the most sophisticated training gear for adventure tourism.

3.10 Strategy Followed by Nation's Adventure Tourism Industry

In recent years, there has been a meteoric rise in what is known as "adventure tourism" all over the world, with travellers travelling to previously uncharted locations. This makes it possible for new locations to portray themselves as being truly unique, making them more enticing to tourists who are looking for uncommon and unparalleled experiences. A National Strategy for Adventure Tourism has been developed in order to expand adventure tourism in the country. The Strategy places a primary emphasis on the development of adventure locations, the promotion of safety in adventure tourism, the development of skills, the construction of capacity, and marketing.

3.11 Future of Adventure Tourism

Adventure tourism is one of the most rapidly expanding sub-industries within the tourism business. It does this by luring high-value consumers, bolstering the economies of local communities, and promoting environmentally responsible behaviours. The expansion of the adventure tourism industry is beneficial not only to the growth of the national revenue but also to the economy of the destinations themselves, as well as to the people and the environment of those destinations. "The tourism industry is always evolving and will never come to an end. For us to have a proper understanding of our tourism industry's past, legacy, boundaries, and potential, we need to have the eyes of an eagle. To achieve the goal of increasing the annual number of tourists, it is necessary to design new tourism packages centered on activities such as extreme sports, festivals, and adventure travel.

4. REVIEW OF LITERATURE

Sourav Mangoch and Deepak Jain (2022) determined that a rising service sector's share of the overall GDP is an indication of economic expansion. The Indian economy's service sector is growing and its percentage of the GDP is rising. The different sub-sectors of the service industry influence its growth. Currently, it is believed that the tourist sector has a significant amount of room to grow. The growth and expansion of this sector will have a significant impact on India's GDP. The study is based on a survey that was conducted in several J&K tourism destinations in India, where primary data was gathered in three categories: prospects, J&K adventure tour operators, and tourists visiting the region. 350 tourists were randomly selected to participate in the primary data collection, which took place during several tourist seasons in both distant and popular tourist destinations. In a similar manner, 50 locals and 40 adventure tour operators were randomly selected to collect primary data. The study's findings showed that the majority of visitors, adventure tour operators, and locals themselves believed that adventure tourism gives locals living in those areas jobs. The survey also demonstrates how satisfied the locals are with their employment because they have worked in the tourism industry for many generations and are continuing to do so. They claim that adventure tourism in the Himalayas will be managed by future generations.

Michal Apollo and Viacheslav Andreychouk (2022) investigated how climbing may play a significant role in supporting measures for economic growth and environmental conservation that aim to improve the quality of life for the community as a whole. It provides a glimpse into the future in order to build sustainable mountain tourism. With its blend of theory and first-hand investigations of the effects of mountain tourism on local communities, this book is energizing reading for scholars studying adventure tourism and human geography. Representatives of the industry, decision-makers in the field, and experts in the subject will all find it to be a fascinating read.

Ajay Singh Katoch (2021) claimed that travel had become a global phenomenon. The hospitality sector is also booming. Asia and Oceania are two increasingly popular travel destinations. Generally speaking, the reasons that support the region are thought to be a growing desire for adventure and a quickly rising interest in pristine nature. This is the ideal time to favourably impact the influx of tourists by promoting new and developing kinds of tourism because India is geographically advantageous and has an enormous amount of natural and cultural tourism resources. Adventure travel is evolving into a variety of travel services over time. A brief survey was conducted in a number of adventure tourist locations in the Himachal Pradesh regions of Kangra, Shimla, and Kullu as part of this study, which focuses on the support and involvement of the hospitality sector in promoting adventure tourism.

MousimeXalxoa (2020) noted that many young people who want to experience the unknown parts of the globe are attracted by adventure tourism whichcontinues to expand and become one of the major and fastest-growing sectors that contributes to the economy of the country. The tourism business can grow in India since it has a wide variety of items to offer, and the country is blessed with a lot of adventurous destinations and chances. Karnataka, a state with a thriving biodiversity that draws visitors from all over the world, provides a wide range of activities. The Western Ghats, which stretch from Maharashtra to Kerala and contain national parks, biosphere reserves, wildlife sanctuaries, and the highest peak in Karnataka after the Karakoram Mountain range in the Himalayas, are one of its main assets. Adventure tourism is all about the pleasure and excitement that tests one's physical boundaries, encouraging risk-taking, achievement, novelty, ego enhancement, and fascinating experiences for those who engage. In order to satisfy customer demand and improve product quality by altering and introducing a small amount of risk along with high-quality equipment, the market should be active.

4.1 Research Gap

There have been numerous studies on international and pilgrimage tourism, but none have been done on adventure visitors. There has been no research on the global trends of adventure tourism, nor has there been any study on growth and development. India as a destination for adventure tourists has not been the subject of any research-based studies. The term adventure tourism and mountaineering practice hasnot been defined clearly, and it appears that different authors are each using their own definition. As tourist destination for adventure, Ladakh, Himachal Pradesh, and Arunachal Pradeshhas not been the subject of any research-based studies.

5. STATEMENT OF THE PROBLEM

There are several adventure tourism resources in India. Although Ladakh, Himachal Pradesh, and Arunachal Pradeshare known for theirmountains, peaks, and stunning stretches of rivers and slopes, which serve as the ideal settings for adventure-related sports, it has been unable to fully realize the potential of their adventure tourism due to a number of factors, including government indifference and a lack of a state-specific policy. Though there are some well-known locations for adventure tourism, such as Dal Lake, Zanskar Valley, Manali, Haridwar, Auli, and many more, these locations are even unable to create the image of world-class adventure tourist destinations. The main issue with the development of adventure tourism in the selected study areas is the obnoxious state of the roads, telecommunication network, and rescue facilities there.

6. SIGNIFICANCE OF THE STUDY

Due to the variety of experiences provided for visitors, adventure tourism has been recognized as one of the most significant segments of the tourism sector. The goal of the study has been to identify and categorize the various aspects of adventure tourism and to investigate the potential economic impacts of this market segment on local, regional, and national economies in addition to the tourism industry. The goal of the current study is to pinpoint all the elements that significantly affect the growth prospects of adventure tourism. For the purpose of creating an authentic policy framework, segmenting adventure tourists poses a difficult issue; these characteristics required to be backed by empirical data that was particular to the states of Ladakh, Himachal Pradesh, and Arunachal Pradesh. Thus, the purpose of the study is to gain a comprehensive understanding of the many factors that affect a visitor's decision to pursue a particular form of adventure tourism.

7. OBJECTIVES OF THE STUDY

- To study the concept of adventure tourism and mountaineering practice and comprehend its business perspective in the selected study areas
- To analyze the factors influencing adventure tourists and tour operators towards growth and development of the sector
- To suggest a strategic model in association with the development and promotion of adventure tourism and mountaineering practice in the selected study areas

8. SCOPE OF THE STUDY

Since theselected states are the main drivers of adventure tourism in the nation, the author has made the decision to do a study of the industry there. By studying adventure tourism in these states, we can learn about the growth, development, and trends in adventure tourism in each, as well as how adventure tourism and mountaineering is perceived as a business. As a result, we can make recommendations based on the study. This study will be useful in identifying the issue with these states' adventure tourism. The planners, travel industry professionals, and tourism academics will all benefit from this study. The findings of this study can serve as a starting point for other research on the same or related topics.

9. RESEARCH METHODOLOGY

This section includes information on the research design, data collection technique, sampling strategy, study area, and statistical methods employed.

Research Design: In order to comprehend adventure tourism and mountaineering practice from a business viewpoint in northern India, the current study employed qualitative and quantitative methods of research approach and took numerous elements into account.

Research Instrument: A structured instrument was created based on the study's goals. The respondents were given a questionnaire with statements and questions about adventure travel and climbing practices from the chosen northern regions, and they were asked to assess the statements and questions. Sustainable marketing policy, Infrastructure facilities, Safety &Security, Tourism business, and Growth & development of tourism are the variables used in this investigation. A five-point Likert scale, from strongly disagree to agree was used to evaluate them.

Data Collection: To gather pertinent data for this study, both primary and secondary data sources were used. The study that formed the basis of the research was carried out in a number of popular tourist areas in Ladakh, Arunachal Pradesh, and Himachal Pradesh. Using a self-made questionnaire created by the researcher, a survey was conducted among tourists, adventure and mountaineering tour operators, and locals living in various tourist areas near Ladakh, Arunachal Pradesh, and Himachal Pradesh to learn about the practice of adventure tourism and mountaineering from a business perspective. For the purpose of this study, secondary data was gathered from a variety of publications, books, websites, government annual reports, etc.

The responders were chosen from among tourists, adventure and mountaineering trip operators, and locals in several tourist destinations near Ladakh, Arunachal Pradesh, and Himachal Pradesh. To get pertinent answers, both the male and female populations were taken into account.

Sample size and sampling strategy: A sample population of 250 respondents was selected to get answers to the questionnaire created specifically for this study. The systematic random sampling method was the sampling strategy advocated for the current study.

Data analysis and statistical tools: The Statistical Package for Social Sciences (SPSS) version 20 and AMOS version 20 were used to enter and evaluate the replies obtained from the respondents. The basic data were analyzed and the present study's hypothesis was tested using statistical methods like one-way ANOVA, SEM, and descriptive statistics.

10. ANALYSIS OF DATA

One –Way ANOVA between Age and Study Variables

H_{01} – The age of the respondents does not have a significant relationship among the factorsof sustainable marketing policy, infrastructure facilities, safety & security, tourism business, and growth and development of tourism.

As can be seen in Table 1, the study variables infrastructure, safety & security, and tourist business are all strongly correlated with age. Since the significant value is less than the 5% threshold for the variables infrastructure facilities, safety & security, and tourism business, the null hypothesis is rejected. As a consequence, it seems that the infrastructure, safety and security, and tourism business differ greatly

Table 1. One –Way ANOVA between age and study variables

Factors	F-Value	Significant	Inference
Sustainable marketing policy	2.309	0.129	NS
Infrastructure facilities	9.599	0.000	S
Safety & security	7.003	0.000	S
Tourism business	16.630	0.000	S
Growth and development of tourism	3.102	0.030	NS

*5% Significant Level; S- Significant/ NS – Not significant

depending on the age of the respondents. The null hypothesis is accepted in terms of sustainable marketing practises and the growth and development of tourism.

One –Way ANOVA between experience and Study Variables

H_{02-} Theexperience of the respondents does not have a significant relationship among the factorsof sustainable marketing policy, infrastructure facilities, safety & security, tourism business, and growth and development of tourism.

As indicated in Table 2, the research variables safety & security and tourism business are all strongly correlated with experience. Since the significant value is less than 5% level for the experience factor, safety & security, and tourism business, the null hypothesis is rejected. The results suggest that depending on the respondents' experiences, safety and security and the tourism business differ greatly. The null hypothesis is accepted in terms of infrastructural facilities, growth and development of tourism, and sustainable marketing strategy.

SEM (Structural Equational Modelling)

Structural equation modelling is a multivariate statistical analysis strategy that is utilized to analyze underlying connections. This procedure is a blend of factor analysis and different regression analysis, and it is utilized to dissect the primary connection between estimated variables and latent constructs.

Hypothesis Discussion

H_{03} – Sustainable marketing policyhas no significant influence on Tourism Business

The null hypothesis is accepted because, as can be seen from the above table, the p-value is bigger than the significant value (0.01). Therefore, a sustainable marketing policy has little to no impact on the tourist business.

H_{04}– Infrastructure facilities have no significant influence on Tourism Business

Table 2. One –Way ANOVA between experience and study variables

Factors	F-Value	Significant	Inference
Sustainable marketing policy	2.275	0.105	NS
Infrastructure facilities	2.603	0.076	NS
Safety &Security	4.732	0.010	S
Tourism business	10.192	0.000	S
Growth and development of tourism	2.442	0.089	NS

*5% Significant Level; S- Significant/ NS – Not significant

Figure 1. SEM analysis
Source-Amos Output

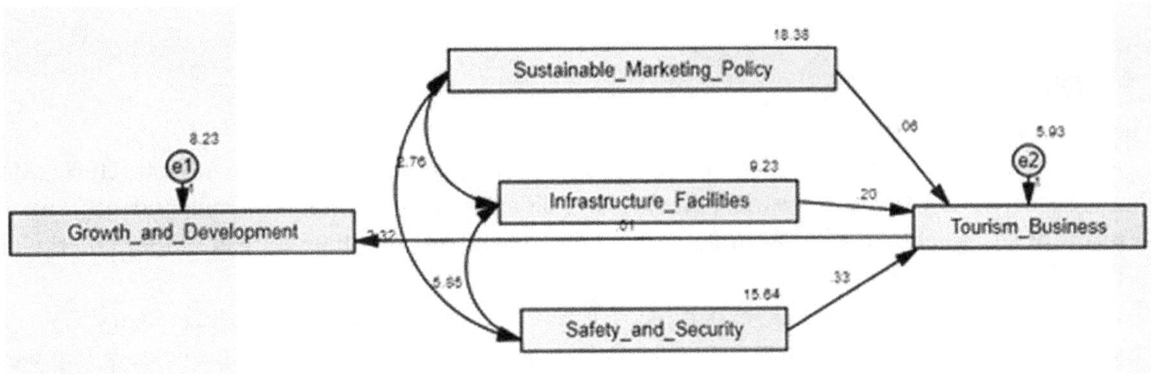

Table 3. Hypothesis testing

Variables Relationship	Estimation	SE	CR	P-Value
Tourism Business<---Sustainable marketing policy	0.056	0.037	1.532	0.126
Tourism Business <--- Infrastructure facilities	0.201	0.059	3.410	0.000
Tourism Business <--- Safety and Security	0.330	0.045	7.375	0.000
Growth and development of tourism <--- Tourism Business	0.014	0.061	9.223	0.003

*@ 1% level

Factors	Observed values
GFI	0.991
CFI	0.986
TLI	0.953
RMSEA	0.058

Table 3 demonstrates that the null hypothesis is rejected since the p-value is smaller than the significant value (0.01). Therefore, infrastructure facilities have a big impact on the tourism business.

H_{05}– Safety and Security have no significant influence on Tourism Business

The null hypothesis is rejected because the p-value in the preceding table is smaller than the significant value (0.01), indicating that safety and security have a substantial impact on the tourist business.

H_{06}– Tourism Business has no significant influence on the Growth and development of tourism

The null hypothesis is rejected because, as indicated in Table 3, the p-value is smaller than the significant value (0.01). Therefore, the tourist business has a big impact on the growth and development of tourism.

The research evaluated the metrics indicating the overall fit absolute goodness of fit (GFI = 0.991), comparative fit indices (CFI = 0.986), Tuker Lewis index (TLI = 0.953), and root mean square error of approximation (RMSEA = 0.058) to determine if the overall model was adequate. Although there are no set numbers for these SEM fit measures, there are general principles that may be used to make a judgment. According to the research, RMSEA may range from 0.06 to 0.08, while GFI, CFI, and TLI values can be more than 0.90. The aforementioned number, therefore, indicates that the model is well-fitted.

11. FINDINGS

- According to one-way ANOVA analysis findings for the age component, the infrastructure, safety and security, and tourist industries, all vary significantly with respect to the respondents' ages. In terms of sustainable marketing strategies and the expansion and improvement of tourism, the null hypothesis is accepted.
- Depending on the respondents' experiences, safety and security and the tourist industry may vary substantially, according to one-way ANOVA analysis findings for the experience component. Regarding physical infrastructure, tourist growth and development, and sustainable marketing tactics, the null hypothesis is accepted.
- The results of the SEM (Structural Equation Modelling) indicate that the RMSEA may vary between 0.06 and 0.08, while the GFI, CFI, and TLI values may exceed 0.90. Therefore, the aforementioned figure suggests that the model is well-fitted.

12. SUGGESTIONS

- There should be new and updated security and safety regulations. Adventure travellers are more likely than regular tourists to visit far-off places and engage in leisure activities that may involve risks. In the age of immediate communication, travellers will research destinations using both online reviews and official travel warnings. Important safety and security updates should be delivered via a destination marketing tool, and real-time communication technologies, such as Twitter, can assist in their dissemination.
- Mountaineering tourism providers must provide trips that promote individual engagement and give them the chance to satisfy their demand for thrills by taking on new challenges and learning new abilities. Additionally, tourism managers must recognize the close connections between mountaineering tourism and recreation and collaborate with these organizations to create top-notch experiences for their customers.
- For both future expansion and guest security, there should be a greater understanding of the social norms that must be upheld while offering Adventure Tourism services. The majority of travel operators offer every adventurous activity, which shows a lack of knowledge. It is important to receive the appropriate training before teaching any adventure activity.

13. STUDY IMPLICATION

This article demonstrates the growing significance of mountainous regions for the tourist industry, while also highlighting the significance of proper management of mountainous settings and the recreational activities that take place there for the viability and continued existence of mountainous regions. Ladakh, Arunachal Pradesh, and Himachal Pradeshare naturally, historically, and socially very profound regions of Northern India; yet, theyare lacking in a real headship in the sphere of economy, tourism development, and overall socio-economic way of life. In spite of the fact that more enabled products and efficiency with regard to key areas such as transportation, energy, health, drinking water, etc. are available, it is necessary to have a proper national-level strategy, planning, and visions rather than the traditional attitude. India's adventure tourism industry requires a first-world level of attitude, research that is knowledge-focused, national plans and policies with futuristic visions, and the mobilization of active youth in order to function well. In conclusion, in order for the Adventure and Mountaineering Tourism Business in selected Northern regions of India (Ladakh, Arunachal Pradesh, and Himachal Pradesh) to be successful, a decentralized system that includes the participation of all inbound and outbound associated personalities in their entirety and that works professionally within a win-win strategy is required.

14. CONCLUSION

In India, adventure tourism is flourishing and is anticipated to be among the most lucrative in the coming years. As a result of their abundant natural resources, the majority of states have identified areas for the growth of adventure tourism. The diversity of mountaineer adventure tourists must be taken into account by climbing tourism suppliers. A happy experience is what mountaineer adventurist aim for, despite the fact that achieving this goal invariably involves dealing with conflicting emotions. Therefore, mountaineer tourist suppliers should work to promote the flow or peak experience. With more people becoming aware of the market, safety is now given first importance and the sector is expanding faster than normal. Adventure Sports have the capacity to transform you, enhance your thinking, and give you the chance to face your anxieties head-on and triumph. Thus, adventure has always held a special place, and India and Adventure have a strong bond that is only growing. People in areas of the world where rural economic development projects are frequently hampered by a lack of investment, knowledge, and market access have major business potential with long sustainability in the adventure tourism industry.

REFERENCES

Ajitha, S. (2004). Tourism - The Other Side. *The Journal of Business Studies*, *2*(1), 87.

Anandan, (2004). Marketing of tourism products. *The Journal of Business Studies,* 2(1), 94.

Apollo, M., & Andreychouk, V. (2022). *Mountaineering Adventure Tourism and Local Communities: Social, Environmental and Economics Interactions*. Edward Elgar Publishing. doi:10.4337/9781802209389

Arunmozhi, T., & Panneerselvam, A. (2013). Types of tourism in India. *International Journal of Current Research and Academic Review*, *1*(1), 84–88.

Bashir, S., & Muthueleckuvan, R. (2018). Effect of mountaineering training on selected physiological variable among moderate altitude inhabitants. *International Journal of Physiology. Nutrition and Physical Education*, *3*(2), 720–722.

Clawson, D. L., & Johnson, M. L. (2004). World Regional Geography: A Development Approach, 8th ed. 36–37.

Kang, S. K., Hsu, C. H., & Wolfe, K. (2003). Family traveller segmentation by vacation decision-making patterns. *Journal of Hospitality & Tourism Research (Washington, D.C.), 27*(4), 448–469. doi:10.1177/10963480030274005

Katoch, A. S. (2021). Adventure Tourism- A Case Study of Himachal Pradesh. [IJCRT]. *International Journal of Creative Research Thoughts, 9*(6).

Katoch, A. S. (2021). Adventure Tourism- A Case Study of Himachal Pradesh. [IJCRT]. *International Journal of Creative Research Thoughts, 9*(6).

Mangoch, S., & Jain, D. (2022). Adventure tourism in northern states of India. *Academy of Marketing Studies Journal, 26*(1), 1–19.

MousimeXalxoa (2020). A Study on Tourist Behavior Towards Adventure Tourism. *Perspectives on Business Management & Economics, 1.*

National Strategy For Adventure Tourism. (2022). Ministry of Tourism vide O.M. No. NT-1401/2/2021. https://tourism.gov.in

Sung, H. H., Morrison, A. M., & O'Leary, J. T. (1996). Definition of adventure travel: Conceptual framework for empirical application from the providers' perspective. *Asia Pacific Journal of Tourism Research, 1*(2), 47–67. doi:10.1080/10941669708721975

Chapter 2
Analyzing Socio-Economic Impact of Adventure Tourism in the Mountain State:
Sikkim, India

Shakuntala Gurung
Sikkim University, India

ABSTRACT

This study focuses on adventure-based tourism development in popular chapter station of Sikkim that affects tourist movement and socio-economic life of the local communities. This chapter begins with the meaning and importance of adventure tourism and its impact on the tourism business, local societies, and environment of the tourism destination of Sikkim. Qualitative interviews were conducted with tourists, tourism operators and local communities. Findings of this study reveal that regulation of protected areas, legislation, and government programs play an important part in the development of sustainable adventure-based tourism in the study area. Nature education, development of business concepts, and safety of the tourists are the important points of findings. The role of the government and local communities are also highlighted here. These findings are used to suggest implications for policymakers and practitioners.

1. INTRODUCTION

Sikkim is a remote Indian state located in the lap of Himalayan Mountains. The state is bordered by Nepal to the west, Tibet to the north and east, Bhutan to the east and West Bengal to the south. Sikkim is the least populous state in India and the second-smallest state after Goa in total area, covering approximately 7,096 km^2. The climate of Sikkim ranges from subtropical to high alpine, and the world's third-highest peak 'Kanchenjunga', is located on Sikkim's border with Nepal. Due to its location in the Himalayas, Sikkim is blessed with varieties of flora and fauna. Sikkim is a popular tourist destination,

DOI: 10.4018/979-8-3693-0823-3.ch002

due to its geography, landforms, biodiversity, scenery and culture. Sikkim's capital and largest city is Gangtok. There are four districts in Sikkim.

Sikkim is the paradise for nature lovers, botanist & plant lovers. For adventure sports, the state provides trekking trails, hiking path and offers dense forest with rhododendron, ethnic monasteries, glaciers, splendid mountain views, high passes and historical silk routes. The Kanchendzonga, the third highest mountain in the world is in this state. Trekking and hiking in Sikkim offers a great opportunity to the adventure tourists to get variety of hill based adventure sports.

This paper examines various types of adventure sports available and their effect on Tourism Business in Sikkim that affects tourist movement and socio-economic life of the local communities. This article highlights the meaning and importance of adventure tourism and its impact on the Tourism business and local economy in the state. Adventure tourism can be beneficial in terms of conservation and economic benefit to the local people. On the other hand, lack of proper planning can be detrimental to the environment, flora, fauna and local people. Nature education, awareness, impact analysis and measuring carrying capacity, visitor management etc. are essential to reduce negative impact on the socio-economic life of the local people, Responsibility of the local bodies; local communities and tourism service providers are also discussed here.

2. OBJECTIVES

The main objectives of the Research paper include the following: -

Figure 1. Sikkim political map
Source: Sikkim political map (www.maps of India.com)

- To understand the concept, development and impact of adventure tourism, especially in Sikkim
- To examine the tourism potential of Sikkim in general and adventure tourism in particular.
- To analyze various prospects and challenges of adventure tourism in Sikkim.
- To study socio-economic impact of Ecotourism in Sikkim

3. METHODOLOGY

To gain insights into the general composition and some of the issues of Adventure Tourism, extended conversations were held with the local people, service providers and tourists in the study area. These types of experiences were chosen because they represent a diversity of issues in the tourism destinations. This study was conducted using self-administered questionnaires. For collecting the data from the primary sources, two sets of structured questionnaire made both for tourists and local people including local service providers. Apart from that, observation, schedules and interview methods were followed for the above said purpose.

4. ADVENTURE TOURISM: CONCEPT AND MEANING

Adventure tourism is a type of niche tourism linking exploration or travel to remote areas, where the traveler may expect the unexpected. It is a type of tourism that involves exploration or travel with perceived risk and potential requiring specialized skills and physical exertion. Priest (1990) said that there must be an element of uncertainty associated with the event. Ewert (1985) and Hall (1992) write that 'risk' plays a primary role in the decision to engage in adventurous activities.

According to the U.S based adventure travel trade association, 'adventure travel may be any tourist activity including two of the following three components a physical activity, a cultural exchange or interaction and engagement with nature'.

According to Canadian Tourism Commission, adventure tourism is 'an outdoor leisure activity that takes place in an unusual, exotic, remote or wilderness destination, involves some form of unconventional means of transportation, and tends to be associated with low or high levels of activity'.

Christiansen (1990) highlighted that, it is the task of the adventure tour operator to provide the client with an adequate perception of risk, while ensuring a high level of safety and security.

Adventure tourism is rapidly growing in popularity as tourists seek unusual holidays, different from the mass tourism and involve some degree of risks. Adventure tourist travel to achieve mental status characterized as rush or flow, resulting from stepping outside of their comfort zone. Mountaineering, trekking, bungee jumping, rafting mountain biking, canoeing, paragliding and rock climbing are some of the examples of adventure tourism.

5. IMPORTANCE OF ADVENTURE TOURISM

Adventure tourism is attracted by adventure seekers or who used to take risk. This form of tourism basically caters to the young to middle age group of tourists who indulge in adventure sports activities like water skiing, scuba diving, snorkeling, paragliding, rock or mountain climbing, river rafting and bungee jumping. Due to the benefit of fresh natural environment there has been an increasing variety of sports activity such as mountaineering, skating, fishing, trekking etc.

The beauty of natural and man-made tourist destinations attracts tourists towards them. Further places of historic and religious significance are also attracted to many tourists. Adventure sports like: water based sports offer water skinning, parasailing, rafting and water scooter driving. Land based sports offer skiing on snow-clad terrain, horse riding, horse racing, hiking, trekking etc. Air-based sports offer paragliding, bungee jumping, parachuting and other such activities related to travel in the air.

6. ADVENTURE TOURISM IN SIKKIM

Sikkim is a small tiny state of India in the lap of Himalayas. It is called the land of majestic splendor because of its high rise mountains, meandering dreams, lush green tea garden and forest. The state is situated between Nepal, China (Tibet), Bhutan & West Bengal. It is the ultimate paradise for botanist & plant lovers. For adventure lovers, the state offers vacant forest, rhododendron- covered hill sides, remote monasteries, glaciers, high passes and magnificent mountain views. Kanchendzonga, the third highest mountain in the world lies in the small state of Sikkim. Trekking in Sikkim offers a great opportunity to get experience of the rich heritage and village lifestyle of north east.

7. TYPES OF ADVENTURE TOURISM IN SIKKIM

Sikkim has an immense potential for adventure tourism related activities. Some of the important adventure activities being promoted in the State are: Trekking, Mountain climbing and Mountaineering, Rock climbing, River crossing, Paragliding, Kayaking, Hang Gliding, White water rafting, Angling and many other types of traditional adventure.

Figure 2. Kangchendzonga and other picks of Sikkim
Kangchendzonga and other picks of Sikkim

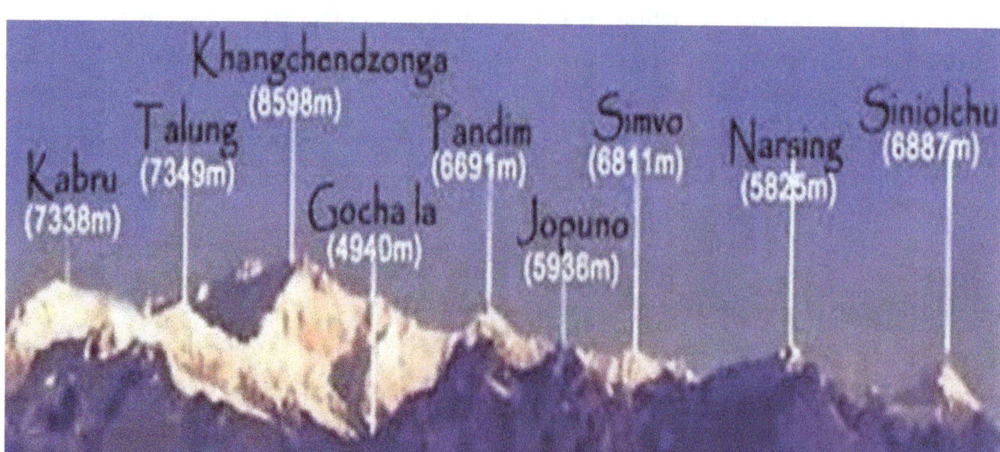

7.1. PARAGLIDING

Paragliding is one of the most important forms of adventure tourism. Travel Agent Association of Sikkim initiated a paragliding at lower Luing Resithang in Gangtok, East Sikkim. The venture was inaugurated in April 2010; but commercially they started flying from 9th November 2010. Sikkim has got many points which are highly suited for the paragliding.

In 2004, department of sports and youth affairs of Sikkim government introduced paragliding in Sikkim. Initially this adventure sports was restricted to Maghey Mela in Jorethang which happened once in a year, witnessing its success in Jorethang a rake was conducted in Gangtok. Paragliding in Sikkim is more challenging and exciting than most other parts of the world. Sikkim Tourism has taken a keen interest in promoting paragliding as sports and as tourist's attractions.

The SPASCS (Sikkim paragliding Adventure sports cooperation society) was established in 2010. The organization comprises of all Sikkim paragliding pilots trained under the Sikkim governments Chief Minister Self Employment Scheme and capacity building program organized by the department of Sports and Youth Affairs.

7.2. MOUNTAINEERING IN SIKKIM

Sikkim is blessed with world famous Himalayas and offers huge scope for adventure sports like mountaineering. Among the variety of adventure sports, mountaineering in Sikkim is gaining immense popularity now-a-days. Jopuno, Tinchenkhang, Lamo Angden and Byrmkhangse are the popular peaks of Sikkim where mountaineering is allowed. One can undertake mountaineering in these peaks with special permits from the concerned authority.

In the southwestern part of Sikkim, in the main Himalayan range, lies the majestic Kanchenjunga, the third largest peaks, rising to a height of 8,585 m (28,169 ft.) above sea level. Besides this other famous mountaineering peaks in Sikkim include- Kabru (7338 m), Talung 7,349 m) and Siniolchu (6,887 m). Mountaineering expeditions require more planning and paperwork as clearance is required from the Indian Mountaineering Foundation, New Delhi. Besides these the local governments also impose royalties on certain peaks.

The state of Sikkim has many mountains, but for the people of Sikkim, Mount Khangchendzonga is more than just the 3rd highest peak in the world it is also the abode of the states guardian deity. To them all of these mountains are sacred. Mount Khangchendzonga literally means five treasures of the eternal snow.

7.3 MOUNTAIN BIKING IN SIKKIM

Sikkim's best adventure product is mountain biking. The rugged and dramatic terrain of Sikkim offers exiting biking trails to the adventure tourists. The beauty of the landscape, flora and fauna and lifestyle of the Sikkim people is best experienced on a bike's seat. There is no specific season for mountain biking. In very time there is different fun and feeling. The mountain bikes available in Sikkim are durable in toughest for hire or for days package. A repair support and all kinds of logistics are provided in the package. The landscape, wilderness, the country side offers an interesting challenge to the mountain bikes who can feel the typical culture & tradition & unique lifestyle of Sikkimese people.

At present there are number of new route that have been introduced in Sikkim, they are

Figure 3. Climbing

- Gangtok – Rumtek – Sang – Sirwani – Temi – Damthang – Rabong – Kewzing – Tashiding – Yoksum – Geyzing – Dentam – Rinchenpongn – Soreng – Somberia – Jorethang – Namchi – Rangpo.
- Gangtok – Phodong – Rangrang – Dikchu – Makha – Sirwani – Temi.
- Gangtok – Rumtek – Sang – Sirwani – Temi – Namchi – Namthang – Rangpo.
- Gangtok – Rumtek – Sang – Sirwani – Timi – Namchi – Jorethang – Melli.

7.4 TREKKING IN SIKKIM

Sikkim is a Trekkers and Adventure lover's paradise. Trekking in Sikkim offers a great opportunity to get a glimpse of the rich heritage and village lifestyle of north east India. Trekking through the snow covered mountains, beautiful orchards, exotic flora and fauna and transparent water streams satisfy the adventure expedition of the trekkers.

Trekking in Sikkim is a wonderful experience a remedy to city stress of daily life and its memories will be cherished lifelong. While trekking one relish the pure air and the silence of the wilderness and feels the awakening of senses blunted by urban living. Goechala Trek, Dzongri Trek, Kasturi Trek, Yambong Singalila Trek, Uttarey Singalila Trek and Kanchendzonga Trek in west Sikkim is a proper well beaten and famous trekking trails. The trekking trails goes through forest of Rhododendrons, Magnolia and silvers with views of mighty Mt. Khanchendzonga and surrounding peak, like Mt. pandim, Jupono, Kabru, Kothang, Thinkchankhang, Thalung etc. One can see the different peaks of Mt. Khanchendzonga range and villages of Nepal on the other side on the clear sunny day. In this particular trekking trail there are many high altitude lakes on the way and it is also known as the lake trek.

7.5 ROCK CLIMBING IN SIKKIM

Rock climbing is a popular adventure sports in Sikkim. Rock climbing is a physically and mentally demanding sport and is one of the best outdoor activities. Free climbing and aid climbing are the two types of rock climbing available in Sikkim. While doing free climbing gears and ropes for safety are provided. However, while undertaking aid climbing a route up, a piece of rock is created with the help of equipment that is kept in the rock for progressing upwards. The training required for those trying out rock climbing can be availed at various locations throughout Sikkim.

Some of the popular and recommended trek routes in Sikkim are:
Monastic Trek
Itinerary: Pemayangtse- Sangacholing-Khecheopalri- Hongri- Sinon- Tashiding- Ralong
Rhododendron Trek
(i)***Itinerary:*** Hilley/Soreng- Barsey- Dentam-Pemayangtse
(ii) Hilley- Barsey- Jorbutey- Kalijar- Chewabhabjang- Uttary
Khangchendzonga Trek
Itinerary: Yuksom- Bkhim- Tsoka- Dzongri- Thangsing/Kokchurong- Lamuni/Chaurigang- Goechala/Rathong glacier and back.
Coronation Trek
Itinerary: Rumtek- Sang- Yangyang- Ravang- Tashiding- Yuksom.
Khedi Trek
Itinerary: Assam Lingzey-Pashing Teng Kha (Base Camp, 1425m)-Chauri Kharka-Do Bato-Khedi- Sela Pass (3150m)-Do Basti-Tal Kharka-Nugang-Passang Teng Kha.
Singalila Trek
Itinerary: i) Uttarey-Chewabhanjang-Dhor(Wahong)-Sikkim Megu-Dafey Bhir-Gomanthang/Boktok
ii)Pelling- Nambu- Chongri- Yambung- Dafeybhir- Gomathang- Tegyapla- Dzongri- Thanghshing- Lamuney- Kokchurong- Tsokha-Yuksom- Tegyapla- Bikbari- Dzongri Tsoka- Yuksom
Samartek Trek
Itinerary: kodong-Dokshing-TsenKhong-Lenchok-Samartek-Pakchong-Kungo-Kuling Nae(Cave) Gungramgampo-Singhik-Dak Bungalow
Rinchenpong/Soreng Trek
Itinerary: Kaluk-Rinchenpong Dak bunglalow-Rigsum Gumpa-Kalud-Sribadam Jhandi Dara-Soreng
Himalayan Trek (South Sikkim)
Itinerary: Namchi(Base camp)-Tendong-Damthang-Ravangla-Maenam-Bhaley Dhunga-Yangyang-Sinchuthang-Bermoik-Parbing-Namchi
Tosar Lake Trek (North Sikkim)
Itinerary: Mangan –Naksuk-Sohar Dhe-Soahar Antsok-Patam-Tosar Lake-Mayong Tar-Kalep Mangan
Meanam Hill Trek (South)
Itinerary: Rabong-Meanam-balleydunga-Ralong-Rabong
Tendong Hill Trek
Itinerary: Dentam-Tendong- Samdruptse - Namchi

Source: Department of Tourism & Civil Aviation, Government of Sikkim

7.6 WHITE WATER RIVER RAFTING IN SIKKIM

Recently, White water rafting is added in adventure sports in Sikkim. Only the Teesta and Rangit rivers offer long stretches which are ideal for safe rafting. Teesta is graded 4 on the International scale. The icy cold waters of the rugged Teesta offers one of the finest rafting stretches in the world. Rangit, a tributary of the Teesta, has more turbulent waters and offers many challenges to more experienced rafters.

The riverbanks of both the rivers have deep forests, with some patches of terraced cultivation, and small villages. Innumerable white sandy beaches offer good campsites. Adequate facilities for overnight River beach camping is also available.

Major River Rafting Routes in Sikkim are:

1. Makha - Sirwani - Bardang - Rongpo (Teesta River)
2. Sikip - Jorethang - Majitar - Melli (Rangit River).

Table 1. Last ten years data of foreign tourists and domestic tourist arrival in sikkim

YEAR	FOREIGN TOURISTS	DOMESTIC TOURISTS	TOTAL
2005	16518	347650	364168
2006	18049	421943	439992
2007	17837	465204	483041
2008	19154	512373	531527
2009	17730	615628	633358
2010	20757	700011	720768
2011	23945	552453	576398
2012	26489	558538	585027
2013	31698	576749	608447
2014	49175	562418	611593

Source: Department of Tourism & Civil Aviation, Government of Sikkim

8. LAST TEN YEARS DATA OF FOREIGN TOURISTS AND DOMESTIC TOURIST ARRIVAL IN SIKKIM

8.1 Socio-Economic Impact of Adventure Tourism in Sikkim

- The USP of Sikkim is its natural beauty. Nature has gifted lush green, environment friendly and undisturbed valleys, peaceful, hospitable and smiling people and a rich cultural heritage to Sikkim. The most peaceful, crime free and total sanitation are the strength of Sikkim.
- 'Tourism' as a major engine for economic growth, employment generation & poverty alleviation, the Government of Sikkim has approved this sector as the main industry of the State. To enhance social and economic benefit, government is also trying to improve the efficiency of Tourism in general and eco-tourism in particular.
- Sustainable development of tourism and encouraging the private sectors to develop tourism related infrastructure & services without disturbing ecology and environment is the primary objective to develop tourism in the state.
- The state government and local bodies and Tourism organizations must work together towards the maintaining sustainability and socio-economic advancement of the local communities.
- Adventure Tourism in remote areas help in elimination of poverty, ending unemployment, creating new skills, enhancing the status of women, improving overall environment and socio-economic life of indigenous people

9. FINDINGS

Major findings of this research paper are:

1. The State is gifted by nature with a variety of adventure based destinations and circuits that enthralled tourists. Sikkim has impressive varieties of Tourism Products.
2. Sikkim is the peaceful place and its uniqueness are: green, neat & clean environment, total sanitation, plastic and pollution free environment, hospitality and beautiful scenic beauty with high altitude landforms.
3. The people are encouraged to preserve the old monuments and heritage sites. Adventure tourism in rural and remote area is encouraged to preserve cultural heritage & handicrafts and to create better economic and employment opportunity in the rural areas.
4. It has been observed that many trekking points in Sikkim is facing problems like, connectivity, safety and security for the international as well as domestic tourist.
5. There are number of mountains that are restricted for expedition in Sikkim which is creating negative impact in adventure tourism in Sikkim.
6. Lack of trained, educated and language tourist guide & tourist driver are the constraint for developing adventure tourism in Sikkim.
7. It is observed that some tour operators & travel agencies charge high rate in very pick season.

10. RECOMMENDATION

Though, the government of Sikkim is trying its level best for promoting & developing adventure tourism in Sikkim. But still it is important to analyze the following recommendations:

1. Development of adventure tourism shall not only benefit the economic growth but it shall bring overall development of the State with high happiness to the local communities.
2. Sikkim government must provide appropriate connectivity in adventure tourism activities such as trekking, Mountaineering, river rafting etc. so that international as well as domestic tourist can feel safe during their adventure trip in the state.
3. Only licensed tour guide & tourist driver should be allowed to accompany the tourists and refresher course to the licensed guide should be conducted to enrich their knowledge, so that the tourist can enjoy the whole journey as well as feel safe.
4. LCD televisions may be installed in the entry point of the state that contains all information about the adventure tourism of Sikkim.
5. Promotion, publicity & marketing of adventure tourism must be taken up in well planned manner at national and international level. Advertisement may be given in various newspaper, television, Magazine, Journals and internet.
6. State government should introduce a legislative framework to regulate adventure tourism and ensure safety & security and health care facilities for the tourists.
7. A section of the state police or other force may be allocated or private security agency may be engaged to act as Tourist Police and special training may be given to them.
8. Government should ensure that neither over-exploitation of natural resources will be permitted nor the carrying capacity of the destination is ignored. No private organizations will be allowed, directly or indirectly to secure short-term gains.
9. Proper cleanliness, sanitation and solid waste management to be implemented to promote sustainability and quality tourism in the state

10. Adventure tourism along with rural tourism, Home Stay and tourism in small settlements, all together may be promoted to get more socio-economic benefit of the indigenous people of the state.
11. Responsible Tourism in the adventure sites along with involving local community with local resources would be the major thrust area for sustainable adventure tourism.

11. CONCLUSION

Adventure tourism is a type of tourism involving exploration or travel with perceived risk and requiring specialized skills and physical hard work. As the topic of this paper is "Analyzing Socio-Economic Impact of Adventure Tourism in Sikkim, India". It can be said that Sikkim has an immense potential for adventure tourism related activities. Some of the land based adventure activities include, rock climbing, trekking, mountaineering, mountain biking. Water based adventure activities in Sikkim are, river rafting, kayaking, canoeing. Air based adventure activities are paragliding, hang gliding, parachuting etc. Trekking, hiking and Mountaineering are the most popular adventure tourism activities in Sikkim along with other adventure activities. To maintain safety of the participants in adventure activities, there is a need to fix minimum acceptable standards in terms of equipment and human resources.

For promoting or developing the adventure tourism in Sikkim government of Sikkim has been taking effort. Sikkim is said to be a virgin state because of unexplored places, so, in future also adventure activities in Sikkim will prosper. As tourist want to go close to the nature and want to experience the nature's beauty, adventure activities can fulfill this experience by exploring the unexplored places.

REFERENCES

Amed, A., Rawat, J. S., & Rai, S. C. (1990). 'An analysis of the Himalayan environment and Guidelines for its Management and Ecologically Sustainable Development. *The Environmentalist, 10*(4), 281–298. doi:10.1007/BF02239722

Andrews, A. (2007). *Introduction to Tourism & Hospitality Industry*. Tata McGraw Hill, West Patel Nagar.

Batta, R. N. (2000). *Tourism & the Environment A Quest for Sustainability*. Indus Publishing Company.

Canter, W. L. (1996). *'Environmental Impact Assessment* (2nd ed.). McGraw Hill Inc.

Ewert, A. (1985). Why people climb: The relationship of participant motives and experience level to mountaineering. *Journal of Leisure Research, 17*(3), 241–250. doi:10.1080/00222216.1985.11969634

Fennel, D. (2008) *Ecotourism*. Rutledge.

Hall, C. M. (1992). Adventure, sport and health tourism. In *Special interest Tourism*. Belhaven Press.

Karpagam, M. (1991). *Environmental Economics: A text Book*. Sterling Publishers Private Ltd.

Priest, S. (1990). *'The semantics of adventure education', Adventure education*. Venture Publishing.

Chapter 3
Assessment of Factors Responsible for Promoting Religious Tourism in the Himalayan Region:
Special Reference to Dev Bhoomi, Uttarakhand

Dillip Kumar Kumar Das
https://orcid.org/0000-0002-6032-9348
University of Burdwan, India

Susanta Ranjan Chaini
https://orcid.org/0000-0002-8509-5306
SOA University, India

ABSTRACT

Religious tourism activities are now the emerging trends in the Himalayan region along with adventure tourism. The Himalayan region is the abode of many Goddesses. Factors responsible towards exploiting and promoting religious tourism in Himalayan region have been carefully analyzed with reference to Dev Bhoomi, Uttarakhand, but many issues still need more attention. This research chapter uses surveying of 150 stakeholders in relation towards assessment of factors responsible towards development of religious tourism in Himalayan region with special reference to Uttarakhand. Research results are analyzed on the basis of six factors influencing the growth and development of religious tourism in Himalayan region: transportation, accommodation, safety and security, environment, economy, and attitude of the local people. The result analysis depicts factors responsible for promoting religious tourism in Uttarakhand as only quite average. This chapter has proposed some important solutions towards promoting religious tourism in the Himalayan region with special reference to Uttarakhand.

DOI: 10.4018/979-8-3693-0823-3.ch003

1. INTRODUCTION

Tourism industry is one of the most prominent sectors in the state of Uttarakhand providing huge economic benefits including better employment opportunity, improving the standard of living and better infrastructure development. Religious tourism is one the oldest type of tourism that prevails in Himalayan region. Dev Bhoomi, Uttarakhand in the Himalayas is very much popular in the world for religious tourism. Uttarakhand is situated in the low Himalayan Mountain range. Uttarakhand state borders Tibet (China) in the North, Nepal in the East, Himachal Pradesh in the North Wet and West and Uttar Pradesh in the South. Uttarakhand is one of the fastest developing states of India. Religious tourists visit shrines, attends religious fairs and festivals and attends various rituals which is a very common practice among all. Religious tourism provides employment opportunities and also livelihood to several local people. In today's arena travel business is contributing generously to a country's GDP particularly for developing countries like India. Religious tourism destinations operate as a dynamic complex system along with independent element and many factors that are responsible for promotion and development of tourism particularly in Uttarakhand. Religious tourism is considered as powerful tool for socio-economic development. Uttarakhand is very much popular and famous for Char-Dham Yatra among religious tourists. Badrinath, Yamunatri, Kedarnath and Gangotri are major four pilgrimage sites of Uttarakhand which are most sought-after destinations among religious tourists. Badrinath is the abode of Lord Vishnu and is one of the best destinations not only among religious tourists but also among all the tourists. Kedarnath is the abode of Lord Shiva and is one of the twelve Jyotirlinga and only one in Uttarakhand. Yamunatri is the place from where sacred river Yamuna originates. Gangotri is the origin point from where the sacred river Ganga flows. Every religious tourist visit Uttarakhand never misses the Char Dham Yatra. To encourage, attract and promote more religious tourism in the state, Government of Uttarakhand has taken several important steps in terms of financial assistance or aid and subsidies. The state is having a strong infrastructure network which is helping to attract more and more religious tourists. It has been observed that more religious tourists are visiting Uttarakhand comparable to any other form of tourism like eco-tourism, agri-tourism and adventure tourism. The state of Uttarakhand is having extensive road and rail network along with two airports to support various kinds of tourism in the state. Foreign direct investment (FDI) to the state of Uttarakhand is steadily increasing from year to year. Mainly the common Annul Yatra time generally starts from the month of April and completes in the month of October when a large number of religious tourists visit the state of Uttarakhand. Especially during winter season, major tourists visit the Garhwal Himalaya for river-rafting, trekking and mountaineering. This is the main reason of blooming of hotels, eateries, Dhabas and way side amenity centers along the highways. It has been found out that there are ample services delivery centres, located within 280 km between Rishikesh and Badrinath. Religious tourism manifests ownership or belongingness and also generates employment as well as income and the same goes well with the local people of the Garhwal Himalaya.

Religious tourism has multiple impacts on local economy as it increases employment opportunities and living standard of the people and thus, there is an increase in the national income (Sati & Kumar 2004). Tourism is a major source of income in the Uttarakhand region as it earns between 15 - 20 per cent of its revenue from taxes and levies imposed on pilgrims each year. Presently in India, tourism industry is the second-largest source of foreign currency behind the gem and jewelry sector. It has been observed that the villages that are located on the major routes to the high shrine pilgrimage centers are now fully dependent on the income from the religious tourists.

2. LITERATURE REVIEW

A study conducted on Tourism as not just a pleasure or entertainment activity, but it is also a revitalizing one **(Joveriya and Mariya, 2019)**. Tourism is considered as one of the greatest service-sector industries in the world, generating employment opportunities or job avenues, particularly for the destination based local people, and also an opportunity for cultural, economic, and social growth on a large scale. Various international organizations related to tourism, like WTO (World Tourism Organization), have highlighted tourism's huge development potential, particularly in developing countries.

A study on Rural Tourism for Uttarakhand, its Significance, Scope, and Challenges **(Rakesh Sajwan, 2019)**. The state of Uttarakhand was founded on November 9, 2000. Uttarakhand has a total area of 53,483 km2, containing 901.92 km2 of the urban area, and 52,581.08 km2 of rural area, with 69.77 percent of the people dwelling in rural regions and just 30.23 percent residing in urban areas.

Tourism sector is one of the fast-growing service sectors in the world. It acts as a powerful instrument for economic growth. The growth of tourism sector has resulted in employment generation, foreign exchange earnings, expansion of infrastructure facilities, capital investment, socio-economic growth, increasing in contribution to GDP and so on. India has evidenced sustainable and inclusive economic growth due to the wide expansion of tourism sector.

Tourism is one of the fastest growing service sectors in the world and a power tool for economic growth as well as development **(Dayanand, K.C & Leelavati, D.S, 2016)**. The growth of tourism industry has manifested the employment generation on a massive scale along with huge foreign exchange earnings.

A dip bathing in river Ganga cleanses the sin and soul which leads to millions of Hindus took bath every year. They also immerse the remains of their loved ones **(Sachdeva, 2016)**.

Religious tourism not only has economic benefits for the local sites visited but also gives more benefit to the people employed in hospitality, transport and selling handicrafts **(Duran- Sanchen et al. 2018).**

Tourists are always in search of better places of tourist's interest to explore. Most of the tourist sites are located in small villages, undeveloped areas, lack of infrastructure etc. In addition to visit of religious or pilgrimage site, tourists also prefer to getaway to holidays most of the time **(Shinde, 2007).**

The major priority of tourists in the region includes pilgrimage and holiday as compared to other activities like societal engagement, official trip or naturopathy **(Rongali, 2017)**.

Pilgrims of all religions visit pilgrimages all over the world **(Stoddard & Morinis, 1997)** and exchange ideas on culture and spirituality with each other **(Chidester & Linenthal, 1995).**

Agenda 2030 has been introduced by United Nations with 17 Sustainable Development Goals (SDGs) that have been formed with global consensus among various experts around the world. It is very much essential for a developing country like India towards attainment of SDGs, with growing populations and limited resources. India's overall development agenda also encompasses the SDGs largely **(Khalid et al. 2018).**

In India, tourism is growing in a faster pace and is being promoted strongly by the government In recent years. Growth and expansion of tourism in a country also improves the projection of Its soft power and image **(Bhonsale 2019).**

India is rich in culture, heritage and a secular country having ancient civilisation. Every religion in India has worship centres and concepts of special congregations on special occasions. The visitors to such places have a greater tendency to return to the destinations within a shorter time frame and hence, have repeat value **(Durán-Sánchez et al. 2018).** For example, different religious community in India has its own religious sites where they visit regularly (Hindus - Char Dham or four adobes at Badrinath in Uttarakhand, Dwarka in Gujarat, Lord Jagannath, Puri in Odisha, Rameswaram in Tamil Nadu; Sikhs visit Golden Temple in Amritsar (Punjab), Gurudwara Hemkunt Sahib (Uttarakhand); Muslims – shrines of Moinuddin Chisti (Rajasthan), Nizamuddin Auliya (New Delhi), Haji Ali (Mumbai) and Christians visit Basilica of Bom Jesus (Goa), St Paul Cathedral (Kolkata) etc. There is a great employment generation occurs in India because of religious tourism. Every state promotes tourism actively, but restrictions at some designated religious sites make it unpopular for tourists of other faiths and hence alternate tourist attractions should be Promoted **(Sultan, 2015).**

Religious tourism is to visit different places of worship, special interest on worshiping God and Goddesses, visiting spiritual sites, having deep interest to study the temple heritage etc. Religious tourism is very popular in all sections of society.

3. OBJECTIVE OF THE STUDY

Based on the literature review and rationale of the study, the following objectives were set to justify the research study which includes the following:

- To study the tourism potentials of the state especially for religious tourism for Dev Bhoomi, Uttarakhand.
- To identify the role of government for the development of religious tourism in the state.
- To analyze major factors responsible for promoting religious tourism in Himalayan region including some measures to promote religious tourism with reference to Uttarakhand.

4. RESEARCH METHODOLOGY

The research study includes both primary as well as secondary method of data collection. The researchers' personal experiences from visits to Uttarakhand, as well as a survey to the experts / different stake holders including hotel owners were mainly taken into consideration for analysing the data. Apart from that various travel agencies and tour operators, transport operators including some government officials related to development of tourism in the study area were consulted who provided the primary information to the researchers through a schedule questionnaire. The authors created the questionnaire based on several factors responsible for promoting religious tourism in the state of Uttarakhand. The secondary data for this research were collected from magazines, newspapers, journals, and various similar other research papers including the statistical report published by state tourism department, including Uttarakhand tourism policy (2018). Mostly primary data has been collected through the state tourism office and a direct interaction with religious tourists. Total 150 tourists (both domestic and foreign) visiting several pilgrimage sites of Uttarakhand state were taken into consideration for collection of data. The random convenience sampling approach was utilized from Uttarakhand pilgrimage sites or destinations. Several factors were considered which are responsible for promoting religious tourism in the state of Dev Bhoomi, Uttarakhand namely commitment towards Gods & Goddesses, spiritual rejuvenation, beliefs for miracle, pilgrimage, beauty of the nature surrounding the pilgrimage destinations, Exploring the interesting stories associated with the pilgrimage sites or destinations, devotion to Gods & Goddesses, desire to interact with priests & local people, desire to partake Prasad or holy food, desire to see the historic value & architecture etc.

5. UTTARAKHAND AS A RELIGIOUS TOURIST DESTINATION

Uttarakhand, which is popularly known as Devbhoomi or 'Abode of Gods' is a site of temples and pilgrimage, endowed with salubrious climate, exotic greenery, rich flora besides its rich culture. Moreover, the beautiful lakes, mountain peaks, mesmerizing scenic landscapes makes Uttarakhand an attractive tourist destination in India and abroad

Situated in the northern Himalayan region, Uttarakhand a northern Himalayan state of Indian subcontinent, tourism is considered to be an important vehicle for economic and social development. The economy of the state has been largely dominated by the services sector including tourism which contributes nearly 40 other destinations in the Uttarakhand are known and regarded as favorite destination for finding peace and sensing spirituality. Tourism and pilgrimage in Uttarakhand add around 27% in gross domestic production of state (GSDP). According PHD Chamber of Commerce and Industries (PHDCCI), tourism sector in Uttarakhand alone contributed around 60% in GDP of state as compared to whole services sector which contributes nearly 51% in the GDP of the state. over 51% to the total GSDP from 2006-07 to 2016-17. Tourism is already a major driver in the economic growth and livelihood promotion in Uttarakhand. it is the place where the world-famous UNESCO cultural heritage 'Kumbh Mela' is celebrated after every three years rotated over four important pilgrimage centres.

6. MAJOR TOURIST DESTINATIONS IN UTTARAKHAND

The above table depicts the tourism resources of Uttarakhand which has been categorised on the basis of several purpose of visit as well as categories of tourism resources available in the state. Depending upon the importance and purpose of visit the authors categorised the tourism resources into religious, adventure, nature and wildlife, health and rejuvenation, rural tourism and sight-seeing etc.

Table 1. Major tourist attractions of Uttarakhand

Theme	Major destinations
Pilgrimage and Festivals	Gangotri, Yamunotri, Kedarnath, Badrinath, Rishikesh, Haridwar, Jageshwar, Baijnath, Piran Kaliyar, Hemkund Sahib, Nanda Devi
Adventure & Watersports	Rishikesh, Auli, Trekking at Shri Hemkund Sahib, Jharipani, Maldevta, Tons Valley, Dhanaulti, Tehri
Nature & Wildlife	Jim Corbett National Park, Rajaji National Park, Binsar Wildlife Sanctuary, Kedarnath Musk Deer Sanctuary, Nanda Devi National Park, Askot Musk Deer Sanctuary, Neel Dhara Pakshi Vihar, Benog Wildlife Sanctuary, Govind Wildlife Sanctuary
Health & Rejuvenation	Rishikesh, Haridwar, Champawat, Pithoragarh and Ramgarh, Jageshwar, Almora, Nainital
Rural Tourism	Mana, Chokhta, Chakrata, Deora, Pallyu, Shaukiyathal, Bageshwar, Chamoli, Almora and Tehri
Sight seeing	Mussoorie, Nainital, Valley of Flowers, Almora, Kausani, Auli

Source: Authors own compilation

7. ROLE OF GOVERNMENT TO PROMOTE RELIGIOUS TOURISM IN THE STATE OF UTTARAKHAND

The main vision of the state government is to promote tourism in the state by establishing different tourist destinations, tourist circuits and clusters with tourist-friendly infrastructure and to enable world-class tourism experiences to both the domestic and foreign tourist in the State.

The famous Char Dham yatra, the major tourist circuit in the state which is the life line for the state tourism industry is the prime attraction mainly for religious tourists. The Uttarakhand government have planned to develop pilgrimage & heritage sites in the state similar to Kedarnath and Badrinath Dham to boost tourism. As per the current Uttarakhand state chief Minister who clearly mentioned that he had extensive discussions with the Prime Minister about the planned development of pilgrimage and religious tourism in the state. The Chief Minister also stressed that devotees who come to visit Char Dham Yatra sometimes have to wait for two to three days for helicopter tickets. "Such arrangements should be made that during this time the travellers can visit the nearby tourist places and will make good use of the time of these travellers on one hand, while on the other hand, the local residents will get employment due to the time spent by them while travelling.

Several government schemes including Pilgrimage Rejuvenation and Spiritual Augmentation Drive (PRASAD)' which was **launched by the Ministry of Tourism in the year 2014-15** with the objective of holistic development of identified pilgrimage destinations in India is already implemented in differs destinations in the state of Uttarakhand. The projects identified under this scheme shall be implemented through the identified agencies by the respective State/ Union Territory Government. Rejuvenation and spiritual augmentation of important national/ global pilgrimage and heritage sites. Community-based development and create awareness among the local communities are the main attraction for growth & development of the pilgrimage sites. There will be an integrated tourism development of heritage city, local arts, culture, handicrafts, cuisine, etc., to generate livelihood particularly in these pilgrimage destinations. Strengthening the mechanism for bridging the infrastructural gaps so that pilgrimage destination becomes more popular.

8. RESULT AND DISCUSSION / MAJOR FINDINGS

The state Uttarakhand is famous for its exotic mountain range, wildlife sanctuaries and national parks, popular trekking paths/ circuits, religious importance, rich flora and fauna including existence of strong regulation for conservation and preservation. The followings are some of the major findings came out during the analysis and discussion with different stakeholders including some experts. These includes:

- The study revealed that religious tourist attractions of Uttarakhand has been scattered in several districts with very few districts like Dehradun, Haridwar, Pauri Garhwal, Tehri Garhwal, Chamoli, Almora and Nainital received maximum number of both foreign and domestic tourist in comparison to the other districts of Uttarakhand.
- The study mentioned that the prime reasons for visiting the state of Uttarakhand is for religious purpose. However, yoga and meditation are found to a significant factor which influence the tourists to visit the state. Hence effort should be made to promote and develop some good yoga and meditation centres in different districts of Uttarakhand.

- Here is no doubt that good roads and other infrastructure facilities are very important especially in a hilly state like Uttarakhand. In this regard, the government efforts are found to be satisfactory. However, to attract more repeat visitors as well as attracting senior citizens including the foreign tourists the government should provide better infrastructure facilities so that tourist should visit the state year around.
- Better tourist traffic management, wayside amenities for the tourists including safety and security are found to the most important influencing factors especially for the religious tourists in the state.
- Religious tourists visiting the state of Uttarakhand are now more interested to avail budget accommodations including homestay facilities at different religious sites.
- In case of foreign tourists' better train and flight connection are found to be more important influential factors to visit the state.
- Lack of co-ordination among different government agencies are found to missing which causes inconvenient to both the foreign and domestic tourists.
- Proper understanding of benefits and costs of tourism in terms of economic, socio-cultural and environmental impacts are found to be missing among the local people as well as local communities.
- Especially in rainy season poor maintenance of roads made accessibility a key constraint for improving quality as well as numbers of tourists to different religious sites of the state.
- It was found that every year during the peak season a large influx of tourist visited mainly the popular char Dham of Uttarakhand resulting in a severe traffic congestion as well as accommodation problem causing a threat to the local environment against a small economic gain mainly gained by the local people/ community mainly break the carrying capacity of the destination.
- It was observed that although a huge number of tourists especially religious tourists visited the state regularly but the medical facility available in the state is very poor. The local people / community including the tourist normally prefer to travel Delhi NCR to get proper treatment which is a major cause of concern for the state in terms of medical facility.

9. SUGGESTIONS AND RECOMMENDATIONS TO IMPROVE RELIGIOUS TOURISM IN UTTARAKHAND

- Being a hilly region, the religious tourist destinations are located in a scattered way inside the state of Uttarakhand. Absence of proper public system sometimes create a big problem for the tourists to reach the destination in time with proper facilities. So, the state government should provide with online public transport application so that the tourist can get conform reservation in bus and shared transport facility to have easy access to the religious destination within a stipulated time frame.
- Need for Improving connectivity to specific religious tourism circuits to promote religious tourism in the state was strongly felt by the tourist.
- Pricing being a sensitive issue so several factors should be taken prior to pricing of religious tourism products in the state.
- Introduce a documentary comprises of all the religious tourist places and advertises it on TV channels of different languages.

- Launch a mobile app which will show distance and time to reach the destination and all about how to opt for the tour. Packages that are available and what all amenities are provided in it.
- Marketing is significant factor now days. It should be advertised on the digital media to increase awareness about the options available for religious tourism in Uttarakhand. Information on accessible tourism destinations and products in Uttarakhand may be collected and published in English as well as major Indian languages for wider diffusion. This will definitely increase the revenue.
- To maintain the carrying capacity of the religious destinations there is a greater need for Proper visitor Management mainly during peak season. As a large number of tourists mainly visit the state and as a result it became very difficult to manage the all tourist in terms of Traffic control, accommodation and other basic amenities which are needed to properly address to maintain the carrying capacity of these religious destinations of the state of Uttarakhand.

Table 2. Tourist arrivals to chardham (in Uttarakhand)

Year	Tourist arrivals to char Dham yatra
2000	1292411
2001	809266
2002	790211
2003	1033424
2004	1040155
2005	1341304
2006	1662935
2007	1942785
2008	2199608
2009	1904239
2010	1942232
2011	2440855
2012	2468838
2013	1273553 (Floods in Kedarnath)
2014	290509 (After effects of floods)
2015	804008
2016	1404707
2017	23,24,158
2018	27,69,508
2019	34,00,000
2020	4,48,000 (Effects of Corona)
2021	5,00,000 (Effects of Corona)
2022	33.46,000
2023	20,000,00 till April 2023 (provisional)

Source: compiled data by authors from https://www.sacredyatra.com

10. TOURIST FLOW TO CHAR DHAM WITH EFFECT FROM 2000 TO 2023

The above table depicts the tourist follow to Char Dham, which is the most important religious circuits of the state Uttarakhand. If we observe the tourist flow from 2000 to 2023, we can say that this circuit is the lifeline for generating economic benefits to the state which is growing at a steady rate except the year 2013-14 where the furious floods created devastation in Kedar Nath and Badrinath. Again, the tourist flow was declined in the year 2020-2022 mainly due the Covid-19 pandemic in our country. So, seeing the trend we can predict that this circuit will act a game changer for promoting religious tourism in the state of Uttarakhand.

11. CONCLUSION

Religious tourism has a big future in Uttarakhand. Uttarakhand is richly gifted with ancient temples and religious festivals. From the above discussion regarding Assessment of Factors Responsible for promoting Religious Tourism in Himalayan region with Special Reference to Dev Bhoomi, Uttarakhand we may concluded that there is an immense tourism potential for promoting religious tourism in the state as the place is receiving a huge number of tourists every year. The measure reasons may be attributed to several factors including cultural and religious sites, natural and scenic landscape, soothing climate, historical and archeological sites, yoga and meditations, cuisines and culinary are some of the major pull factors. The state no doubt attracts both domestic and international tourists in large numbers, however, they are confined to some limited destinations inside the state only. Keeping in mind of these factors, the government should effectively promote and market these untapped tourist destinations of Uttarakhand so that economic development can be maximized through tourism business. Apart from that preparation of master plan is urgently needed including encouragement given to different stakeholders especially the travel agencies and tour operators to participate and international and national events to promote and market in a bigger way to achieve economic sustainability. A good infrastructure facility should be provided to each and every destination year around as well as educating the local people/ communities regarding the benefits and cost of tourism to achieve environmental and economic sustainability including profit making should be the need of the hour for promoting religious tourism in the state of Uttarakhand.

REFERENCES

Dayananda, K. C., & Leelavathi, D. S. (2016). Tourism Development, Economic and Employment Growth in India. *IOSR Journal of Humanities and Social Science (IOSR-JHSS), 21*, 35-40.

Duran-Sanchen. (2018). *Religions*, *9*(9), 249. doi:10.3390/rel9090249

Gaurav, R. & Kumar, S. (2016). *Prospects and Problems of Tourism Industry in Uttarakhand* (Vol. 4, pp. 277–284). Prashant Gautam, Sandeep Walia.

Joshi, Pursotam & Saxena, S. (2019) Religious tourism in Uttarakhand. *JETIR1907C81 Journal of Emerging Technologies and Innovative Research (JETIR).* www.jetir.org

Joveriya & Mariya. (2019). Problems and prospects of tourism industry in Uttarakhand. International Journal of Geography. *Geology and Environment*, *1*(1), 10–16.

Joveriya & Mariya. (2019) Problems and prospects of tourism industry in Uttarakhand. *International Journal of Geography, Geology and Environment, 1*(1), 10-16, 2706-7483

Khalid, Shahid, M., Natasha, Bibi, I., Sarwar, T., Shah, A., & Niazi, N. (2018). 2018, A Review of Environmental Contamination and Health Risk Assessment of Wastewater Use for Crop Irrigation with a Focus on Low and High-Income Countries. *International Journal of Environmental Research and Public Health, 15*(5), 895. doi:10.3390/ijerph15050895 PMID:29724015

Rongali, A. (2017). *Issue 11* [Year - 2017]. Vol. 8). Visiting Purposes of Tourists in Uttarakhand: A Comparative Study of Kumaun & Garhwal Regions, CASIRJ. ISSN.

Sajwan, R. (2019), Rural tourism in Uttarakhand, Importance, Scope and Challenges. *JETIR, 6*(3), 2349-5162.

Sharma, N., & Kalotra, A. (2016). Hospitality Industry in India: A Big Contributor to India's Growth. *International Journal of Emerging Research in Management &Technology*, *5*(6), 202–210.

Stoddard & Morinis. (1997). *The Geographic Contribution to Studies of Pilgrimage: Introduction to Sacred Places, Sacred Spaces*. University of Nebraska – Lincoln.

Sufi, T. (2015). Indian Hotel Industry: Past, Present and Future. Journal of Hotel management. ResearchGate.

Sujith, T. S., & Jisha, M. K. (2017). Tourism in India: Opportunities & Challenges . *Imperical Journal of Interdisciplinary Research, 3*, 201–2033.

The Travel & Tourism Competitiveness Report. (2015). World Economic Forum.

Chapter 4
Challenges and Recommendations for Fostering Pilgrimage Tourism in the Himalayan Region of India

Gaurav Bathla
CT University, India

Ashish Raina
https://orcid.org/0000-0001-5812-5920
CT University, India

Varinder Singh Rana
https://orcid.org/0000-0001-6158-9195
City University, Ajman, UAE

Amit Kumar
https://orcid.org/0000-0002-6915-5495
Central University of Haryana, India

ABSTRACT

This chapter delves into the realm of pilgrimage tourism in Himalayan region of India, aiming to shed light on the multifaceted aspects of this religious and cultural phenomenon. Through a comprehensive investigation, the study strives to achieve three primary research objectives. Firstly, it seeks to discern the demographic profiles and travel patterns of pilgrims traversing the sacred landscapes of India. Secondly, it endeavours to uncover the myriad challenges and hardships encountered by tourists embarking on their spiritual quests, encompassing issues ranging from logistical hindrances to cultural disparities. Lastly, drawing upon the findings from the preceding objectives, the chapter aspires to present a set of well-informed recommendations geared towards the enhancement of the pilgrimage tourism experience in Himalayan regions of India. This chapter presents valuable insights to propel the evolution and enrichment of pilgrimage tourism in culturally diverse and spiritually vibrant nations.

DOI: 10.4018/979-8-3693-0823-3.ch004

1. INTRODUCTION

Tourism is defined as the movement of people from one location to another, and it is an economic, social, and cultural phenomenon. It is a transient movement of individuals to another location or destination. Tourism can be classified into four types based on its purpose: particular interest, pleasure, business, and education. Outbound, inbound, and domestic tourism are the three categories of tourism. According to Page and Hall (2014), the genesis of Indian tourism policy began with an examination of the global advancement of tourism. It has enormous potential for creating needed job opportunities, national integration, and foreign exchange revenues. In terms of the many types of tourism in India, pilgrimage tourism has grown significantly in recent years. People from all around the world travel India to worship with their specific faiths. The practice of traveling for the sake of one's faith dates back to prehistoric times. Major and smaller pilgrimage centres for Hinduism, Buddhism, Jainism, and Sikhism are all spread out over the country. People also frequently visit churches, mosques, temples and other religious buildings, including Sufi centres. In fact, religious pilgrimage has traditionally been the primary inducement for the vast majority of India's domestic travellers. All of our country's most important religious buildings are clustered on the hills or along the banks of major rivers. Income for both governments and private businesses can be directly attributed to the tourism industry. Both macro and micro goals can be traced back to the Ministry of Tourism. The national gross domestic product (GDP), foreign exchange, and employment are all macro targets of indicators. Tourists can experience underwater nature tourism, beach tourism, nature tourism, religious tourism, and culinary tourism, just to name a few. There has been a rise in the number of people seeking out "spiritual" or "religious" vacations in recent years. According to Gladstone (2013), pilgrimage tourism is a trip to a temple, shrine, or other destinations significant to the tourists' faiths and beliefs. It is also related to religious tourism. During the 2017 fiscal year, the government of India earned approximately 51,587 INR in Foreign Exchange Earnings due to pilgrimage tourism (Reader, 2013). Additionally, the income of the tourists and hotel industry increased by 5.1% in 2017 (Jafari and Scott, 2014). Religious tourism, as defined by Shinde (2018), consists of trips to holy sites that serve both spiritual and touristic purposes. While on the subject of expanding one's spiritually, Haq and Jackson (2009) define a religious tourist as someone who travels to a new and different setting in order to do just that. Pilgrimages serve as complex spiritual trip phenomena with an external place that endures and transcends (Singh, 2006). According to Wilson et al. (2013), spiritual tourism consists of vacation packages that combine opportunities for religious observance and pilgrimage with leisure and social pursuits.

2. THE HIMALAYAN REGIONS IN INDIA

The Himalayas in India are nestled amidst the awe-inspiring grandeur of the world's tallest mountain range lies the Himalayan region, a geographical marvel that has captivated the imagination of travellers, adventurers, and spiritual seekers for centuries. Stretching across the northern borders of India, this majestic expanse is not only a sanctuary for pristine natural beauty but also a sacred haven for Hindu pilgrimages of profound significance. The Himalayas, often referred to as the "Abode of the Gods," hold a special place in the hearts of million people, as the region is intricately woven into the tapestry of India's spiritual heritage. In this chapter, we will embark on a journey to explore the mystical Himalayan region and delve into the profound Hindu pilgrimages that draw devotees and seekers alike to its rugged and spiritual landscapes.

3. HIMACHAL PRADESH AND HIMALAYAS

The name of the state of Himachal Pradesh comes from the Himalayas. Himachal Pradesh is mountainous land between 30° 22' and 33° 12' north latitude and 75° 47' and 79° 4' east longitude. Kashmir, Punjab, Haryana, Uttar Pradesh, and Tibet border it. Except for a few sub-mountainous regions bordering Punjab and Haryana, the state is mountainous. Different locations are 350–7000 meters above sea level. Geophysical factors vary greatly across the state, affecting climate and rainfall. Physiographical, the state is Himalayan. The Shivaliks (outer Himalayas), Inner Himalayas (mid-mountains), and Alpine zone (larger Himalayas) can be topographically split from south to north.

Himachal Pradesh has changed in size and governance as well as culturally. It began as a Chief Commissioners Province in 1948 and became a state of the Indian Union in 1971. Four districts—Chamba, Mahasu, Mandi, and Sirmaur—formed it initially. This region underwent the harshest feudal exploitation under princes. The monarchs did not see the need to develop their lands by using their natural riches (Singh, 1985). Himachal has developed well since its founding. The state's size and administrative structure between 1948 and 1971 affected its development. Himachal's feudal economy and institutional framework hindered socio-economic growth. Thus, the state started development behind the rest of the country.

Himachal Pradesh's tourism, agricultural, textile, and hydroelectric power export industries raise capital. Himachal Pradesh, in north India, signifies snowy mountains. Dev Bhumi—Gods' own home—is another name for it. Himachal Pradesh has a high per-capita income. Himachal Pradesh has many Hindu pilgrimage sites. Himachal Pradesh has 2,000+ temples. These temples are magnificent to draw tourists from all parts of the world and many of these have ancient traditions attached to the places.

4. PILGRIMAGE IN INDIA

Even though religion has become a politically charged word in recent years in India, it is the glue that holds the entire country together, from Kashmir to Kanyakumari. This article does not cover the complexities of religion and its function in India; however, it does discuss the economic face of religion and how it can be used to eliminate economic imbalance across the country. India is a country that has a startling number of diverse religions, including Hinduism, Islam, Christianity, Buddhism, Sikhism, Jainism, and Persian, to name a few. Despite this, the people who embrace these various religions coexist peacefully. In addition, adherents of many religions travel to the holy sites of those of other religions, which is something that is both remarkable and unexpected. Even though India has a diverse population that respects one another's differences, the culture is quite complicated. For researchers interested in the field of behavioural science, it is always a wonderful place to work. People in India follow a variety of religions, customs, and traditions, and have vastly different ways of life, tastes, and preferences. Due to the fact that this diversity presents a constant challenge for marketers, many multinational corporations send their global managers to India for training because the country serves as a potential training ground. Marketers also focus on encasing the religious belief of the population by targeting religious tourism locations to market their locations. Locations such as Kumbamela at Varanasi, Allahabad, Rajmundry, and other locations like these are some of the hottest locations for the marketers to target their potential consumers. These destinations attract a large number of pilgrims from around the nation and bring together a diverse group of potential customers in one spot. As a result, they provide businesses the opportunity to reach a large number of people at a low cost per promotion. It is possible to bridge the economic gap between different regions of India's population by visiting one of the many holy sites that dot the country from north to south, east to west, and everywhere in between. The dissemination of money to the appropriate locations is contingent on the travels of domestic tourists to a variety of destinations. For example, Tirupathi is a well-known holy site in the Indian state of Andhra Pradesh that attracts visitors from all over the country and the world. It is considered to be the most luxurious pilgrimage venue after the Vatican. Tirupathi is responsible for providing work possibilities for a large number of people and is home to a sizeable population. It does not contain any of the city of Tirupathi's core competencies, which can be defined as the capacity of the location to produce employment opportunities suitable for people possessing particular skill sets. However, thanks to the renowned Balaji Temple, it operates like a multinational corporation and generates billions of dollars in revenue each year.

5. LITERATURE REVIEW

According to one definition, a pilgrimage is "a journey deriving from religious motives, externally and internally for spiritual aims and interior enlightenment." This is a frequent religious and cultural phenomenon in human civilizations and an important aspect of the world's major faiths (Barber, 1993).

A rise in the practice of pilgrimage is constantly being seen all across the world (Digance, 2003; 2006). A form of population mobility known as the pilgrimage is described here. The political, economic, social, and cultural repercussions of pilgrimages are significant, and they even have the potential to influence worldwide trade and health. The act of pilgrimage requires movement through space and is also an important topic because of the breadth and depth of its implications for that space. Pilgrimage not only promotes additional mobilities such as trade, cultural exchanges, political movements, and the less desired spread of sicknesses and epidemics, but for some nations and cities, it has been the principal economic activity (Barber, 1993; Vukoni, 2002).

In general, there are misunderstandings regarding the similarities that exist between pilgrimage and tourism from the points of view of religion, the pilgrims themselves, the tourism sector, and researchers (Olsen and Timothy 2006). In his book titled "Religion and Tourism," published in 2011, Stausberg presented a comprehensive analysis of the intersection of religion and tourism from the point of view of religious studies. The word "pilgrimage," which derives from the Latin word "peregrinus," can be interpreted in a variety of ways, including the journey of a foreigner, a wanderer, an exile, or a traveller, as well as that of a newcomer or stranger. This is despite the fact that the term "pilgrimage" typically refers to the religious journey of a pilgrim, most often to a holy shrine or holy places.

According to Nicolaides (2016), if there are gaps in marketing communication, insufficient infrastructure, and other areas, this could discourage the development of pilgrimage and religious tourism, which are both a massive industry and a niche market. As long as there is a need to fulfil and go beyond the expectations of visitors to holy sites, there will be a constant demand for change (Raj, 2007).

Religiously inspired tourism is a global phenomenon that dates back to the origins of religion and is observed across various religious faiths (Lanczkowski, 1982). Scholars have examined the correlation between religion and tourism from many angles (Collins-Kreiner, 2018; Terzidou et al. 2017).

In his work, Bremer (2005) identifies three distinct perspectives through which scholars examine the intersections of religion and tourism. These perspectives include the spatial approach, which focuses on the coexistence of pilgrims and tourists in shared physical spaces but with distinct behavioural patterns. The historical approach examines the historical relationship between religious forms of travel and tourism. Lastly, the cultural approach explores pilgrimage and tourism as contemporary practises within a post-modern society. In the study conducted by Millán-Vázquez de la Torre et al. (2016), the concept of religious tourism is examined from two perspectives. Firstly, it is viewed as tourism driven by religious motivations, either exclusively or partially, which aligns with the traditional understanding of this phenomenon as proposed by Rinschede (1992). Secondly, the researchers consider tourism as a modern-day spiritual quest, drawing on the perspective put forth by Sharpley (2009).

The phenomenon of religious tourism exhibits strong connections with various other forms of tourism, such as holiday, cultural, social, and group tourism. Consequently, it is also influenced by the concept of seasonality, as discussed by Collins-Kreiner (2018), Raj et al. (2015), Rinschede (1992), and Timothy and Boyd (2006). Throughout history, religious journeys have consistently served several purposes, even when religious motivations appeared to be the primary driving force. Nevertheless, it appears that religious motivation holds less significance in contemporary society compared to ancient ones (Rinschede, 1992).

The distinction between tourists and pilgrims, who are both participants in the religious tourism business, is a highly contested topic among scholars in this field. The concept of pilgrimage is frequently characterised as "a religiously motivated journey to a sacred location, undertaken by individuals seeking spiritual enlightenment and internal comprehension" (Barber, 1993, p. 1). This definition implies that a pilgrimage involves the travel of a devout follower of a particular religion to a revered religious site (Turner, 1973; Turner & Turner, 1978). According to Collins-Kreiner (2010b) and Morinis (1992), pilgrimage can be defined as the act of embarking on a journey to a specific location that represents significant value, profound meaning, or serves as a fundamental aspect of the traveler's identity.

According to Hyde and Harman (2011, p. 1343), in a contemporary society characterised by a growing secularisation, a significant number of individuals who do not identify with any religious affiliation engage in pilgrimages to locations that hold profound personal significance for them. Secular pilgrimages encompass various expeditions to the burial sites and memorials of renowned individuals, iconic sports arenas, or locations of political importance (Digance 2006 & Margry 2008). According to Hyde and Harman (2011), the concept of pilgrimage extends beyond its religious connotations, challenging the traditional paradigm that associates pilgrimage solely with religious aspects (Collins-Kreiner, 2010a, 2010b). Two other forms of pilgrimage must be acknowledged: the religious and the secular.

In addition, there is a significant demand for the consideration of pilgrimage tourism as it relates to the promotion of sustainable development. According to Cannell (2013), the significance of pilgrimage tourism in India is rather high due to the fact that the people of India have a strong cultural and religious identity. As a result, it is essential to both preserve the existing infrastructure and ensure the safety of the location.

6. PROBLEM STATEMENT AND RESEARCH GAP

India is home to a variety of religions, and as a result, it is home to a significant number of pilgrimage sites that are known all over the world. Religious tourism in India is a unique sub segment of the country's tourism industry and offers a diverse range of options within the component. India has the potential to develop into a location that is attractive from a business perspective throughout the course of time. There are not many studies done on the challenges and opportunities facing India's pilgrimage tourism industry (Vijayanand & Pujari 2012).

Very little research has been done in the context of a comprehensive approach to determining how pilgrims feel about religious tourism at various pilgrimage destinations on India. Also, the demographic and travel patterns involved in determining the level of satisfaction experienced by religious tourists, and analysing the challenges that religious tourists are facing at such religious destinations.

7. OBJECTIVES OF THE STUDY

The study aims:

1. To identify demographic profiles and travel patterns of pilgrims visiting the Himalayan regions of Himachal Pradesh.
2. To find any difficulties faced by tourists during pilgrimage tourism in the Himalayan regions of Himachal Pradesh.
3. To recommend ways to create better pilgrimage tourism experience in the Himalayan regions of Himachal Pradesh.

8. RESEARCH METHODOLOGY

After consulting adequate number of published literature, a structured questionnaire was framed to meet the research objectives. The framed questionnaire was validated by two senior academicians in the field of tourism research. The researcher selected five temples in the vicinity that are published to have visitors from around the nations in the previous literatures. These temples were selected from the various districts of Himachal Pradesh. The temples selected for the study were Jwalamukhi Temple at Kangra, HP, Chamunda Devi Temple at Dharamshala, HP, Manimahesh Temple at Bharmaur,Manimahesh, HP, Chintpurni Temple at Una, HP and the Naina Devi Temple at Bilaspur, HP. The researcher collected the data from the research areas with the help of interviews and questionnaires distributed in hard copies. A total of 40 questionnaires were distributed and collected with data personally by the researchers. Total 200 filled questionnaires were collected from the respondents (40 from each temple visitors/devotees). Convenient sampling techniques were used to collect the primary data by the researchers. The data was collected between the time period of March 2022 and August 2022. The data collected was analysed to meet the objectives of the current study.

9. ANALYSIS OF DATA

Table 1 represents the distribution of data in the different study areas. The researcher has collected data from total 200(N) numbers of visitors to the religious destinations in Himachal Pradesh. The sample

Figure 1. Expenditure of infrastructure in the tourism sector across India from financial year 2016 to 2021, with estimates until 2023
(Source-Statistica.com)

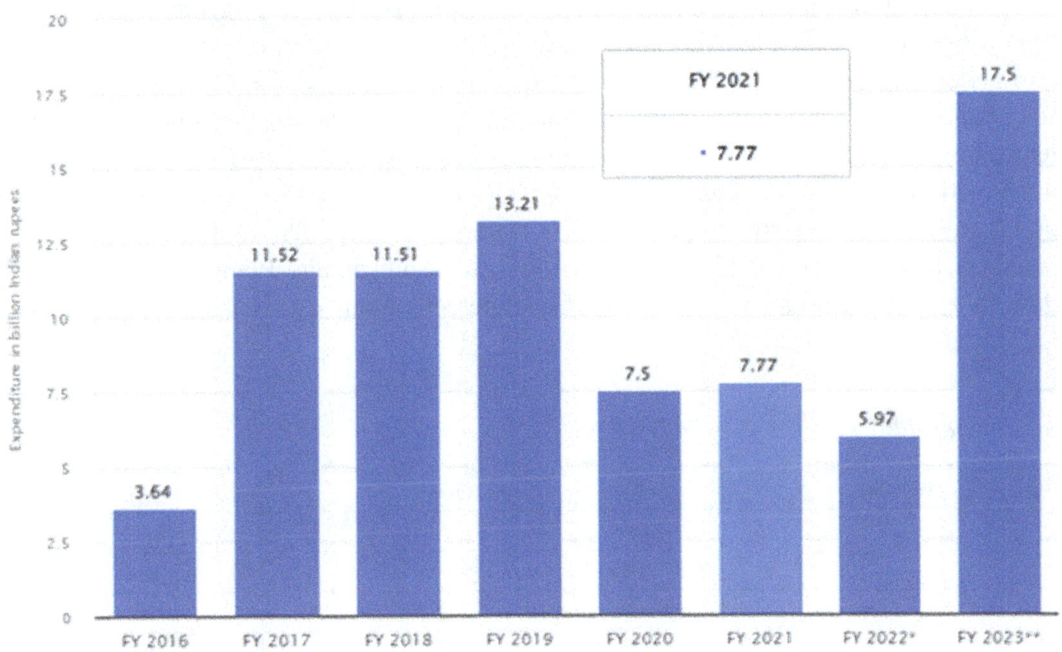

size was decided to be 40 responses from each study area, making it 20% of the total data from each selected location.

In the table 2 the demographic profiles of the responding populations are described. The data is resented to analyse various demographic aspects of the visitors to the selected temples in the Himachal Pradesh. These visitors collectively create pilgrimage tourism in Himachal Pradesh. The responding population was analysed to be more in terms of male respondents. From the total number of respondents (200), the data reported only 31% (62) females. Also, the maximum number of responding population was observed in the age group above 40 years and 13.5% (27) the responding population was below 20

Table 1. List of temples taken into research study

S. No.	Name of the Temple	Location	n	% of total sample
1	Jwalamukhi Temple	Kangra, HP	40	20
2	Chamunda Devi Temple	Dharamshala, HP	40	20
3	Manimahesh Temple	Bharmaur,Manimahesh, HP	40	20
4	Chintpurni Temple	Una,HP	40	20
5	Naina Devi Temple	Bilaspur, HP	40	20
Total (N)			200	100

years of age. It is also observed that maximum number of visiting population was bearing graduation as their educational qualification and 59.5% of the respondents were married. Seeking information on the occupation of the responding population from the study areas were working in the public sector (41%) followed by respondents working in the private sector (35.5%).

Table 3 outlining the itinerary details of pilgrims visiting Himalayan pilgrimage sites presents several noteworthy findings. First and foremost, it highlights that a significant portion of pilgrims, 46.5%, choose to undertake this spiritual journey with their families, underscoring the profound significance of familial bonds in the context of pilgrimage. Additionally, 34.5% of pilgrims opt for a blend of family and friends, revealing a desire for both personal and communal spiritual experiences. While 15.5% travel exclusively with friends, 3.5% embark on this sacred journey as solo travelers, reflecting the diversity of personal preferences. In terms of group size, a majority, 62%, prefer smaller groups with less than three members, fostering a more intimate connection with the pilgrimage. Planning-wise, 39.5% dedicate three days to a week for preparation, while an equal percentage invest weeks to months, highlighting

Table 2. Demographic profiles of respondents

Variable	Features	n	%
Gender	Male	138	69
	Female	62	31
Age	Below 20 years	27	13.5
	21 - 30 years	53	26.5
	31 - 40 years	41	20.5
	Above 40 years	79	39.5
Educational qualifications	No education	5	2.5
	High Schooling	34	17
	Graduate	116	58
	Post Graduate and above	45	22.5
Marital status	Married	119	59.5
	Single	81	40.5
Occupations	Private Sector	71	35.5
	Public Sector	82	41
	Self-employed	32	16
	House lady/House men	13	6.5
	Others	2	1
Annual household income	Less than Rs.100000	32	16
	Between 100001 - 500000 INR	129	64.5
	Between 500000 - 1000000 INR	31	15.5
	Above Rs.1000000	8	4
Residence	From Himachal Pradesh only	71	35.5
	From another state in India	106	53
	NRI's/Foreigner	23	11.5

the varying degrees of commitment and readiness among pilgrims. Moreover, a significant 77.5% use online booking platforms, showcasing the convenience of digital resources. Transportation-wise, 44% choose their own vehicles, and accommodation preferences vary, with 45% favouring simpler options like Dharamshalas, Cottages, or Dormitories, while 29.5% opt for the luxury of star-categorized hotels, and 25.5% prefer cost-effective non-categorized hotels. These findings collectively offer valuable insights into the diverse choices and priorities of pilgrims on their transformative journeys to the Himalayas.

The table 4 measuring the satisfaction level of pilgrims visiting Himalayan pilgrimage sites reveals several key findings. Firstly, when assessing the condition of roads, a significant 78.5% of pilgrims expressed satisfaction levels ranging from "Satisfied" to "Highly Satisfied," indicating generally favorable road conditions on their journey. Similarly, in evaluating traffic conditions, 49.5% of respondents reported being "Satisfied" or "Highly Satisfied," suggesting manageable traffic flow. However, when it comes to signals and signboards, nearly 42.5% of pilgrims expressed dissatisfaction to some degree, indicating room for improvement in this aspect of infrastructure. Interestingly, for tolls and taxes, the majority of pilgrims (89.5%) were either "Neutral" or "Satisfied," highlighting a reasonable level of acceptance. Administrative control received mixed feedback, with 66.5% expressing varying degrees of satisfaction. When assessing value for money on travel, a majority (68.5%) considered their expenditure satisfactory. Lastly, regarding overall satisfaction, 83.5% of pilgrims reported being either "Neutral" or "Satisfied," with only 8% expressing dissatisfaction, suggesting that the majority of pilgrims left their Himalayan pilgrimage with a sense of contentment. The mean satisfaction scores for each category

Table 3. Itinerary details

Construct	Category	n	Percentage
Travel type	With Family	93	46.5
	With Friends	31	15.5
	With Family and Friends both	69	34.5
	Solo Traveller	7	3.5
Group Size	less than 3 members	124	62
	3 to 6 members	67	33.5
	more than 6 members	9	4.5
Tour planning time	Within three days	43	21.5
	Three days to a week	79	39.5
	Weeks to Months	78	39
Mode of booking	Online	155	77.5
	(Offline)Travel Agent	45	22.5
Mode of Travel	By Indian Railways	19	9.5
	By bus	68	34
	By Taxi	25	12.5
	By own transport	88	44
Type of stay	Star categorized hotels	59	29.5
	Non Categorized hotels	51	25.5
	Dharamshalas/Cottages/Dormitories	90	45

underline this trend, with "Condition of Roads," "Toll and Taxes," and "Value for Money on Travel" receiving relatively higher satisfaction scores compared to other aspects, showcasing the overall positive sentiment among the pilgrims.

10. CONCLUSION

In conclusion, this study has provided a comprehensive analysis of various aspects related to pilgrims visiting religious destinations in Himachal Pradesh, with a particular focus on Himalayan pilgrimage sites. The findings have shed light on the distribution of data across different study areas, the demographic profiles of the responding population, their travel itineraries, and their satisfaction levels during the pilgrimage.

One significant observation is the dominance of male respondents in the study, comprising a majority of the responding population. This indicates a potential gender disparity among pilgrims visiting these sites, which warrants further exploration and consideration in the context of pilgrimage tourism development.

Furthermore, the study highlighted that a substantial portion of pilgrims prefer to undertake their spiritual journey with their families, emphasizing the importance of familial bonds in pilgrimage experiences. The diversity in group size preferences and planning durations showcases the individuality and varying levels of commitment among pilgrims.

The study also provides valuable insights into the preferences of pilgrims in terms of transportation and accommodation choices. The prevalence of online booking platforms for planning and the popularity of self-owned vehicles for transportation underscore the influence of technology and personal convenience in modern pilgrimage practices.

In terms of satisfaction levels, it is encouraging to note that a significant majority of pilgrims reported overall satisfaction with their pilgrimage experience. The positive feedback on road conditions, traffic, tolls, and taxes, as well as value for money on travel, suggests that infrastructure and services have been generally well-received by the pilgrims.

Table 4. Satisfaction level of pilgrims

	Highly Dissatisfied		Dissatisfied		Neutral		Satisfied		Highly Satisfied	
	n	%	n	%	n	%	n	%	n	%
Condition of Roads	36	18	2	1	62	31	95	47.5	5	2.5
Condition of traffic	54	27	46	23	36	18	25	12.5	39	19.5
Signals and Signboards	12	6	61	30.5	24	12	58	29	45	22.5
Toll and Taxes	6	3	12	6	116	58	63	31.5	3	1.5
Administrative Control	19	9.5	47	23.5	41	20.5	84	42	9	4.5
Value for money on Travel	7	3.5	112	56	44	22	25	12.5	12	6
Overall satisfaction	3	1.5	9	4.5	21	10.5	154	77	13	6.5
Mean	19.6		41.2		46.28		72		18	

However, there are areas for improvement, particularly in the provision of signals and signboards, as a notable percentage of respondents expressed dissatisfaction in this regard. Additionally, administrative control received mixed feedback, indicating the need for further examination and potential enhancements in this aspect.

In summary, the findings of this study offer valuable insights for stakeholders involved in pilgrimage tourism in Himachal Pradesh. The study's comprehensive analysis of demographic profiles, travel itineraries, and satisfaction levels provides a foundation for informed decision-making and the enhancement of the pilgrimage experience in the region. As pilgrimage tourism continues to play a crucial role in the cultural and economic landscape of Himachal Pradesh, these insights can guide efforts to ensure that pilgrims have a fulfilling and satisfying journey to the Himalayan pilgrimage sites.

11. LIMITATIONS OF THE STUDY

1. The research is restricted to specific five Hindu holy temples in Himachal Pradesh.
2. The research is limited to domestic Hindu pilgrims alone.
3. Due to limited data, the survey instrument's sample faced a number of difficulties. A one-time measurement for data collection, a limited questionnaire, and the timing of the survey all contributed to this constraint.
4. However, it is hoped that these limits would suggest and inspire the development of new avenues and standards for future research.

REFERENCES

Barber, R. (1993). *Pilgrimages*. The Boydell Press.

Collins-Kreiner, N. (2018). Pilgrimage-Tourism: Common Themes in Different Religions. *International Journal of Religious Tourism and Pilgrimage*, *6*, 3.

de la Torre, M.-V., Genoveva, L. P.-N., & Martínez-Cárdenas, R. (2016). Factores que Determinan el Crecimiento del Turismo en Destinos Religiosos. [RCS]. *Revista de Ciencias Sociales*, *22*, 85–97.

Digance, J. (2006). Religious and Secular Pilgrimage: Journeys Redolent with Meaning. In *Tourism, Religion and Spiritual Journeys* (pp. 52–64). Routledge.

Gladstone, D. L. (2013). *From pilgrimage to package tour: Travel and tourism in the third world*. Routledge. doi:10.4324/9780203689363

Haq, F., & Jackson, J. (2009). Spiritual journey to hajj: Australian and Pakistani experience and expectations. *Journal of Management, Spirituality & Religion*, *6*(2), 141–156. doi:10.1080/14766080902815155

Hyde, K. F., & Harman, S. (2011). Motives for a Secular Pilgrimage to the Gallipoli Battlefields. *Tourism Management*, *32*(6), 1343–1351. doi:10.1016/j.tourman.2011.01.008

Jafari, J., & Scott, N. (2014). Muslim world and its tourisms. *Annals of Tourism Research, 44*, 1–19. doi:10.1016/j.annals.2013.08.011

Lanczkowski, G. (1982). *Die heilige Reise-Auf den Wegen von Gottern und Menschen*. Herder Verlag.

Lois-González, R., & Santos, X. (2015). Tourists and Pilgrims on their Way to Santiago. Motives, Caminos and Final Destinations. *Journal of Tourism and Cultural Change, 13*(2), 149–164. doi:10.1080/14766825.2014.918985

Margry, P. (2008). *Shrines and Pilgrimage in the Modern World: New Itineraries into the Sacred*. University of Amsterdam Press. doi:10.5117/9789089640116

Page, S. J., & Hall, C. M. (2014). *The geography of tourism and recreation: Environment, place and space*. Routledge.

Raj, R., Griffin, K., & Blackwell, R. (2015). Motivations for Religious Tourism, Pilgrimage, Festivals and Events. In *Religious Tourism and Pilgrimage Management: An International Perspective* (pp. 103–117). Cabi. doi:10.1079/9781780645230.0103

Rinschede, G. (1992). Forms of Religious Tourism. *Annals of Tourism Research, 19*(1), 51–67. doi:10.1016/0160-7383(92)90106-Y

Sharpley, R. (2009). Tourism, Religion and Spirituality. In T. Jamal & M. Robinson (Eds.), *The SAGE Handbook of Tourism Studies* (pp. 237–253). SAGE Publications Ltd. doi:10.4135/9780857021076.n14

Shinde, K. (2018). Governance and Management of Religious Tourism in India. *International Journal of Religious Tourism and Pilgrimage, 6*(1), 7.

Singh, R. P. B. (2006). Pilgrimage in Hinduism: historical context and modern perspectives. In D. J. Timothy & D. H. Olsen (Eds.), *Tourism and Religious Journeys* (pp. 220–236). Routledge.

Terzidou, M., Scarles, C., & Saunders, M. (2017). Religiousness as Tourist Performances: A Case Study of Greek Orthodox Pilgrimage. *Annals of Tourism Research, 66*, 116–129. doi:10.1016/j.annals.2017.06.009

Turner, V. (1973). The Center out There: Pilgrim's Goal. *History of Religions, 12*(3), 191–230. doi:10.1086/462677

Turner, V., & Turner, E. (1978). *Image and Pilgrimage in Christian Culture: Anthropological Perspective*. Basil Blackwell.

Wilson, J. A. J., & Grant, J. (2013). Islamic marketing – a challenger to the classical marketing canon? *Journal of Islamic Marketing, 4*(1), 7–21. doi:10.1108/17590831311306327

Chapter 5
Ecological Impacts of the Himalayan Region in West Bengal

Achintya Barman
Pailan College of Management and Technology, India

Monami Barman
Bhaban's Netaji Subhash Chandra Bose Vidyaniketan, Haldia, India

Asik Rahaman Jamader
https://orcid.org/0000-0002-6938-5901
Pailan College of Management and Technology, India

ABSTRACT

The establishment of protected areas, the promotion of responsible tourist behavior, and the adoption of sustainable tourism practices: the effectiveness of these measures will be assessed based on their contribution to biodiversity conservation, local community empowerment, and overall environmental sustainability. The chapter will provide recommendations for the future management of mountain tourism in the Himalayan region and beyond. These recommendations will include the need for stronger regulations and enforcement mechanisms to ensure that tourism activities do not harm delicate ecosystems. Additionally, it will emphasize the importance of community involvement and education in promoting sustainable tourism practices and raising awareness about the value of biodiversity. In conclusion, mountain tourism in the Himalayan region and beyond has the potential to promote economic growth and cultural exchange. However, it is crucial to acknowledge and address the ecological impacts associated with this form of tourism.

1. INTRODUCTION

The growth of mountain tourism in the Himalayan region and beyond has captured the imagination of travelers seeking unique and breathtaking experiences in nature (Gansser, 1974). The allure of snow-

DOI: 10.4018/979-8-3693-0823-3.ch005

capped peaks, rugged landscapes, and vibrant cultures has led to a surge in tourist arrivals, bringing both opportunities and challenges.While mountain tourism presents economic benefits and opportunities for cultural exchange, it also raises concerns about its ecological impacts and sustainability. As visitor numbers increase, so does the pressure on delicate ecosystems and biodiversity (Singh et al., 1987). Habitat destruction, pollution, and disturbances to wildlife populations have become pressing issues that necessitate further examination and action.This introduction sets the stage for an exploration of mountain tourism and its ecological impacts in the Himalayan region and beyond. It highlights the need to understand and address the environmental consequences of this form of tourism in order to promote a more sustainable approach(Mitra et al., 2020).

By conducting a comprehensive review of existing literature and analyzing data, this study aims to shed light on the key ecological impacts of mountain tourism. The emphasis will be on understanding the specific challenges faced by the Himalayan region and exploring broader insights that can be applied to similar mountainous areas worldwide. Furthermore, this study will investigate the initiatives and policies that have been implemented to mitigate the ecological impacts of mountain tourism and promote sustainability (Shekhar et al., 2010). It will evaluate the effectiveness of measures such as the establishment of protected areas, the promotion of responsible tourist behavior, and the adoption of sustainable tourism practices.

Ultimately, this research seeks to provide recommendations for future management of mountain tourism in the Himalayan region and beyond ***(Shroder et al., 1998).*** These recommendations will draw upon the findings of the study and emphasize the importance of stronger regulations, community involvement, and education to ensure the preservation of mountain ecosystems and their rich biodiversity.In conclusion, understanding and addressing the ecological impacts of mountain tourism is crucial for the long-term sustainability of this industry. By examining the specific case of the Himalayan region and expanding insights to other mountainous areas, this study aims to contribute to the development of more environmentally responsible and sustainable practices in mountain tourism(Bhutia, 2014).

2. ECOLOGICAL IMPACTS OF HIMALAYAN REGION IN WEST BENGAL

Mountain tourism, while offering breathtaking experiences and a connection with nature, also imposes significant ecological impacts on the fragile mountain ecosystems. The following are some key ecological impacts associated with mountain tourism and explained the proper decision support system which is mention in figure 1:

- ***Habitat Destruction***: The construction of infrastructure, such as hotels, roads, and ski resorts, often leads to the destruction and fragmentation of natural habitat. This loss of habitat can result in the displacement and disruption of wildlife populations, including endangered species.
- ***Pollution:*** Increased tourist activities in mountain regions contribute to various forms of pollution. Air pollution from transportation and energy use, water pollution from sewage and waste disposal, and soil pollution from improper waste management practices all have detrimental effects on the environment and its ecological balance.
- ***Erosion and Land Degradation***: Human activities associated with mountain tourism, such as deforestation, land clearing, and excessive hiking or skiing, can cause erosion and land degradation.

Figure 1. Ecological Impacts and decision support system
Source: https://www.mdpi.com/2225-1154/9/8/121#

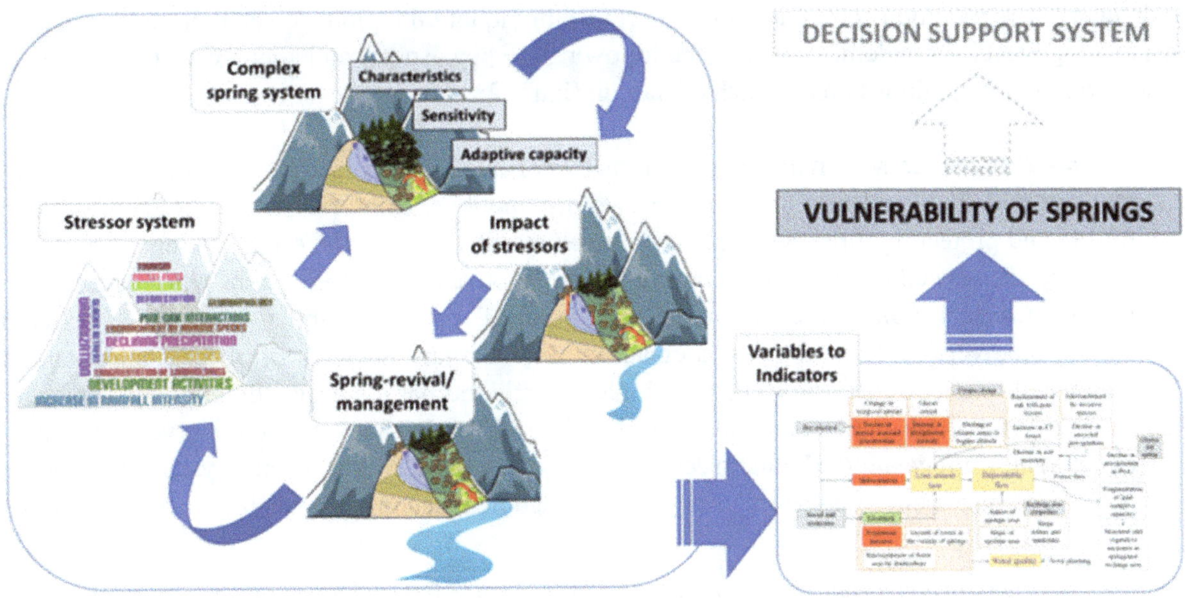

This can result in destabilized slopes, decreased soil fertility, and alterations in natural drainage patterns.

- ***Disturbance to Wildlife:*** The influx of tourists can lead to noise pollution, disturbance of sensitive breeding or resting areas, and increased interactions between humans and wildlife. These disturbances can disrupt natural behaviors and stress wildlife populations, potentially leading to long-term negative effects.
- ***Fragmentation of Habitats:*** The expansion of infrastructure and the establishment of tourist facilities can fragment natural habitats, isolating wildlife populations and disrupting migration patterns. Fragmentation can reduce genetic diversity, limit access to essential resources, and increase the vulnerability of species to threats.
- ***Overuse of Natural Resources:*** The demand for resources associated with mountain tourism, such as water, energy, and food, can exceed the capacity of local ecosystems. Overuse of natural resources can deplete water sources, strain local ecosystems, and lead to conflicts with local communities that depend on these resources.

It is important to note that these impacts can vary depending on the specific characteristics of each mountain region and the management practices in place. However, addressing these ecological impacts is crucial for preserving the integrity and long-term sustainability of mountain ecosystems and their biodiversity(***Rawat et al., 2028).***

3. BIODIVERSITY CONSERVATION IN HIMALAYAN REGIONS

Biodiversity conservation is of paramount importance in mountain regions, given their unique ecosystems and high levels of endemic species. The following are key aspects and approaches to biodiversity conservation in mountainous areas which explain in figure 2:

- ***Species Protection:*** Mountain regions often harbor a wide range of plant and animal species, many of which are endemic or have restricted distributions. Conservation efforts focus on identifying and protecting threatened and endangered species through the establishment of protected areas, species-specific conservation programs, and captive breeding programs when necessary.
- ***Habitat Conservation***: Preservation and restoration of mountain habitats are critical for maintaining biodiversity. This includes safeguarding key habitats, such as forests, wetlands, and alpine meadows, as well as protecting ecological corridors that connect different habitat patches. Restoration projects may be undertaken to rehabilitate degraded habitats and improve connectivity.
- ***Ecosystem Management***: Mountain ecosystems are complex, interconnected systems that require holistic management approaches. Adaptive management techniques, such as ecosystem-based

Figure 2. Biodiversity conservation is of paramount importance in mountain regions
Sources: https://www.mdpi.com/2071-1050/14/6/3342

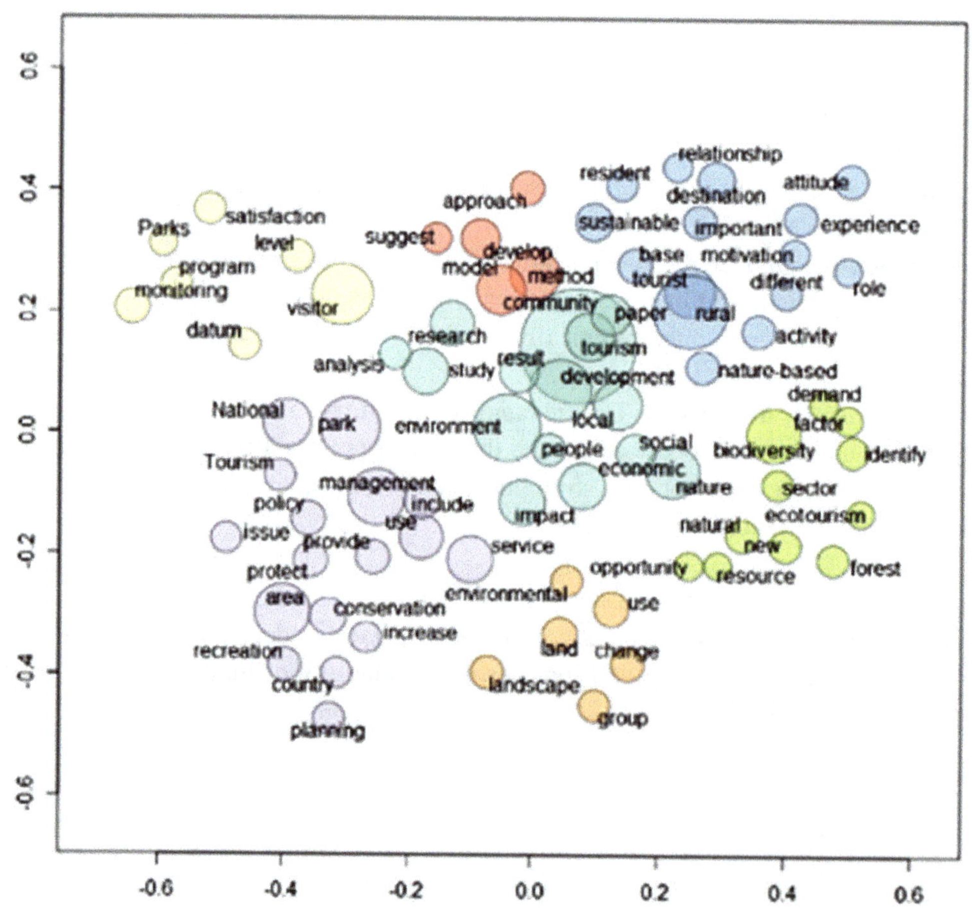

management and landscape conservation, aim to ensure the long-term health and functioning of these ecosystems while considering the needs of both wildlife and local communities.

- ***Community Involvement***: Engaging local communities in conservation efforts is essential for sustainable mountain biodiversity conservation. Collaborative management models that involve indigenous and local communities in decision-making processes, traditional ecological knowledge, and nature-based livelihoods can help foster a sense of shared responsibility and promote effective conservation outcomes.
- ***Sustainable Tourism Practices***: Promoting sustainable tourism practices can contribute to biodiversity conservation in mountain regions. This includes minimizing the ecological footprint of tourist activities, promoting responsible behavior among visitors through education and awareness campaigns, and supporting community-based tourism initiatives that encourage the protection of local biodiversity.
- ***International Cooperation:*** Mountain ecosystems often span multiple countries, requiring transboundary cooperation for effective conservation. Collaborative initiatives, such as biodiversity corridors, transnational protected areas, and joint research and monitoring programs, facilitate information sharing, collaboration, and coordinated conservation efforts among countries.
- ***Climate Change Adaptation***: Mountains are particularly vulnerable to climate change, which can have profound impacts on biodiversity. Conservation efforts need to incorporate climate change adaptation strategies, such as identifying and protecting climate refugia, facilitating species migration, and promoting climate-resilient ecosystem management practices.

By implementing these approaches and collaborative efforts, it is possible to conserve mountain biodiversity and ensure the long-term survival of unique species and ecosystems. These conservation efforts not only have ecological significance but also contribute to the sustainability and resilience of mountain communities and their livelihoods***(Dey et al., 2017).***

4. SUSTAINABLE TOURISM PRACTICES

Sustainable tourism practices are essential for mitigating the negative environmental and social impacts of tourism and ensuring the long-term sustainability of destination areas. In the context of mountain tourism, the following practices are crucial for promoting sustainability which is explain in figure 3:

- ***Conservation of Natural Resources:*** Sustainable tourism practices in mountain regions include minimizing the consumption of water, energy, and other natural resources. This can be achieved by promoting energy-efficient accommodation, water-saving measures, waste reduction, and recycling programs.
- ***Responsible Waste Management:*** Proper waste management is crucial to prevent pollution and ensure the integrity of mountain ecosystems. Effective waste management involves minimizing waste generation, promoting recycling and composting, and ensuring proper disposal of hazardous materials.
- ***Low-Impact Recreational Activities:*** Encouraging low-impact recreational activities in mountain regions reduces damage to natural habitats and wildlife populations. This can involve promoting

Figure 3. Sustainable tourism practices
Source:https://www.researchgate.net/figure/Relationship-between-sustainable-tourism-ecotourism-and-some-of-the-main-forms-of_fig1_283215188

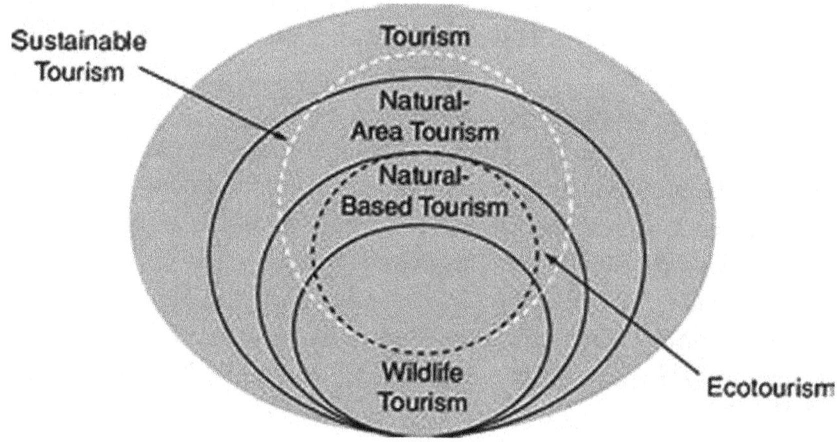

designated hiking and biking trails, regulating off-road vehicle use, and providing education on responsible behavior, such as not feeding or approaching wildlife.

- ***Sustainable Transportation:*** Sustainable transportation options, such as public transportation, cycling, and pedestrian-friendly infrastructure, should be encouraged to reduce carbon emissions and traffic congestion. Mountain destinations can also promote the use of electric vehicles or implement shuttle services to reduce the impact of private car use.
- ***Eco-friendly Accommodations:*** Sustainability can be promoted through the development and operation of eco-friendly accommodations. This can include using renewable energy sources, implementing energy-efficient designs, reducing water consumption, and adopting sustainable practices for waste management and food sourcing.
- ***Preservation of Cultural Heritage:*** Sustainable tourism practices in mountain regions should also focus on the preservation and promotion of local cultural heritage. This can be achieved through the respectful engagement with local communities, supporting traditional livelihoods and cultural practices, and ensuring that tourism development respects and benefits local people.
- ***Education and Awareness:*** Promoting awareness and educating both tourists and local communities about the importance of sustainable tourism practices is vital. This can involve information campaigns, interpretive signage, visitor codes of conduct, and educational programs that emphasize the value of conservation and responsible behavior.
- ***Collaboration and Stakeholder Engagement:*** Sustainable tourism cannot be achieved in isolation. Collaboration among governments, tourism operators, local communities, and conservation organizations is essential. Engaging all relevant stakeholders in planning, decision-making, and monitoring processes creates a sense of ownership and encourages shared responsibility in achieving sustainable outcomes.

By implementing these sustainable tourism practices in mountain regions, it is possible to mitigate environmental impacts, enhance local communities' livelihoods, and preserve the unique cultural and natural heritage of these areas for future generations(***Ghosh & Ghosal, 2020).***

5. POLICIES AND REGULATIONS

Policies and regulations play a crucial role in mitigating the ecological impacts of mountain tourism and promoting sustainability. Governments, authorities, and tourism organizations can implement various measures to regulate and guide tourism activities. Some key aspects related to policies and regulations in mountain tourism include the following:

Protected Areas and Conservation Designations; Governments can establish protected areas, national parks, and other conservation designations in mountain regions to safeguard fragile ecosystems and biodiversity. These areas should have clear regulations regarding visitor access, permitted activities, and conservation goals. ***Zoning and Land Use Planning***; Zoning and land use planning are essential tools to manage tourism development in mountain regions. Governments can designate specific areas for tourism infrastructure, ensure the protection of environmentally sensitive zones, promote sustainable transportation, and regulate development in ecologically significant areas. ***Environmental Impact Assessments***; Mandatory environmental impact assessments can help evaluate the potential ecological consequences of proposed tourism projects in mountain regions. These assessments should be thorough, transparent, and considerate of the unique characteristics and vulnerabilities of mountain ecosystems. ***Tourism Carrying Capacity; Setting*** and enforcing tourism carrying capacity limits is crucial for managing visitor numbers in mountain regions. Governments can establish guidelines for the maximum number of tourists that can be accommodated sustainably, considering factors such as habitat sensitivity, infrastructure capacity, and the preservation of ecological processes. ***Regulation of Infrastructure Development***; Governments can regulate the development of infrastructure, such as hotels, roads, ski resorts, and cable cars, in mountain regions to ensure sustainability. This includes enforcing guidelines on building codes, land use, and environmental standards to minimize impacts on water resources, wildlife habitats, and scenic beauty. ***Waste Management and Pollution Control;*** Policies and regulations should be in place to regulate waste management and pollution control in mountain regions. This includes requiring tourist facilities to adopt waste reduction and recycling practices, promoting sustainable wastewater treatment systems, and enforcing regulations to minimize pollution from transportation and recreational activities. ***Licensing and Accreditation;*** Governments can mandate licensing and accreditation systems for tour operators and accommodations in mountain regions. These systems can ensure compliance with sustainability standards and best practices, such as responsible tourist behavior, waste management, resource conservation, and engagement with local communities. ***Collaboration and Enforcement;*** Effective policies and regulations require collaboration among various stakeholders, including government agencies, tourism organizations, local communities, and enforcement agencies. Collaboration helps ensure compliance, enforcement of regulations, and continuous monitoring and evaluation of the effectiveness of policies(***Datta & Behera, 2022).***

It is essential that policies and regulations in mountain tourism are formulated and implemented in a participatory and transparent manner. Regular updates, monitoring, and evaluation processes are necessary to address emerging challenges and adapt to changing environmental, social, and economic conditions. Continuous engagement with stakeholders and learning from best practices worldwide can further enhance policy effectiveness and promote sustainable mountain tourism***(Samal et al., 2003).***

6. COMMUNITY INVOLVEMENT AND EDUCATION

In figure 4 Community involvement and education are crucial aspects of promoting sustainable practices and mitigating the ecological impacts of mountain tourism. Engaging local communities and providing them with knowledge and resources not only empowers them but also fosters a sense of ownership and responsibility towards their natural and cultural heritage. The following are key components of community involvement and education in mountain tourism:

Local Participation and Decision-making; Involving local communities in decision-making processes related to tourism development ensures that their perspectives and needs are considered. This can be achieved through forums, consultations, and participatory planning approaches that enable local community representatives to contribute to the development and implementation of tourism policies and strategies. ***Cultural Preservation and Revitalization;*** Mountain communities often have rich cultural traditions and practices that are intertwined with the natural environment. Supporting the preservation and revitalization of local cultural heritage can enhance the authenticity and sustainability of mountain tourism. This can involve cultural events, traditional craftsmanship initiatives, and educational programs that promote cultural exchange and respect. ***Sustainable Livelihood Opportunities;*** Providing sustainable livelihood opportunities for local communities reduces their dependence on environmentally harmful

Figure 4. Community involvement and education
Source:https://www.researchgate.net/figure/Level-of-community-involvement-across-different-typologies-of-community-management_fig1_283768465

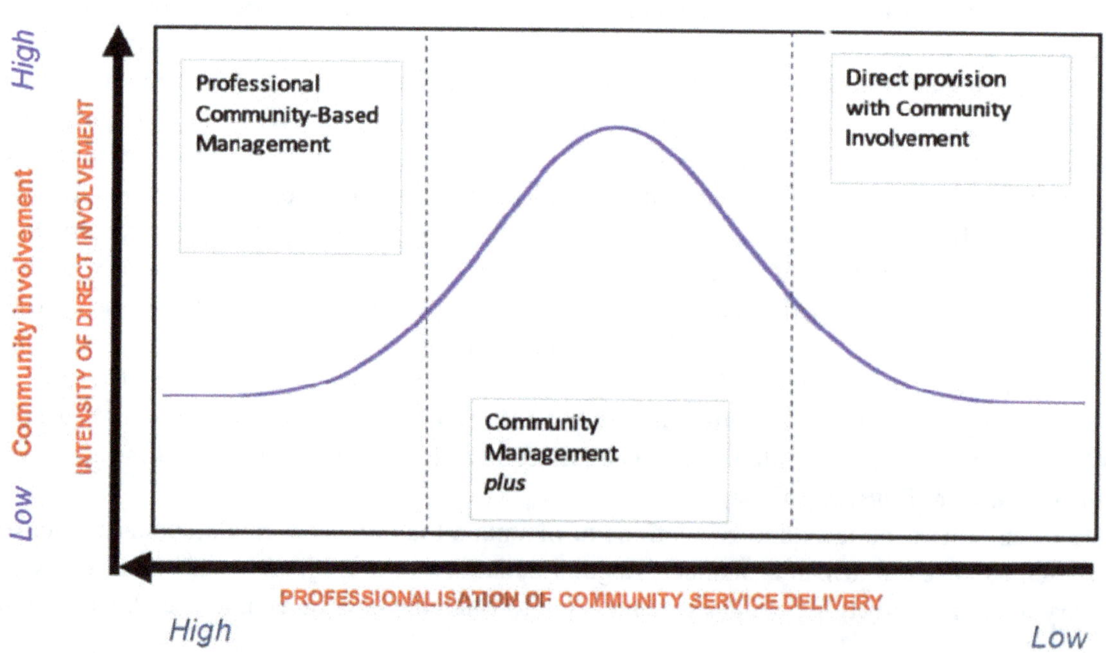

practices and creates incentives for conservation. This can include supporting community-based tourism initiatives, promoting local products and services, and facilitating capacity-building programs that enable local entrepreneurs to engage in sustainable tourism ventures. ***Education and Awareness;*** Education and awareness programs are essential for promoting responsible behavior among tourists and locals alike. These programs should focus on the ecological value of mountain ecosystems, the importance of conservation, and the sustainable use of resources. They can be conducted through interpretive signage, visitor centers, school programs, and community-led initiatives. ***Environmental Training and Skill Development;*** Providing training and skill development opportunities related to sustainable practices in mountain tourism can empower local communities and enhance their capacity to participate in the tourism industry responsibly. This can include environmental stewardship workshops, eco-guide training, sustainable agriculture techniques, and alternative energy solutions. ***Community-based Monitoring and Surveillance;*** Engaging local communities in monitoring and surveillance activities can help identify and address potential ecological impacts of tourism. This can involve training community members to collect data on wildlife populations, water quality, and visitor behavior, and involving them in collaborative monitoring programs with local authorities and conservation organizations. ***Benefit-sharing and Revenue Redistribution;*** Ensuring that local communities benefit directly from tourism revenue is crucial for their continued support and engagement in conservation efforts. Mechanisms such as revenue-sharing arrangements, community development funds, or community-based tourism enterprises can help redistribute the economic benefits of tourism***(Ghosh & Ghosal,2021).***

By implementing these community involvement and education measures, mountain tourism can be transformed into a more inclusive and sustainable industry. The active participation of local communities, along with their understanding and appreciation of the value of conserving their natural and cultural heritage, contribute to the long-term preservation of mountain ecosystems and the well-being of local residents.

7. FUTURE DIRECTIONS AND RECOMMENDATIONS

As mountain tourism continues to grow in popularity, it is crucial to prioritize ecological sustainability to ensure the long-term viability of these fragile ecosystems. To move towards a more sustainable future, the following recommendations can be considered:

- **Strengthen Policies and Regulations:** Governments and tourism authorities should review and update policies and regulations to address emerging challenges and promote sustainability in mountain tourism. This includes stricter enforcement of existing regulations, incorporating ecological indicators into planning processes, and encouraging the adoption of international best practices.
- **Foster Collaboration and Partnerships**: Collaboration among governments, tourism operators, local communities, and conservation organizations is key. Foster partnerships that promote knowledge exchange, resource sharing, and joint initiatives for sustainable tourism development. Engaging with indigenous and local communities and respecting their traditional ecological knowledge is also crucial for collaborative decision-making processes.
- **Invest in Research and Monitoring:** Continued scientific research on the ecological impacts of mountain tourism and the effectiveness of conservation measures is essential for evidence-based

decision making. Monitoring programs should be established to assess the ecological health of mountain ecosystems and track the impacts of tourism over time. This research and monitoring can guide sustainable management practices and inform policy development.

- ***Promote Sustainable Tourism Certification and Labeling***: Encouraging tourism operators in mountain regions to obtain certifications or labels that endorse sustainable practices can help consumers identify and choose responsible tourism options. These certifications should include criteria related to environmental management, community involvement, and cultural preservation.
- ***Support Community-based Tourism Initiatives:*** Investing in community-based tourism initiatives can empower local communities and ensure that tourism benefits are shared equitably. These initiatives should emphasize sustainable resource management, cultural preservation, and direct economic benefits for local residents. Governments and tourism organizations can provide training, marketing support, and financial incentives to help communities develop their tourism enterprises.
- ***Strengthen Visitor Education and Awareness:*** Enhancing visitor education and awareness programs is essential for promoting responsible behavior and minimizing the ecological footprint of tourists. This includes providing information on local ecosystems, cultural practices, sustainable tourism guidelines, and opportunities for meaningful engagement with local communities and conservation initiatives.
- ***Integrate Climate Change Adaptation Strategies***: Given the vulnerability of mountain regions to climate change, integrating climate change adaptation strategies into tourism planning and management is crucial. This includes identifying climate resilient practices, evaluating the potential impacts of climate change on tourism activities, and promoting carbon offsetting initiatives for tourist transportation.
- ***Encourage Sustainable Investment and Funding Mechanisms:*** Governments and international funding institutions should promote sustainable investment and funding mechanisms for tourism development in mountain regions. These mechanisms can provide financial support for sustainable infrastructure, conservation initiatives, and community-based tourism enterprises. They can also incentivize private sector investment in sustainable practices.

By adopting these recommendations, mountain tourism can be transformed into a responsible and sustainable industry that preserves the ecological integrity, cultural heritage, and economic vitality of these unique regions***(Mehta et al., 2020).*** It will require a collaborative effort from all stakeholders to achieve a harmonious balance between tourism development and ecological conservation in mountainous areas.

8. CASE STUDIES

8.1 Bhutan

Bhutan, a small landlocked country located in the Eastern Himalayas, has gained recognition for its sustainable tourism initiatives. The country has embraced a unique approach called "High Value, Low Impact" tourism, focusing on minimizing environmental impacts and promoting cultural preservation. Bhutan has established a strict tourism policy that requires all visitors to obtain a visa and pay a daily fee, ensuring a controlled number of tourists. Additionally, the country invests in conservation efforts,

Figure 5. Bhutan
Source: https://www.worldatlas.com

promotes community-based tourism, and requires sustainable practices from tour operators, building a strong foundation for sustainable mountain tourism ***(Upreti, 1999).***

8.2Switzerland

Switzerland is known for its stunning alpine landscapes and has taken steps to promote sustainability in mountain tourism. The Swiss government has implemented stringent regulations on building and tourism infrastructure development to prevent excessive construction that would damage the natural environment. The country encourages the use of public transportation in mountain regions, such as trains and cable cars, to minimize carbon emissions ***(Gansser, 1981)***. Switzerland also promotes sustainable skiing practices by adopting measures to reduce energy consumption in ski resorts, increase snow efficiency, and preserve the natural habitat of mountain wildlife ***(Molnar, 1984)***.

8.3 Costa Rica

While not located in mountainous regions, Costa Rica showcases how sustainable tourism practices can benefit mountain areas indirectly. Costa Rica has significantly invested in eco-tourism, emphasizing the conservation of its diverse ecosystems, including its cloud forests and volcanoes. Sustainable practices implemented in coastal areas have positively influenced neighboring mountain regions by creating awareness about conservation and promoting sustainable practices ***(Schelling, 1992)***. The country's eco-lodges and adventure activities attract tourists interested in experiencing nature in a responsible way. Costa Rica's success in sustainable tourism has helped promote similar practices in mountain regions globally ***(Lall, 1981)***.

Figure 6. Switzerland
Source: https://print.hpeasystart.com

Figure 7. Costa Rica
Source: https://www.worldatlas.com

8.4 Hokkaido, Japan

Hokkaido, the northernmost island of Japan, is known for its picturesque mountain landscapes and vibrant cultural heritage. The region has implemented sustainable tourism practices to minimize environmental impacts. For example, ski resorts in Hokkaido have started adopting renewable energy sources, efficient snow-making techniques, and waste reduction strategies. Furthermore, community-based tourism initiatives have been developed, allowing travelers to engage with local communities, learn about traditional practices, and support local economies ***(Valdiya, 1980)***. These efforts have helped preserve the natural beauty of the region and contribute to the local community's well-being. These case studies highlight the importance of sustainable tourism practices in mountain regions and demonstrate successful strategies implemented by different countries to maintain ecological integrity, preserve cultural heritage, and promote responsible tourism ***(Acharjee& Barat, 2010).***

9. CONCLUSION

In conclusion, mountain tourism offers unique and breathtaking experiences for travelers while also posing significant ecological impacts. The growth of this industry in the Himalayan region and beyond calls for urgent attention to ensure sustainable development and conservation of mountain ecosystems. Through this chapter, we have explored the ecological impacts of mountain tourism, the importance of biodiversity conservation, and the need for sustainable tourism practices. We have examined policies and regulations, community involvement, education, and the economic and cultural impacts of mountain tourism. Furthermore, we have provided recommendations and case studies highlighting best practices and lessons learned.It is clear that addressing the ecological impacts of mountain tourism requires a collaborative effort involving governments, tourism authorities, local communities, and visitors. Strengthening policies and regulations, promoting sustainable tourism practices, and fostering community involvement and education are essential steps towards achieving ecological sustainability in mountain tourism.

Additionally, considering the threats posed by climate change to mountain ecosystems, integrating climate change adaptation strategies into tourism planning and management is crucial. By adopting sus-

Figure 8. Japan
Source: https://www.worldatlas.com/

tainable practices, preserving biodiversity, and collaborating with stakeholders, mountain tourism can not only benefit local communities and economies but also contribute to the conservation of these fragile ecosystems for future generations.It is our hope that the information and insights provided in this book will inspire policymakers, stakeholders, and tourists to prioritize ecological sustainability in mountain tourism development. The future of mountain tourism lies in striking the right balance between economic growth, cultural preservation, and ecological conservation. By working together and implementing the recommended strategies, we can ensure that mountain tourism continues to offer unforgettable experiences while preserving the natural wonders that make these regions so special.

REFERENCES

Acharjee, M. L., & Barat, S. (2010). Impact of fishing methods on conservation of ichthyofauna of river Relli in Darjeeling Himalaya of West Bengal. *Journal of Environmental Biology*, *31*(4), 431–434. PMID:21186715

Bhutia, S. (2014). Economic development and environmental issues in Darjeeling Himalaya of West Bengal, India: A theoretical perspective. *International Journal of Humanities and Social Science Invention*, *3*(7), 42–47.

Datta, P., & Behera, B. (2022). Assessment of adaptive capacity and adaptation to climate change in the farming households of Eastern Himalayan foothills of West Bengal, India. *Environmental Challenges*, *7*, 100462. doi:10.1016/j.envc.2022.100462

Dey, T., Pala, N. A., Shukla, G., Pal, P. K., & Chakravarty, S. (2017). Perception on impact of climate change on forest ecosystem in protected area of West Bengal, India. *Journal of Forest and Environmental Science*, *33*(1), 1–7. doi:10.7747/JFES.2017.33.1.1

Gansser, A. (1974). Himalaya. *Special Publication - Geological Society of London*, *4*(1), 267–278. doi:10.1144/GSL.SP.2005.004.01.15

Gansser, A. (1981). The geodynamic history of the Himalaya. *Geodynamic Series*, *3*, 111–121. doi:10.1029/GD003p0111

Ghosh, M., & Ghosal, S. (2020). Determinants of household livelihood vulnerabilities to climate change in the Himalayan foothills of West Bengal, India. *International Journal of Disaster Risk Reduction*, *50*, 101706. doi:10.1016/j.ijdrr.2020.101706

Ghosh, M., & Ghosal, S. (2021). Climate change vulnerability of rural households in flood-prone areas of Himalayan foothills, West Bengal, India. *Environment, Development and Sustainability*, *23*(2), 2570–2595. doi:10.1007/s10668-020-00687-0

Lall, J. S. (1981). *The Himalaya aspects of change*. Oxford University Press.

Mehta, P., Sekar, K. C., Bhatt, D., Tewari, A., Bisht, K., Upadhyay, S., Negi, V. S., & Soragi, B. (2020). Conservation and prioritization of threatened plants in Indian Himalayan Region. *Biodiversity and Conservation*, *29*(6), 1723–1745. doi:10.1007/s10531-020-01959-x

Mitra, S., Roy, A. K., & Tamang, L. (2020). Assessing the status of changing channel regimes of Balason and Mahananda River in the Sub-Himalayan West Bengal, India. *Earth Systems and Environment*, *4*(2), 409–425. doi:10.1007/s41748-020-00160-y

Molnar, P. (1984). Structure and tectonics of the Himalaya: Constraints and implications of geophysical data. *Annual Review of Earth and Planetary Sciences*, *12*(1), 489–516. doi:10.1146/annurev.ea.12.050184.002421

Rawat, D. S., Dash, S. S., Sinha, B. K., Kumar, V., Banerjee, A., & Singh, P. (2018). Community structure and regeneration status of tree species in Eastern Himalaya: A case study from Neora Valley National Park, West Bengal, India. *Taiwania*, *63*(1).

Samal, P. K., Palni, L. M. S., & Agrawal, D. K. (2003). Ecology, ecological poverty and sustainable development in Central Himalayan region of India. *International Journal of Sustainable Development and World Ecology*, *10*(2), 157–168. doi:10.1080/13504500309469794

Schelling, D. (1992). The tectonostratigraphy and structure of the eastern Nepal Himalaya. *Tectonics*, *11*(5), 925–943. doi:10.1029/92TC00213

Shekhar, M. S., Chand, H., Kumar, S., Srinivasan, K., & Ganju, A. (2010). Climate-change studies in the western Himalaya. *Annals of Glaciology*, *51*(54), 105–112. doi:10.3189/172756410791386508

Shroder, J. F. Jr, & Bishop, M. P. (1998). Mass movement in the Himalaya: New insights and research directions. *Geomorphology*, *26*(1-3), 13–35. doi:10.1016/S0169-555X(98)00049-X

Singh, J. S., & Singh, S. P. (1987). Forest vegetation of the Himalaya. *Botanical Review*, *53*(1), 80–192. doi:10.1007/BF02858183

Upreti, B. N. (1999). An overview of the stratigraphy and tectonics of the Nepal Himalaya. *Journal of Asian Earth Sciences*, *17*(5-6), 577–606. doi:10.1016/S1367-9120(99)00047-4

Valdiya, K. S. (1980). The two intracrustal boundary thrusts of the Himalaya. *Tectonophysics*, *66*(4), 323–348. doi:10.1016/0040-1951(80)90248-6

Chapter 6
Ethnic Tourism and Its Prospects in the Eastern Himalayas:
With Special Reference to Northeast India

Trinayanee Devi Das
Assam University, India

ABSTRACT

The tourism sector has seen great development in recent years and expanding its forms to target different segments of tourists, and Ethnic tourism is one such form, which has been a common strategy used by many nations to promote economic and cultural growth and aid in the preservation of ethnic heritage. The eastern Himalayan region includes Bhutan, southern, central, and eastern Nepal along with Northeast India. Each region has different characteristics and features, which have contributed to the region's rich ecological and cultural diversity. The Northeastern region of the Eastern Himalayas is dominated by many indigenous communities which can be distinguished on traditional values, lifestyle, and customs. Ethnic tourism in the northeastern region has a considerable number of positive impacts which can enhance the value of the present tourism scenario of the region. Thus, this study shall focus on the prospects of this special interest tourism form, while addressing the need for improvement of planning strategies for the growth of ethnic tourism.

1. INTRODUCTION

The promotion of ethnicity as a tourist appeal and as a means of helping ethnic groups generate money and acquire foreign currency has grown. Many nations have used ethnic tourism to promote local economic growth by using their cultural variety. Ethnic tourism has also grown to be a potent tool for preserving ethnic groups' sense of self. It offers a huge opportunity for ethnic communities to promote their diverse cultures and heritages and to rejuvenate their customs, languages, and sense of cultural pride. Ethnic tourism has been a common strategy used by many nations to promote economic and cultural growth and

DOI: 10.4018/979-8-3693-0823-3.ch006

aid in the preservation of ethnic heritage. Ethnic tourism may have positive economic and social effects, but it may also have a severe negative effect on the traditional cultures, lifestyles, and sense of identity of ethnic communities. The growth of ethnic tourism has been supported by economic advantages. Although most of them are not ethnic members, the government and tourist business owners are the key forces behind the development of ethnic tourism. Their administrative and economic engagement in the tourism industry has a significant impact on how ethnic culture is staged, packaged, and represented in tourist destinations. The decisions made by the government and entrepreneurs affect the authenticity of attractions rather than the resource producers, the ethnic minority. The commercialization of ethnic culture, the creation of cultural events, and the creation of other tourist attractions are all utilized to serve commercial and political objectives. Because they have little influence over tourist resources and activities, minorities are frequently economically and politically sidelined or disadvantaged. The majority of minorities that work in tourism only have low-wage occupations.

Van den Berghe (1992) defined ethnic tourism as the type of travel where the main attraction is the exoticism of the local culture. Three groups—tourists, tourees (natives who actually make a show of themselves), and intermediaries (who facilitate contacts between tourists and tourees and offer culinary services—are divided into labor to carry out this complicated system of interethnic connections). According to Harron and Weiler (1992), ethnic tourism is described as travel that is largely driven by the desire to have direct, authentic, and at times personal interaction with individuals who have a different ethnicity or cultural background than the tourists. Direct interaction with the host culture and environment was stressed, generally through trips to local houses and villages to witness and take part in ceremonies, rituals, dances, and other traditional activities.

Ethnic minorities typically encourage the growth of ethnic tourism in their area, and they especially appreciate ethnic tourism's role in fostering economic development. The majority of people of color are optimistic about the socio-cultural effects of ethnic tourism on their neighborhood, although a small percentage sees certain drawbacks, including increasing cultural integration and harmed folk practices and regional festivals/cultural events.

Ethnic Tourism offers unique opportunities for both traveler and the communities they visit. Let's explore the significance of ethnic tourism;

- **Cultural Enrichment:** Ethnic tourism influences visitors to get involved closely with the different ethnicities and its forms. It gives an opportunity to discover specific customs, music, art, cuisines, and languages. Cross-cultural understanding and respect may fostered by the cultural enrichment.
- **Economic Benefit:** For local communities, ethnic tourism may bring revenue from the selling of its components. Like, visitors frequently buy locally created souvenirs, which help the community in earning monetary value. The community gains from the creation of jobs in the tourist industry.
- **Preservation of Traditions:** Ethnic communities have a reason to conserve and promote their roots since tourists are interested in learning about their traditions and customs. This may result in the revival of extinct customs and languages, ensuring their transmission to next generations.
- **Sustainability:** Many ethnic tourism initiatives emphasize sustainability and responsible tourism practices. This can help to protect the natural environment and cultural settings, ensuring that they remain intact for years to come.
- **Cross-Cultural Exchange:** Ethnic tourism fosters a meaningful exchange of ideas and experiences between tourists and local communities. Travelers gain insights into different customs of lifestyles, leading to promote mutual respect and understanding.

- **Marketing Opportunities:** With the rise of social media and digital marketing, ethnic tourism destinations have the chance to reach broader audiences. Effective promotion can attract tourist globally, boosting the local economy.

Ethnic tourism offers intriguing possibilities for both the communities hosting them and tourists looking for unique experiences of culture. It may promote economic growth, cultural preservation, and intercultural understanding when done responsibly and sustainably.

1.2 Objectives of the Study

- To find out the different ethnic communities living in the Northeastern Region
- To study the prospects of Ethnic Tourism in the Eastern Himalayas especially in Northeast, India.
- To analyze the factors influencing the growth of ethnic tourism in the study area

2. METHODOLOGY

The study on the topic "*Ethnic Tourism and its Prospects in the Eastern Himalayas, With Special Reference to Northeast, India*" has been created based on the secondary resources. The relevant data were gathered from a variety of research papers and articles that were published in different periodicals. Apart from, few information's from some official websites was also taken into account for the study.

2.1 Himalayas and the Eastern Himalayas

The Eastern Himalayas Region encompasses Bhutan, Northeastern India, and southern, central, and eastern Nepal. Sitting at the juncture between Asia and the Indian subcontinent, the region is geologically young and characterized by extreme altitudinal variations, which contribute to its rich biological diversity. (CPRF, 2005)

In terms of economics, sociocultural, and the environment, the Himalaya is a valuable worldwide resource. The Himalaya is a very popular travel destination due to the variety of its physiographic landscapes, climates, and environmental settings. Additionally, the area has a wide variety of cultural mosaics, ancient civilizations, archaeological resources, and other historical places. The area is renowned for its warmhearted people who live up to the Upanishad's slogan Atithi Devo Bhava, which translates as "The Guest is God." Despite these resources and attractions, the Himalayan region's tourist industry is still in its infancy, contributing less to the economy than the average for the world due to a number of physical, social, economic, and political difficulties. The Himalayan region is very sensitive to both natural and human-induced environmental and socioeconomic changes occurring at local, national, regional, and global dimensions due to its young age as one of the world's youngest mountain ranges. These changes include, but are not limited to, political unrest, population increase, and natural calamities. The Himalayas are distinct from other mountains due to their antiquity, particular and dynamic geophysical processes, cultural variety, underdevelopment, and isolation. The Himalayas are distinguished by their fragility, marginality, and variety (Jodha, 2001). A concentrated effort should be made to focus on tourism and sustainable development because of the significant implications this uniqueness has for socio-economic development in general and tourist development in particular.

Figure 1. Himalayas and the Eastern Himalayas
Source: CPRF (Critical Ecosystem Partnership Fund, Prepared by WWF- US, Asia Program, 2005)

The Eastern Himalayan area's ethnic tourism provides visitors with a distinctive and immersive journey into the many cultures, customs, and lifestyle of the indigenous ethnic groups who call this spectacular region home. This type of tourism not only offers tourists an authentic cultural encounter, but helps in the preservation and long-term growth of these thriving communities. The Eastern Himalayan region's ethnic tourism is described in further detail below:

- **Cultural Diversity:** The Eastern Himalayan region is a melting pot of ethnic diversity, with numerous indigenous communities coexisting for centuries. These communities include Tibetans, Bhutanese, Sikkimese, Nepalese, Adivasis, Mongoloid tribes, Assamese and various other hill tribes and sub-tribes. Each group has its own distinct language, customs and traditions.
- **Festivals and Rituals:** The regions calendar is dotted with numerous colorful festivals. Bhutan's Paro Tsechu, Nepal's Dashin, Sikkim's Losar are a few ones. One can witness these celebrations up-close which often revolve around religious and agricultural events and community gatherings.
- **Monasteries and Spiritual practices:** The region is steeped in Buddhism, and one can explore ancient monasteries like the Potala Palace in Tibet (UNESCO World Heritage Site), Taktshang (Tigers Nest) in Bhutan, Rumtek in Sikkim and many more. Witnessing the divine rituals of each ones would be a profound spiritual insight for the visitors allowing them to be rich spiritually and educationally.
- **Gastronomic Delight:** Food is an integral part of culture, and Eastern Himalaya region offers a diverse culinary delights. Starting from Butter Tea in Tibet; fiery Bhutanese dishes like Ema

Datshi, Sikkimese Phagshapa, Momo's and Thukpa and others unique flavours, the exploration of the culinary landscape provides an intimate glimpse into the lives of the communities.

- **Art and Crafts:** Local artisans create exquisite handicrafts in the region. Tourists can purchase and experience traditional craft like Thangka paintings, Bhutanese mask making, hand-woven textile of Sikkim, bamboo and terracotta of Assam are a few of them.

In addition to the above, the eastern Himalayas are home to renowned trekking routes. While Bhutan's Druk Path Trek provides unspoiled nature and cultural exchanges, the Everest Base Camp trek passes via a Sherpa village. Physical difficulties and cultural exposure are also present in these journeys. The preservation of indigenous cultures is significantly enhanced by tourism. Communities frequently take delight in presenting to the outside world their cultural treasures, which also promotes the continuance of practice. With the passage of time and changing perspectives, many ethnic tourism initiatives are now focusing on sustainability in order to reduce their negative environmental effects while ensuring that the communities benefit economically from tourism, enhancing the host's well-being and preserving cultural authenticity.

Ethnic tourism in the Eastern Himalayas essentially offers a comprehensive cultural experience by allowing tourists and visitors to interact with the ethnic communities there, fostering a profound understanding of the rich and diverse cultures that coexist in this remarkable region of the world.

2.2 Northeast and its Ethnic Communities

Northeast India, one of the main geographical parts of the Eastern Himalayan domain, is a geographic region with an extraordinary diversity in terms of both geography and culture. It consists of eight states and is renowned for its beautiful scenery, distinctive culture, and extensive history. Arunachal Pradesh, Assam, Manipur, Meghalaya, Mizoram, Nagaland, Sikkim, and Tripura are the eight states in the area. With over 200 different ethnic groups, this region's amazing cultural fusion and rich natural history, which offers visitors an unmatched experience, are what genuinely set it apart from the mainland. The Northeast India is a treasure of natural wonders; including the living root bridge in Meghalaya, wildlife of Assam, the lake beauty of Sikkim, hill road and valleys of Arunachal Pradesh and Nagaland and many more. Apart from the natural wonders, the region offers and extensive view on its cultural heritage as well. The cultural assimilation of different ethnic communities offers distinctive traditional dances, music, arts & crafts and the culinary delights. Here are the details of ethnic groups in more details, along with the state profile:

I. **Arunachal Pradesh:** Arunachal is home to some 30 communities and their 47 sub-groups. The list of Scheduled Tribes includes 25 tribes. Of the remaining five communities, Meyor, Mikir, Mishing and Lisu, have been given the facilities due to the Scheduled Tribes, but are not recognised as such by the state government. The fifth, i.e., Chakmas, is regarded as a non-tribal community. Going by the settlements, the state is divided into five cultural zones. The first is the Tawang and Kameng regions, where the tribes settled are Buddhist Monpas, Sherdupens, Akas, Mijis and Khowas. The second zone lying to the east of the first covers the area of East Kameng and Lower and Upper Subansiri. The tribes settled there are Nyishis, Banginis, Sulungs, Apatanis, Nas, Tagins Mikirs and Hill Miris. The third zone covers some parts of the Upper Subansiri and the East and West Siang districts, where the Adis, a major tribe, dominate. The Adis have 15 sub-tribes, namely the

Ashing, Bori, Karko, Milang, Padam, Pangi, Ramo, Tangam, Bokar, Gallong, Komkar, Minyong, Pailibo, Pasi and Shimon. The Dibang valley and Lohit regions constitute the fourth cultural zone, where the major tribes include Khamtis and Mishmis. The fifth zone spreads over the Tirap and Changlang regions and includes tribes like Noctes, Wanchos, Tangsas and Singphos.

II. **Assam:** Assam's physiography and people landscape are both vibrant and diverse. It has served as an intersection for several ethnic and cultural streams. Assam's population has been enriched by the influx of individuals from several ethnic groups, including Austro-Asiatics, Dravidians, Tibeto-Burmans, Mongoloids, and Aryans. Assam is home to the Bodos, Dimasas, Karbi, Tai-Ahom, Mishing, Rabha, Deori, and Tiwa ethnic groups.

III. **Manipur:** The people of Manipur are grouped into three main ethnic communities - Meiteis those inhabiting the valley and 29 major tribes in the hills dividing into two main ethno-denominations, namely Nagas and Kuki-Chins. Under the Meiteis, Bamon and Meitei Pangans are also included. All speak Meiteilon or otherwise known as Manipuri to the outsiders. In addition to Meiteis, the valley is also inhabited by Nepalis, Bengalis, Marwaris and other Indian communities. At present several people from the hill have also migrated and settled in the valley. The Naga group consists of Zeliangrong (composed of three related tribes, namely, Rongmei or Kabui, and Liangmei and Zemei or Kacha Nagas), Tangkhul, Mao, Maram, Maring and Tarao. The Chin-Kuki group consists of Gangte, Hmar, Paite, Thadou, Vaiphei, Zou, Aimol, Chiru, Koireng, Kom, Anal, Chothe, Lamgang, Koirao, Thangal, Moyon and Monsang. In recent times, several Chin-Kuki communities have identified themselves as Nagas e.g. Anal, Kom, Thangal, etc. depending on socio-economic and geo-political advantages to the tribes. The term Chin is used for the people in the neighboring Chin state of Myanmar whereas Chins are called Kukis in the Indian side. Other groups like Paite, Zou, Gangte, and Vaiphei identify themselves as Zomi and have distanced themselves from the name, Kuki. Thadous remain the major Kuki population in this Chin-Kuki group while Hmar identify closer to the Mizo or Lushei group. It should be noted that all the different ethnic groups are of the same Mongoloid group, and have very close similarities in their culture and traditional habits.

IV. **Meghalaya:** Meghalaya is mainly the homeland of three main tribes, the Khasis, the Jaintias and the Garos, who follow a clan system. The Garos inhabit Western Meghalaya, the Khasis in Central Meghalaya and the Jaintias in Eastern Meghalaya. The Khasis, Jaintias known to be one of the earliest ethnic group of settlers in the Indian sub-continent, belonging to the Proto Austroloid Monkhmer race. The Garos belonging to the Bodo family of the Tibeto – Burmese race were said to have migrated from Tibet. A unique feature about the people of Meghalaya is their matrilineal society. The lineage and inheritance follow the mother's line.

V. **Mizoram:** The majority of the Mizos, who compose the majority of the population of Mizoram, are Hmar, Lushai (or Lusei), Paite, Lai (Pawi), Mara (Lakher), and Ralte. Numerous sub-clans exist within these groupings; for instance, the Hmars are divided into the Thiek, Faihriem, Lungtau, Darngawn, Khawbung, Sote, and other sub-clans. People of the Mizo and related ethnic groups may be found in Bangladesh, Burma, and other Northeastern region. The other tribal indigenous groups of Mizoram are the Bru (Reang) and Chakma. Additionally, a sizable population of Nepalis is dispersed throughout the state.

VI. **Nagaland:** Nagaland is inhabited by 16 major ethnic communities along with a number of sub-tribes. Ao, Angami, Chang, Konyak, Lotha, Sumi, Chakhesang, Khiamniungam, Kachari, Phom, Rengma, Sangtam, Yimchungru, Kuki, Seliang and Pochury are the major ones. Each community is distinct and unique in character in terms of customs, language and attire. The colorful and intricately designed costumes and ornaments that are traditionally worn can help distinguish each tribe and subtribe.

VII. **Sikkim:** The people of Sikkim consist of three ethnic groups, namely Lepcha, Bhutia and Nepali communities with different shades that intermingle freely in Sikkim to form a homogeneous mixture of Sikkimese culture, encompassing many customs and ways of life while managing to maintain their own identity.

VIII. **Tripura:** Communities who are considered to be the original inhabitants consist largely of Tripuri, Reang, Jamatia, Noatia, Lushai, Uchai, Chaimal, Halam, Kuki, Garo, Mog and Chakma, who have migrated to the region at different points of time in the ancient past. The Tripuris constitute the largest community in the state.

2.3 Prospects of Ethnic Tourism in the Region

The Northeastern region of India holds significant potential for ethnic tourism. One can witness and experience the diverse and distinct traditional and cultural assets of the region. Here are some prospects for ethnic tourism in Northeast:

- Northeast India is culturally diverse. This diversity is a unique selling point for ethnic tourism, attracting visitors interested in exploring traditions and lifestyle. One of the major distinct views is that the Matrilineal Societal System followed in the state of Meghalaya.
- Each ethnic group of the region has its distinctive traditional clothing. For example, Dokhona for Bodo, Chador Mekhlea for Assamese, Jainsem for Khasis, the Nagas have Shawls etc.
- The region's hosts celebrate different festivals. Although majority of festivals are agricultural events, the way of celebration along with practices and techniques are apart from one community to another. Bhangni Mutairenai in Tripura, Solung' celebrated by the Adis in Siang and Dibang valley of Arunachal Pradesh, Chomkan celebrated by the Karbis, Behdeiñkhlam festival' of Jaintia Hills, Meghalaya are some.
- Dance and music are also an integral part of Northeast's ethnic communities. Bagurumba, (Bodo's traditional dance), Baidima (Dimasa's dance form), Pung Cholom (Manipuri's classical dance form), Cheraw (Bamboo Dance) and Chai lam from Mizoram are few from the Northeast, India.
- Northeastern cuisine is distinctive and relatively unexplored by the mainstream tourist. Each state has its own culinary specialties. Although Rice is a staple food to the region, there are plenty of other food options as well, namely, Jadoh from Meghalaya, Zutho from Nagaland, Thukpa from Arunachal Pradesh, Pitha from Assam, Churpi from Sikkim, Eromba from Manipur etc.
- Talking about the ethnic tourism prospects, the entire region of Northeast is known for its skilled artisans who produce excellent handicraft and handlooms, including handwoven textile, bamboo and cane product, terracotta, pottery, traditional paintings, and hand crafted traditional jewelry.
- Other prospects, which holds a major significance in each communities is folklore. Few examples are, the legend of Manas and Manas Kund (Assam), the tale of U Soso Tham (Meghalaya), the legend of Khonoma (Nagaland), the origin of Ziro Valley (Arunachal Pradesh), the story of Pakhangba (Manipur), are some worth mentioning.

In conclusion, the region has a wide range of opportunities owing to its rich cultural variety, delectable cuisine, distinctive festivals, and other significant encounters. Ethnic tourism must, however, be approached with consideration and respect for local customs and beliefs. A special cultural consideration must be included throughout the planning phase since exotic ethnic culture is seen as ethnic tourism's key draw and its most vulnerable resource.

3. FACTORS INFLUENCING THE GROWTH OF ETHNIC TOURISM

The above figure illustrates the relations among the three stated variable; Independent variable (factors), Moderate variable (intervening) and the Dependent variable (end product). Independent variable such as community involvement, advance technological options, components of tourism which can be denoted as 5 A's (Attraction, Accommodation, Accessibility, Amenities and Activities) can have a direct influence on the growth of ethnic tourism. Similarly, when the moderate variable factors (such are tourism policy, different marketing strategies, political scenario, rules and regulations) are effectively implement in the ethnic tourism, they will influence the growth of ethnic tourism scenario. And lastly comes the dependent variable (tourist satisfaction, livelihood of locals, community development) are certainly dependent on both independent variables and moderate variables.

Figure 2. Illustrates the relations among the three stated variable; Independent variable (factors), moderate variable (intervening) and the dependent variable (end product)
Source: Author

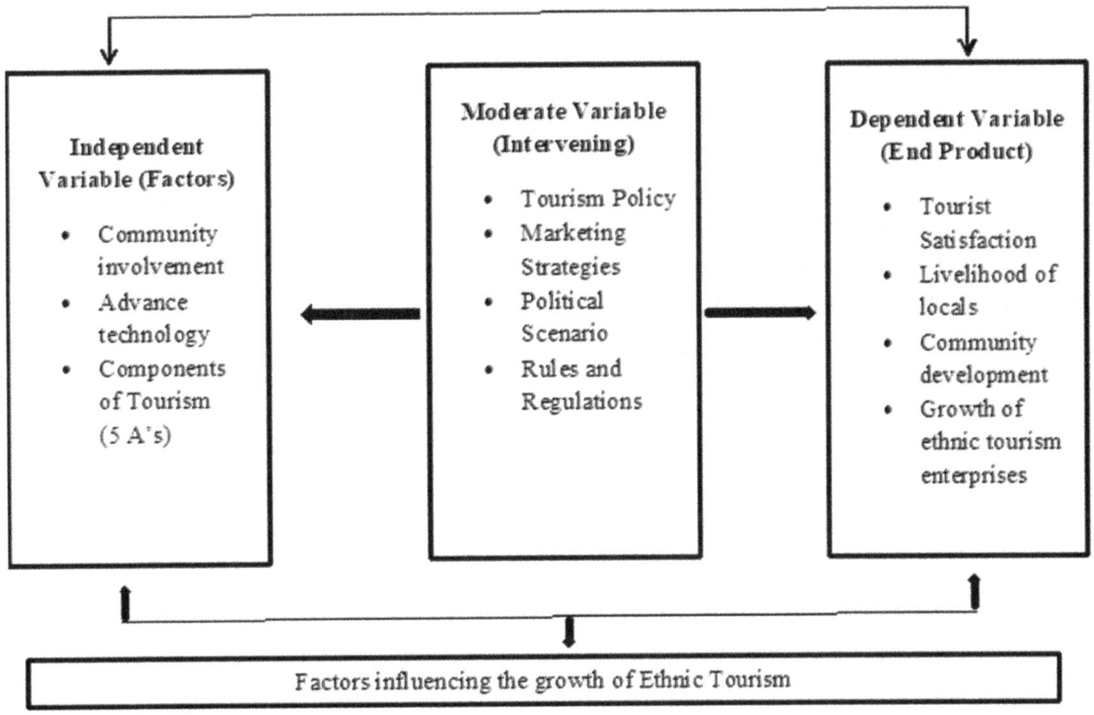

Irrespective of any destination, the different factors stated above have a close relation with one another. As the industry is dynamic in nature, the growth of ethnic tourism in the said region would be shaped by a multitude of interconnected factors. Tourists possess diverse preferences and interests, ranging from natural surroundings to experiences, necessitating destinations to offer a wide array of attractions. Along with the cultural richness and diversity of different ethnic group residing in the Northeastern region, their uniqueness can be a significant draw for tourist interested in experiencing new traditions, customs, and lifestyle which are different from their own. Celebrations, festivals, food & cuisine play a crucial role in attracting tourists who seeks for ethnic tourism experience. Taking a note on Hornbill Festival, the largest ethnic festival of Nagaland is the one which stand on the ground of ethnic tourism in the Northeast, India.

Globalization, which will soon allow travellers to visit distant places with ease, will force locations to compete not only locally but also globally. A destination's growth has been significantly impacted by advanced technology and has transformed travelling. India's northeast is no more an isolated region, yet the entire region has developed its own position in the global tourist industry. However, the ability to protect, celebrate, and promote the region's cultural heritage while providing the necessary infrastructure and support to ensure a positive and immersive experience for visitors interested in exploring and engaging with the heritage of Northeast, India, is crucial to the development of ethnic tourism in the area.

3.1 Challenges and Issues

- To maintain a balance between promoting ethnic pride and using ethnicity as tourism attraction while also protecting the community's values and identities.
- The commodification and marketing of ethnicity for tourism also creates a variety of issues, from the preservation of ethnic communities to the construction of ethnic identities and the changing of ethnic values.
- The conflict between the minority's thirst for contemporary living and the tourist's desire for an authentic cultural experience. While ethnic tourism effectively encourages minorities to pursue modernity and economic progress, it also creates a conundrum for the industry's own growth. Tourists desire authentic experiences of ordinary village life and traditional representations of minority cultures. However, as minority communities' architecture and lifestyles encounter significant transformation, folk villages are losing their exotic appeal.
- The region frequently experiences political unrest and social instability, and at times struggles to encourage ethnic tourism due to guest safety concerns.
- Encouraging responsible tourism practices among tourists is yet become difficult to undertake. At times, the influx of tourists can strain local resources and ecosystems. So, to ensure long-term viability of the destinations, ensuring each tourist to be in sustainable manner can be challenging at times for the host.

Overall, promoting ethnic tourism requires careful planning, collaboration with local communities, responsible tourism practices, and a commitment to preserving and celebrating the destination's cultural heritage while addressing any challenges and issues that may arise, regardless of the destination.

3.2 Approaches to Address the Challenges

- Implement cultural sensitivity training for professionals working in the tourism sector to ensure they are aware of and respectful of the traditions and cultures of the region. Establishing heritage preservation initiatives to save cultural landmarks, artifacts, and practices while facilitating access to responsible tourism.
- Enforce laws and rules that control the commercialization of cultural practices to make sure they adhere to regional values and traditions. Providing tourists with cultural orientation and educational values will also encourage respectful behaviour and an awareness of the destination's standards.
- Involving host communities in decision-making and profit-sharing related activities. Community based tourism models can empower locals and ensure economic benefits are distributed more equitably.
- To reduce negative effects and encourage balanced tourism growth, more effective and efficient planning should taken into account.To create a balance between conflicting stakeholder interests and between economic growth, environmental preservation, and social justice, various stakeholders' viewpoints should be taken into consideration. The government should conduct timely research and consultations to learn how minorities, business owners, and tourists view and feel about the growth of ethnic tourism.

4. CONCLUSION

Ethnic tourism encourages cultural progress and changes in traditional life while giving minority people the chance to enhance their incomes and showcase their identity. Entrepreneurs in the tourism industry are often well aware of the influence of economic growth and cultural change, particularly how acculturation has posed a threat to the unique ways of minority living. The improvement of local lives and the promotion of ethnic minorities' economic independence are two goals that are frequently associated with economic growth through tourism. Although elderly individuals are worried about cultural changes and support the preservation of traditional culture, ethnic minorities typically favor economic prosperity brought on by tourism.

In conclusion, ethnic tourism in Northeast India offers both distinctive chances and challenges to cherish and promote its cultural heritage across the globe. Ethnic tourism in the area has enormous potential, and it stands out for its diverse culture, breathtaking scenery, and distinctive traditions. It is essential to proceed cautiously, emphasizing cultural sensitivity and community engagement, in order to effectively capitalize on the advantages of ethnic tourism in this area. Northeast can provide travellers with meaningful and engaging experiences by finding a balance between presenting the many cultures and maintaining their originality. Along with, strategic planning and a commitment to responsible tourism, with a collaborative effort with the host; Northeastern region of India can emerge as a model for how ethnic tourism can positively impact on destinations growth, benefiting both tourist and host equally.

REFERENCES

Apollo, M. (2016). The clash–social, environmental and economical changes in tourism destination areas caused by tourism the case of Himalayan villages (India and Nepal). *Current Issues in Tourism Research*, *5*, 6–19.

Apollo, M., Andreychouk, V., Mostowska, J., Jin, Z., Maciuk, K., Rettinger, R., & Jones, T. E. (2021). Himalayan nature-based tourism. Potential, regional diversity, nature conservation and touristic load. *International Journal of Conservation Science*, *12*(4).

Bao, J., & Sun, J. (2003). The community participation in tourism planning. *Planners*, *19*(7), 32–38.

Briassoulis, H. (2002). Sustainable tourism and the question of the commons. *Annals of Tourism Research*, *29*(4), 1065–1085. doi:10.1016/S0160-7383(02)00021-X

Chistyakova, O. V. (2020). Ethnic tourism as knowing other. *RUDN Journal of Philosophy*, *24*(4), 720–729. doi:10.22363/2313-2302-2020-24-4-720-729

Critical Ecosystem Partnership Fund. (2005). *Ecosystem profile- Eastern Himalayan Region*. Critical Ecosystem Partnership Fund.

Critical Ecosystem Partnership. Fund. (2007). *Eastern Himalayas region*. Critical Ecosystem Partnership Fund.

Harron, S., & Weiler, B. (1992). Review: Ethnic tourism. In B. Weiler & C. M. Hall (Eds.), *Special interest tourism* (pp. 83–92). Belhaven.

Iseh, G. N., Yarhere, E. O., Ovharhe, O. J., & Apugo, O. E. (2021). Cultural and environmental factors influence on tourists' choice of destination in Delta State, Nigeria. *European Journal of Hospitality and Tourism Research*, *9*(2), 8–15.

Ives, J. D. (2004). *Himalayan perceptions: Environmental change and the well-being of mountain peoples*. Routlege: Pxon.

Jodha, N. S. (2001). *Life on the edge: Sustaining agriculture and community resources in fragile environments*. Oxford University Press.

Konwar, J. (2014). Ethnic tourism- problems and prospects for community development in North East India. Mosaic: A collection of research articles (ed.) by Dibru College. Dibrugarh, Assam, India. 09–22.

North East India People, history and culture. (2017). National Council of Educational Research and Training.

Nyaupane, G. P., & Timothy, D. J. (2022). *Tourism and development in the Himalaya: Social, environmental, and economic forces. Routledge cultural heritage and tourism services*.

Saha, G. K. (2015). Tourism sector in the north eastern states-challenges and opportunities. *International Journal of Developmental Research*, *5*(11), 6077–6081.

Van den Berghe, P. L. (1992). Tourism and the ethnic division of labor. *Annals of Tourism Research, 19*(2), 234–249. doi:10.1016/0160-7383(92)90079-5

Yang, L. (2007). *Planning for ethnic tourism: Case studies from Xishuangbanna, Yunnan, China*. [Doctoral Dissertation, University of Waterloo].

Chapter 7
Exploring the Thrills: Adventure Tourism in the Indian Himalayan Region

Rinzing Lama
Indian Institute of Tourism and Travel Management, India

ABSTRACT

Adventure tourism in the Indian Himalayan Region is a dynamic and rapidly growing sector that offers travellers unique opportunities to explore the breathtaking landscapes and cultural diversity of this region. This chapter provides a comprehensive overview of adventure tourism in the Indian Himalayan. The chapter begins by discussing the regions of Indian Himalayas and its drainage systems. Furthermore, the chapter delves into the Indian Himalayan region and adventure activities. The chapter also discuss about public and private agencies for promotion of adventure tourism in Indian Himalayan Region, which play a pivotal role in ensuring the safety and sustainability of adventure tourism activities. Later, the chapter will discuss about various adventure training institute in Indian Himalayan Region. Lastly, the chapter explores future prospects and emerging trends in adventure tourism within the Indian Himalayan Region. It discusses the role of changing traveler preferences, and the importance of community engagement in shaping the industry's trajectory.

1. INTRODUCTION

Adventure tourism stands as a thrilling and dynamic segment within the travel industry, inviting participants to immerse themselves in exhilarating and often physically challenging activities set in natural or remote environments. This form of tourism is centered around experiences that push the boundaries of comfort and safety, encompassing activities like hiking, trekking, rock climbing, white-water rafting, zip-lining, and wildlife safaris, among others.

At its core, adventure tourism is about the pursuit of excitement, adrenaline rushes, and establishing a profound connection with nature. It involves purposefully engaging in outdoor pursuits that pose challenges, requiring participants to apply tenacity, skills, endurance, and problem-solving abilities to

DOI: 10.4018/979-8-3693-0823-3.ch007

counter potential dangers. Beyond the thrill, adventure activities impart valuable life skills such as co-ordination, decision-making, teambuilding, and leadership. In essence, adventure is a unique blend of entertainment and education, epitomized by the term "edutainment," motivating organizations, schools, corporate offices, and colleges globally to engage their employees, partners, colleagues, and students in numerous Adventure Tourism activities.

Adventure Tourism often entails exploration of remote and pristine destinations, fostering a sense of adventure, and creating lasting memories. Safety assumes paramount importance in this domain, with experienced guides and specialized equipment ensuring that participants can enjoy their activities with minimal risk. This form of tourism appeals to individuals who crave excitement, revel in testing their physical and mental limits, and harbor a passion for the outdoors. Moreover, it plays a crucial role in the conservation of natural environments by promoting responsible and sustainable practices. Whether it's trekking in the Himalayas, diving in the Great Barrier Reef, or embarking on a wilderness expedition in Alaska, adventure tourism offers a diverse range of experiences for those seeking a sense of adventure in their travels.

Adventure tourism in India emerges as a vibrant and diverse sector capitalizing on the country's stunning natural landscapes, rich cultural heritage, and an array of thrilling activities. The Indian Himalayas, spanning union territories and states like Ladakh, Jammu and Kashmir, Himachal Pradesh, Uttarakhand, Sikkim, and Arunachal Pradesh, provide unparalleled opportunities for trekkers, mountaineers, skiers, thrill-seekers, and nature enthusiasts. This region, renowned for its majestic peaks, lush valleys, and rich cultural diversity, is a haven for adventure seekers.

The adventure tourism in the Indian Himalayan region is a captivating blend of adrenaline-pumping activities, cultural immersion, and breathtaking landscapes, making it an increasingly popular choice for travelers seeking excitement and unique experiences in this diverse nation. Emphasizing responsible and sustainable travel, experienced guides, safety measures, and a growing eco-consciousness ensure that visitors can enjoy these adventures while preserving the natural beauty and cultural heritage of the country.

India's adventure tourism has gained both domestic and international recognition, proving increasingly popular among global travelers. The Indian Himalayan region is globally renowned for its trekking and mountaineering opportunities, attracting domestic and international trekkers and mountaineers to areas like Sikkim, Garhwal, Kumaon region of Uttarakhand, Ladakh, and Himachal Pradesh. Popular trekking and mountaineering routes such as Triund Trek, Kheerganga Trek, Friendship Peak, Stok Kangri, Markha Valley Trek, Roopkund Trek, Kedarnath Trek, Flower of Valley Trek, Goecha La Trek, Kanchenjunga Base Camp Trek, and Dzongri Trek have gained acclaim worldwide.

India offers numerous rivers for white water rafting, including the Ganges, Beas, Zanskar, and Teesta rivers, with Rishikesh recognized as a global destination for river rafting. Destinations like Bir Billing in Himachal Pradesh have gained international recognition for paragliding, with adventure festivals and events such as the Bir Billing Paragliding Pre-World Cup attracting participants and spectators globally. Beyond adventure activities, the rich cultural heritage and diverse traditions of the Indian Himalayan region contribute to its international tourism appeal. Travelers often combine adventure activities with cultural exploration, making adventure tourism in the Indian Himalayan region a captivating blend of adrenaline-pumping activities, cultural immersion, and breathtaking landscapes—an increasingly popular choice for travelers seeking excitement and unique experiences.

1.1 Himalayas and its region in India

The term "Himalaya," originating from Sanskrit words 'hima' (snow) and 'alaya' (abode), translates to the "abode of snow." Forged through a colossal collision between Eurasia and the northward-drifting Indian plate around 50 million years ago, the Himalayas stand as the world's youngest and highest mountain range, harboring ten of the planet's loftiest peaks. This singular entity of immense physical dimension extends over India's northern boundaries, showcasing alpine meadows, river valleys, lakes, cold deserts, inter-mountain valleys, deep gorges, snowfields, glaciers, and alluvial plains.

The Himalayan ecosystem holds paramount importance for the ecological security of the Indian landmass, strategically spanning the entire northern boundary. It significantly influences weather patterns across South Asia. As the world's highest mountain chain, the higher Himalaya hosts the most extensive and vital glacier systems globally. The Indian Himalayan Region is also noteworthy for its high forest cover, encompassing over 41.5% of its geographical area—a third of India's total forest cover, with almost half classified as 'very good' forest cover.

Administratively, the Indian Himalayan Region sprawls across 2 Union Territories and 11 states, covering a vast geographical area of over 5 lakh km2. The mountain ranges traverse a west-eastern route from the Indus river to the Brahmaputra river system, elevating from low plains to over 8000m above sea level. Within India, the Himalayas are categorized into three longitudinal belts: Greater Himalaya (Himadri), Lesser Himalaya (Himachal), and Outer Himalaya (Siwaliks), each marked by great bends that give rise to lower mountain ranges and hills.

The Himalayas play a pivotal role in North India's drainage system, serving as the source of rivers like the Indus, Ganga, and Brahmaputra. This drainage system is further sub-divided into three subsystems—Indus System, Ganga System, and Brahmaputra System—providing an ideal setting for water-based adventure tourism. Adventure enthusiasts find the Indian Himalayan Rivers, such as the Indus, Ganga, and Brahmaputra and their tributaries, perfect for thrilling activities like angling, kayaking, white-water rafting, and river expeditions. These adventures not only provide an adrenaline rush but also offer an opportunity to connect with nature and experience the region's rich cultural diversity.

Internationally and domestically, the Himalayan regions of India beckon adventure seekers. Adventure tourism here offers a unique blend of physical challenges, cultural enrichment, and awe-inspiring natural beauty. With some of the planet's highest peaks, rivers, lush valleys, and vibrant culture, the Indian Himalayas invite individuals to test their limits, connect with nature, and create lasting memories. It is a journey where you can explore stunning landscapes, embark on thrilling river expeditions, and immerse yourself in the enchanting allure of this majestic mountain range.

1.2 Indian Himalayan Region and Adventure activities

Adventure Tourism is a kind of tourism that entails visiting far-flung, unusual and fascinating locations in order to partake in thrilling and adventurous activities. It focuses on giving travelers who want excitement, challenges and a change from typical tourist activities distinctive and heart-pondering experiences. One of the key benefits of adventure tourism is that it allows travellers to connect with new cultures, landscapes and people in a firsthand manner. It provides an immersive experience in a real-time environment, fostering a deeper understanding and appreciation of different cultures and natural surroundings. Additionally, participating in adventure activities can boost individual's confidence, happiness and overall sense of well-being. As a result, the demand for adventure tourism is steadily increasing worldwide.

Adventure tourism in the Indian Himalayas is a unique blend of adrenaline-inducing activities, cultural exploration, and the awe-inspiring beauty of the world's mightiest mountain range. Adventure tourism in the Indian Himalayas offers a thrilling experience. As this region is incredibly diverse and attracts adventure enthusiasts from around the world. The Himalayan region in India is a paradise for adventure tourism.

1.3 Here are some popular activities and destinations:

1. **Trekking:** The Indian Himalayas are a trekker's paradise; it offers a plethora of trekking trails that cater to all levels of adventurers for beginners, seasoned trekkers and hardcore professionals. Explore the stunning popular trails like the Goechala Trek, Kanchendzonga Base Camp Trek, Varsey Trek, Chadar Trek, Roopkund Teak, Markha Valley, Hampta Pass, Beas Kund, Har Ki Dun and Valley of Flowers Trek take you through pristine wilderness, dense forests and high altitude meadows to remote villages offering adventure, stunning vistas and breathtaking trekking experiences.
2. **Mountaineering:** The beautiful Indian Himalayan region remains unrivaled when it comes to the thrilling experience of mountaineering. For those with a passion for climbing, peaks like Stok Kangri, Nanda Devi, Kamet, Kedar Dome, Hanuman Tibba, Friendship Peak, Frey Peak, Mr. Jupono Peak and Mt. Kabru provide challenging opportunities. Indian Himalayas offer options for both seasoned mountaineers and novices from around the globe.
3. **White Water Rafting and Kayaking:** The roaring rivers originating in the Himalayas like the Ganges, Beas, Zanskar, Subansari and Teesta offer thrilling white water rafting and kayaking opportunities and experiences. These adventures lake you through picturesque gorges and rapids of varying difficulty levels attract both beginners and experts.
4. **Skiing and Snowboarding:** During the winter months, destinations like Gulmarg, Auni, Yumthang and Manali become hubs for skiing and snowboarding enthusiasts. The powdery snow and panoramic views create a dream setting for winter sports.
5. **Paragliding:** Soaring through the Himalayan skies while taking in the stunning landscape provides a bird's eye view of the stunning valleys and mountains, which provides an unforgettable experience. Ladakh, Bir Billing in Himachal Pradesh, Nainital in Uttarakhand and Bulbule Dara in Gangtok is a world famous destination for paragliding enthusiasts. These places also organize national and international Paragliding competitions every year.
6. **Mountain Biking:** Enjoy mountain biking in the scenic trails of Himalayas. Challenging terrain and scenic routes make the Indian Himalayas perfect for mountain biking adventures in places like Ladakh, Lachen, Lachung, Auli and Spiti Valley.
7. **Wildlife Safaris:** The Indian Himalayas are home to diverse wildlife, including snow leopards, brown and black bear, butterflies, tigers, elephants, one-horned rhinos and various bird species. Explore the rich biodiversity of the Himalayan region in national parks like Jim Corbett, Kaziranga, Kanchendzonga National Park and Hemis. Wildlife safaris in regions like Ladakh, Assam and Uttrakhand offer a chance to spot these incredible creatures.
8. **Climbing:** Test your climbing skills on natural rock formations in Ladakh, Himachal Pradesh, Uttarakhand and Sikkim. Conquer frozen waterfalls and ice formations during the winter season in Indian Himalayas.
9. **Motorcycle Tours:** Embark on epic motorcycle journeys like the Manali-Leh, Sikkim, Arunachal Pradesh and Uttarakhand Highway.

10. **Camping:** Spend nights under the starry Himalayan sky by camping in serene locations. Experience camping amidst stunning landscapes at locations like Spiti Valley, Teesta riverside and Zanskar.
11. **Yak Safaris:** A trip on the back of Yak, moving through the craggy trails, in the midst of angelic beauty of Indian Himalayan region. In mountain region of Ladakh and Tsongmo Lake in Sikkim, you can go on a unique yak safari to explore remote areas.
12. **Cultural Experiences:** Adventure tourism in the Indian Himalayas is not just about physical challenges. It also involves cultural immersion with opportunities to visit ancient monasteries, interact with local communities and savor traditional Himalayan cuisine.
13. **Photography:** Capture the stunning landscape, diverse flora, fauna and vibrant culture through your lens. The stunning vistas in the Indian Himalayas provide endless opportunities of photography.
14. **Yoga and Spiritual Retreats:** The Himalayas have been a spiritual haven for centuries and associated with spiritual practices. Many travelers come here for yoga retreats and meditation sessions, seeking inner peace amidst the natural beauty.

Before engaging in any adventure activity in the Indian Himalayas, it is essential to consider your weather conditions, fitness lever, acclimatization to high altitude and safety precautions. Always follow local guidelines and regulations to ensure a memorable and safe adventure in their majestic region. Always practice responsible tourism to ensure both your enjoyment and the preservation of this pristine environment.

1.4 Promotion of Adventure Tourism in Indian Himalayas

The promotion of adventure tourism in the Indian Himalayan region relies on the collaborative efforts of both public and private agencies. Both public and private agencies have to work together to showcase the region's natural beauty, culture and adventure opportunities.

i. Public Agencies (State and National Government)
   - **Infrastructure Development:** Both government agencies invest in building and maintaining essential infrastructure like roads, airports and trekking routes to improve accessibility to the Himalayan region, making it more attractive to adventure tourists. Government also invest in developing and maintaining infrastructure that supports adventure tourism, including trekking trails, camping sites, adventure facilities and base camps for mountaineers.
   - **Regulations and Safety Standards:** They establish and enforce safety standards and regulations for adventure activities such as trekking, mountaineering and rafting to ensure the well-being of tourists.
   - **Tourism Marketing:** Public agencies promote adventure tourism in the Himalayas through marketing campaigns, both domestically and internationally to attract adventure enthusiasts.
   - **Conservation and Environmental Protection:** Governmental bodies work to protect the fragile Himalayan ecosystem and preserve the natural beauty of the region, promoting responsible adventure tourism practices to minimize environmental impact.
   - **Marketing and Promotion:** Public agencies often lead marketing campaigns to promote adventure tourism in the Indian Himalayas, both domestically and internationally. They highlight the region's natural beauty and adventure opportunities. They also highlight lesser-known and off-the-beaten path destinations within their states to distribute tourist traffic more evenly and reduce the impact on popular spots.

- **Adventure Tourism Training:** They may offer training programmes and certification for adventure guides and operators design and offer adventure tourism packages that cater to various levels of adventure enthusiasts, including trekking, mountaineering and white water rafting.
- **Licensing and Permits:** They issue permits and licenses for adventure activities, ensuring that tour operators and guidelines meet certain standards and qualifications.
- **Emergency Services:** Government agencies often coordinate emergency services, including search and rescue operations to assist adventure tourists in case of accidents or emergencies.
- **Tourism Websites and Information Centers:** They maintain official tourism websites and information centers to provide travelers with up-to-date information on attraction, accommodation and activities in the Indian Himalayan states.
- **Local Craft, Cultural Festival and Events:** Sates organize cultural festivals, fairs, handicrafts, art forms and events to showcase the unique traditions, music, dance and cuisine of the Indian Himalayan. Many states organize adventure festivals and events such as trekking expeditions, rock climbing competitions and river rafting championships. These events attract adventure enthusiasts and showcase the region's potential.
- **Community Involvement:** They engage with local communities to involve them in adventure tourism activities, providing opportunities for local guides, homestays and cultural experiences for tourists.
- **Collaboration with Private Sector:** State governments collaborate with private adventure tour operators, hospitality businesses, NGOs to jointly promote and develop adventure tourism in the region.

ii. Private Agencies (Adventure Tour Operators, Hospitality, NGOs)

- **Tour Packages:** Private tour operators design adventure packages that cater to various levels of adventure seekers from beginners to experts, offering activities like hiking, trekking, rock climbing and paragliding.
- **Accommodation and Services:** Private hospitality businesses provide accommodations ranging from hotels and lodges to campsites and homestays, ensuring that adventure tourists have comfortable places to stay and access to the necessary services.
- **Adventure Activities:** They offer a wide range of adventure activities like rafting, skiing, zip-lining and wildlife safaris, providing tourists with thrilling experiences in the Indian Himalayan region.
- **Safety Measures:** Private adventure companies prioritize safety by providing trained guides, high quality equipments and adherence to safety protocols during adventure activities.
- **Marketing and Promotion:** Private agencies invest in marketing and promotion efforts to attract adventure enthusiasts through digital marketing, social media and collaboration with influencers.
- **Cultural Experiences:** Some private agencies incorporate cultural experiences into adventure tours, allowing tourists to interact with local communities, learn about their traditions and immerse themselves in the culture of the Indian Himalayan region.
- **Customized Adventure:** They often customized adventure experiences allowing tourists to tailor their trips to their preferences, interest and skill levels.

- **Environmental Responsibility:** Many private adventure tourism business practice eco-friendly and sustainable tourism, minimizing their impact on the environment and promoting responsible tourism

Both public and private agencies play vital role in the promotion of adventure tourism in the Indian Himalayas. Their combined efforts help showcase the region's breathtaking landscapes and diverse adventure opportunities while ensuring the safety and sustainability of these activities. For instance, Ministry of Tourism, Government of India is promoting Adventure Tourism as niche tourism product in India. They are promoting Adventure tourism as 365 days destination and attracts adventure enthusiast with specific interest.

The Ministry of Tourism, Government of Tourism has opened 340 mountain peaks for domestic and international adventure enthusiast desirous of obtaining Mountaineering Visa ('MX') for trekking and mountaineering. These peaks are located in the union territories and states of Sikkim (10 peaks), Ladakh and Jammu and Kashmir (133), Himachal Pradesh (91 peaks) and Uttarakhand (106). Apart from this Ministry of Tourism, Govt. of India declared and promoted year 2018 as the "year of Indian Adventure". Adventure tourism was also promoted as major component of Incredible India Campaign.

National organization Adventure Tour Operators Association of India (ATOAI) and Ministry of Tourism has launched the "Indian Adventure Tourism Guidelines (Version 2.0) on Safety and Quality norms for adventure tourism in India. Later Ministry of Tourism, Govt. of India has formulated a National Strategy for Adventure Tourism in 2022. National Strategy for adventure tourism has been prepared. The strategy focuses on developing adventure destinations, promoting safety in adventure tourism, skill development, capacity building and marketing.

By working collaboratively, public and private agencies can effectively promote adventure tourism, offering tourists thrilling experiences while ensuring safety, sustainability and access to the natural and cultural wonders of a region. These contribute to the economic development of the region while preserving its natural and cultural heritage.

1.5 Adventure Training Institutes in Indian Himalayan Regions

Indian Himalayan region is well known for its adventure training institutes that offers courses and programs related to adventure activities. These institutes provide training to enthusiasts and professionals seeking to engage in various adventure pursuits.

Here are some well-known training institutes in the Indian Himalayas:

1. **Himalayan Mountaineering Institute (HMI), Darjeeling:** Tenzing Norgay, one of the first men to reach Mount Everest's summit, founded HMI in 1954, HMI offers mountaineering and adventure courses in the scenic Darjeeling region.
2. **Nehru Institute of Mountaineering (NIM), Uttarakhand:** NIM is located in Uttarkashi, NIM is one of the most prestigious mountaineering institutes in India, established in 1965. It offers a wide range of mountaineering and adventure courses, including basic and advanced mountaineering, rock climbing, and skiing.
3. **Atal Bihari Vajpayee Institute of Mountaineering and Allied Sports (ABVIMAS), Himachal Pradesh:** ABVIMAS is situated in Manali and established in 1961, provides training in moun-

taineering, rock climbing, skiing and other adventure activities. It also conducts adventure sports events and competitions.

4. **Jawahar Institute of Mountaineering and Winter Sports, Jammu and Kashmir:** The Institute was established in 1983 at Aru, near Pahalgam,J&K with a view to provide numerous adventurous activities at one place in close proximity of each other.
5. **Indian Institute of Skiing and Mountaineering (IISM), Jammu and Kashmir:** IISM is located in Gulmarg and established in 1968, IISM specializes in skiing, snowboarding, mountaineering and other adventure courses. It offers training in skiing, snowboarding, and adventure tourism.
6. **Regional Mountaineering Centre (RMC), Himachal Pradesh:** RMC of Atal Bihari Vajpayee Institute of Mountaineering and Allied Sports is based in Dharamshala and offers courses in mountaineering, rock climbing, and trekking.
7. **National Institute of Mountaineering and Allied Sports (NIMAS) Arunachal Pradesh:** NIMAS established in 2013, offers a variety of adventure courses including mountaineering, rock climbing, river rafting and paragliding in the remote and beautiful region of Arunachal Pradesh.
8. **Garhwal Mandal Vikas Nigam (GMVN) Adventure Sports Training Institute, Uttarakhand:** GMVN established in 1976, offers adventure courses in activities such as trekking, river rafting, and paragliding.
9. **Indian Himalayan Centre for Adventure and Ecotourism (IHCAE), Sikkim:** Located in Chemchey and established in 2006. IHCAE prominent mountaineering institutes in India. It offers a wide range of adventure, ecotourism and mountaineering courses, including basic and advanced mountaineering, rock climbing, methods of instructor and search and rescue.
10. **Himalayan Institute of Adventure Sports (HIAS), Himachal Pradesh:** HIAS offer adventure-training programs in activities like rock climbing, trekking and skiing in Himalayan region.

Training institutes in the Indian Himalayan region play a vital role in cultivating the passion of adventure enthusiasts, supporting professionals, and inspiring aspiring adventurers. These institutions serve as crucial hubs, providing individuals with the essential skills and knowledge necessary for safe and responsible participation in a wide range of adventure activities.

Offering diverse courses such as the Basic Mountaineering Course, Advance Mountaineering Course, Search and Rescue Course, Method of Instruction, Basic Adventure Course, Elementary Adventure Course, Skiing Course, Paragliding Course, Rafting Course, and more, these training centers immerse participants in the breathtaking landscapes of the Indian Himalayas.

Through these courses, individuals not only acquire technical expertise but also develop a deep understanding of the environments they explore. This comprehensive preparation ensures that adventurers are well-equipped to navigate the challenges presented by the diverse terrains of the Indian Himalayas while fostering a profound appreciation for the stunning natural beauty of the region.

Beyond these training institutes, the Indian Mountaineering Foundation (IMF) stands out as a prominent and nationally recognized organization devoted to promoting mountaineering, climbing, and adventure activities in the Indian Himalayan region. Established in 1957 and headquartered in New Delhi, the IMF's primary objective is to encourage and support mountaineering, rock climbing, and other adventure pursuits in India.

Offering various training programs and certification courses related to climbing and mountaineering, the IMF collaborates with affiliated mountaineering institutes and organizations. The foundation organizes and sponsors mountaineering expeditions to peaks in India and abroad, and as a member of the International Mountaineering Federation (IMF), it coordinates and funds expeditions worldwide.

The IMF actively supports climbing competitions and adventure events, aiming to nurture talent and foster a keen interest in mountaineering and adventure sports. It provides affiliation to numerous adventure clubs, mountaineering institutes, and adventure organizations, including the Nehru Institute of Mountaineering, Himalayan Mountaineering Institute, Adventure Tour Operators Association, and IITTM Adventure Club.

Establishing safety standards and regulations for mountaineering activities in India, the IMF ensures that mountaineers and climbers adhere to safety protocols and guidelines. The foundation also publishes books, journals, and literature related to mountaineering, adventure activities, and explorations.

Representing India in international mountaineering and climbing organizations, the IMF advocates for Indian mountaineers and showcases their achievements globally. The Indian Mountaineering Foundation has played a pivotal role in cultivating a vibrant culture of mountaineering and adventure activities in India. It continues to provide essential training, support, and opportunities for both seasoned mountaineers and aspiring climbers to explore and conquer challenging peaks and mountains.

1.6 Future prospect of Adventure Tourism in Indian Himalayan Region

The future prospects of adventure tourism in the Indian Himalayan region are promising, but they also come with challenges and considerations.

The following are some key factors:

1. **Growing Demand:** The appetite for adventure tourism is on the rise, both among domestic and international travelers. As more people seek unique and thrilling experiences, the Indian Himalayas offer an enticing destination.
2. **Diversification of Activities:** The Indian Himalayas are diversifying their adventure offerings beyond traditional mountaineering. Activities like trekking, paragliding, river rafting, and cultural immersions are gaining popularity in Ladakh, Himachal Pradesh, Uttarakhand and Sikkim.
3. **Infrastructure Development:** Improved infrastructure, including better roads, airports, and accommodations, will make the Himalayan region more accessible to adventure tourists, expanding the potential for growth in Indian Himalayan region.
4. **Environmental Conservation:** To sustain adventure tourism, there is a growing emphasis on responsible and sustainable practices to protect the fragile Himalayan ecosystem. Conservation efforts will be integral to the region's future.
5. **Local Community Involvement**: Engaging local communities in adventure tourism can lead to economic benefits and cultural exchanges. Homestays and community-based tourism initiatives will likely increase in Ladakh, Himachal Pradesh and Sikkim.
6. **Government Support:** Continued support from government bodies, both at the state and national levels, will be crucial. This includes safety regulations, funding for infrastructure, and promotion of the region.
7. **Safety Standards:** Strict adherence to safety standards and regulations for adventure activities will be essential to maintain the trust of tourists and ensure their safety.

8. **Marketing and Promotion:** Effective marketing campaigns, both by government agencies and private operators, will be essential in attracting adventure tourists. Digital marketing and social media will play significant roles.
9. **Adventure Training:** The availability of quality adventure training institutes and courses will help produce skilled guides and operators, enhancing safety and the overall adventure experience.
10. **Global Recognition:** As Indian mountaineers and adventurers achieve recognition on the global stage, the Indian Himalayas will continue to draw international attention and climbers to climb Indian Himalayan peaks.
11. **Challenges:** Climate change and environmental degradation pose threats to the region. Unplanned development, overcrowding, and waste management need careful attention.
12. **Infrastructure Management:** Managing the balance between infrastructure development and environmental preservation is crucial. Sustainable practices and eco-friendly initiatives will be necessary.

The future of adventure tourism in the Indian Himalayan region is bright, driven by increasing demand, improved infrastructure, and a diverse range of adventure activities. However, sustainable practices, environmental conservation, safety standards, and responsible tourism will be vital to ensure the long-term success and preservation of this magnificent destination.

2. CONCLUSION

In conclusion, adventure tourism in the Indian Himalayan region stands out as a captivating and rapidly expanding sector within the broader tourism industry. The extraordinary natural beauty, cultural richness, and diverse adventure opportunities in this region make it an increasingly popular destination for travelers in search of thrilling and immersive experiences.

The Indian Himalayas offer a plethora of adventure activities, ranging from trekking and mountaineering to white-water rafting, paragliding, wildlife safaris, and more. This diversity caters to adventure enthusiasts with varying interests and skill levels. The Himalayan landscapes, adorned with towering peaks, pristine lakes, lush forests, and vibrant flora and fauna, create a breathtaking backdrop for these activities, drawing adventure enthusiasts from around the world.

Beyond the adrenaline-pumping experiences, adventure tourists have the unique opportunity to immerse themselves in the rich cultures and traditions of the Himalayan communities. There is a growing emphasis on responsible and sustainable tourism practices, aimed at protecting the fragile Himalayan ecosystem and preserving the region's natural beauty for future generations.

Looking ahead, the future of adventure tourism in the Indian Himalayas appears promising, fueled by a surge in demand, improved accessibility, and the diversification of activities. This trend not only provides economic opportunities for local communities and the tourism industry but also allows visitors to forge profound connections with nature and culture.

In summary, adventure tourism in the Indian Himalayan region offers an exhilarating blend of natural beauty, cultural richness, and thrilling experiences. As it continues to evolve and grow, the implementation of responsible tourism practices and conservation efforts will be vital in safeguarding the Himalayas' unique allure for generations to come.

REFERENCES

Baishya, K. (2017). Adventure Tourism in India. *EPRA International Journal of Economic and Business Review.*, *5*(10), 50–55.

Birkett, B., & McGrady, C. (Eds.). (2000). *Classic Treks: The 30 Most Spectacular Hikes in the World.* Bulfinch.

Buckley, R., & Pandey, P. (2017). Trekking tourism in the Indian Himalayas: A case study of environmental impacts and management strategies. *Journal of Outdoor Recreation and Tourism*, *19*, 40–50.

Chaudhary, M., & Lama, R. (2014). Community Based Tourism Development in Sikkim of India- A study of Darap and Pastanga Villages. *Transnational Corporations Review*, *6*(3), 228–237. doi:10.5148/tncr.2014.6302

D'souza, T. B., & Puttaraju, S. (2000). Examining the influence of trekking in nature of adaptive response to stress in individuals. *International Journal of Indian Psychology*, *8*(1), 913–923.

Hackman, J. R., & Wagerman, R. (2005). Tourism and Nature. *Academy of Management Review*, *30*(2), 269–287. doi:10.5465/amr.2005.16387885

Hall, C., & Weiler, B. (Eds.). (1992) Special Interest Tourism. Belhaven, London. Himalayan Mountaineering Institute. About us from https://hmidarjeeling.com/origin-and-growth/

Joshi, P. (2020). Safety and risk management in adventure tourism: A case study of mountaineering expeditions in Himachal Pradesh. *International Journal of Adventure and Sports Tourism*, *3*(1), 45–62.

Joshi, P. (2020). Safety and risk management in adventure tourism: A case study of mountaineering expeditions in Himachal Pradesh. *International Journal of Adventure and Sports Tourism*, *3*(1), 45–62.

Kamble, U. P. (2019). Overview of adventure sports tourism in India. International Journal of Physicology. *Nutrition and Physical Education*, *4*(2), 155–158.

Smith, S. L. J., & Gupta, M. (2016). Community-based ecotourism and sustainable development in Ladakh, India. *Journal of Ecotourism*, *15*(3), 204–220.

Tiwari, A., & Verma, S. (2018). Marketing strategies for promoting adventure tourism in the Indian Himalayan region. *Journal of Travel & Tourism Marketing*, *35*(6), 726–740.

UNWTO. (2014). *Global Report on Adventure Tourism*. UN. https://www.e-unwto.org/doi/pdf/10.18111/9789284416622

Weber, K. (2001). Outdoor adventure tourism: A review of research approaches. *Annals of Tourism Research*, *28*(2), 360–377. doi:10.1016/S0160-7383(00)00051-7

Chapter 8
Factors Responsible for Selecting Government-Owned Accommodations in the Himalayan Region of West Bengal

Dillip Kumar Kumar Das
https://orcid.org/0000-0002-6032-9348
University of Burdwan, India

Ushashi Mitra
https://orcid.org/0000-0001-7927-6875
University of Burdwan, India

ABSTRACT

In the world of tourism, accommodation plays a very crucial role. Accommodation is the key to the success of tourism industry, any development in the tourism Industry is quite not possible unless the problem of accommodation has been fixed. The hotel is actually home away from home for travelers. Tourists generally choose accommodation on the basis of few things like their financial ability, the standard of living, geographical location, environment, available amenities, safety, and security. Arriving in a new location with family or alone always gives tension to the traveler because they are new to the place and not aware of local issues. In this chapter, the authors will discuss the factors that influence any tourist to choose government owned accommodation, to find out why tourists are psychologically attracted towards government owned accommodation. In this current study, the authors select the accommodations provided by Govt. of West Bengal with special reference to the Himalayan region and tried to examine the reasons for which the government owned accommodation attracts the tourists in this region.

DOI: 10.4018/979-8-3693-0823-3.ch008

1. INTRODUCTION

Accommodation plays an important role in tourism industry, as it provides tourists an opportunity to stay for a length of time to enjoy the beauties of that location. Today, the accommodation does not mean only bed and breakfast but lot about amenities or the service they provide. Hotel industries are offering a wide range of services for the tourists to facilitate. Accommodation plays a base role for tourists to explore the destination, experience local food, participating in adventurous activities, witnessing cultural events, shopping from local markets etc. which any way generates the revenue of the locals by creating employment opportunities.

Pre-independence British's formed a few summer houses in hill stations of India to cope up with summer heat, Calcutta and Shimla respectively used to be the capital for the British rulers. From then the Himalayan hill stations of India slowly became the interesting for the tourists. In the early 19th century, northern district of West Bengal situated in the foothills of eastern Himalayas, within close range from Calcutta, Darjeeling became the potential destination for the British for spending summer and relaxing in soothing cool weather. Today, Darjeeling is known as 'Queen of Hills'. Natural beauty, breathtaking view of Kanchenjunga, flora fauna and the British colonial architecture made this place a perfect destination for national and international tourists.

Apart from Darjeeling, its adjacent places like Kalimpong and Kurseong became very popular destination for the tourists. Govt. of West Bengal established few accommodation units with such amenities to facilitate the tourists.

1.1 List of the Accommodation at the Foothill of Himalayas Offers by the Govt. of West Bengal

1. **Darjeeling Tourism Property**: This accommodation unit is situated near famous mall road of Darjeeling. View of the Kanchenjunga from the rooms of this property is phenomenal.

Infrastructure of this property currently offers:

- 9 Bedrooms (2 large, 7 standard)
- 1 restaurant
- 1 tavern
- 1 lawn

2. **Kurseong Tourism Property**: This unit is situated in the road side of Darjeeling, provides rooms and restaurant for the travellers.

Infrastructure of this property currently offers:

- 10 Bedrooms (2 large, 8 standard)
- 1 restaurant
- 1 tavern

3. **Morgan House Tourism Property**: This property is a heritage property that tourism of Bengal offers. In 1930s Mr. Morgan (a jute tycoon of British East India Company) & Mrs. Morgan constructed this mansion to as their summer home. Post independence government took control of this majestic property. Morgan House is a great piece of British colonial architecture with Victorian mansion, gabled roof, large glass windows, towering chimneys and the ivy plant over the building took a place in travellers bucket list.

Infrastructure of this property currently offers:

- 16 Bedrooms (6 in main building, 10 in annexed building)
- 2 Restaurant
- 3 Lawn

4. **Hill Top Tourism Property:** Near to Dr. Grahams home at Kalimpong, Hill-Top tourism Property is another beautiful stay. A peaceful wooden old-fashioned bungalow with a view of Kanchenjunga makes this property a weekend destination.

Infrastructure of this property currently offers:

- 8 Bedrooms
- 1 Restaurant
- 1 Lawn

All these properties provide amenities like, TV, free wi-fi, laundry, room service, restaurant etc.

1.2 The Objective of Study

The major objective of the study is to identify the determining factors that influence the tourists (both domestic and international) to choose govt. owned accommodations in the Himalayan region of West Bengal.

2. LITERATURE REVIEW

According to Pertiwi, P. R. & Sulistyawati, A. S.(2020) "Factors of Influence In Choosing Accommodation -" choose Mas Village, An Artistic Heritage Village In Ubud – Bali in their research paper mainly focussed on the influencing factors for choosing accommodation especially by the foreign tourists. The study revealed that four things tourist mainly considered for choosing accommodation namely location, cleanliness, atmosphere, good value and local touch & security.

According to Baruca, P. Z., & Civre, Z. in (2012) in their study on 'How do guests choose a hotel?'- the authors mentioned that location plays as a primary factor to choose any hotel followed by Price and Past experience are the next most important factors in hotel selection.

According to Gunasekaran, N., & Anandkumar, V. (2012) in their article on "Factors of influence in choosing alternative accommodation: A study with reference to Pondicherry, a coastal heritage town" stated in their paper that there are four factors namely, homely atmosphere, value for money, guest – host relationship and local touch mainly responsible for choosing an accommodation facility. Out of four, value for money and perception played a major concern on choosing alternative accommodation.

According to Chu, R. K. S., & Choi, T. (2000) an importance-performance analysis of hotel selection factors in the Hong Kong hotel industry: a comparison of business and leisure travellers. Studied 26 numbers of hotels. They revealed in their study that business or leisure travellers both gives stress on eight important factors: Service Quality, Business Facilities, Value, Room and Front Desk, Food and Recreation, and Security.

2.1 Methodology

The data for the proposed study has been collected from both primary and secondary sources. The primary data has been collected through a questionnaire. The sample size was restricted to 100 tourists (80 domestic & 20 foreign tourists) who visited or willing to visit Government properties in Himalayan Region. The questionnaire comprised of 15 questions mostly related to Five-factors (i.e, Security, Reliability, Geographic location, Money and service). Apart from that, data also collected from respective locals of the study area, engaged staff and Officials engaged in respective Departments.

3. RESULT AND DISCUSSION

After analysing the data, the following major 5 (Five) factors came out which are mainly responsible for choosing government owned accommodation in Himalayan region of West Bengal, these includes the following:

1. **Safety and Security-** Before arriving in an alien city, every human think about their security which is also very essential to care about. In an unknown place, new people, unaware of the local situation, seeking safety is a Natural Phenomenon, hesitation is justified. On that point Govt. Accommodation plays a major role, it is very easy to trust any Govt. owned service than any other source. In our survey, **96%** of people voted that Safety & Security influenced them for choosing Govt. Accommodation.

Figure 1. Safety factor

Figure 2. Reliability factor

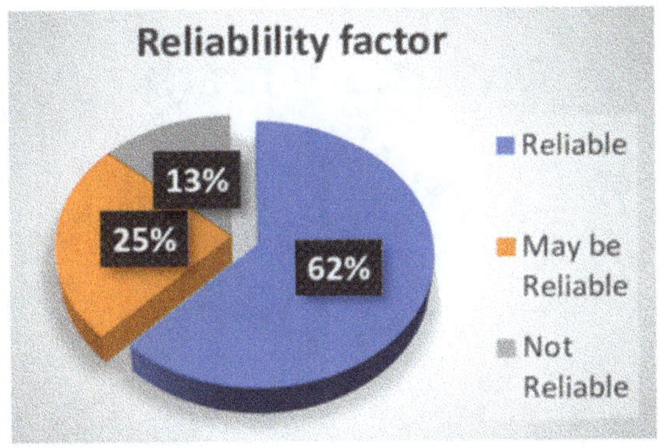

2. **Reliability-** technology doesn't always bring goodness, malpractice of using technical tools is very known nowadays. Fraud and unlawful practices are raising day by day. Paying money online is a big dilemma, on one hand, it is a very useful mode of transaction in the other hand it is doubtful, reasons account hacking, weird money transactions, scams, or fraud. Generally, people try to stay away from these frauds. In Tourism, fraud websites, unrealistic deals, fake bookings are recurrent. It's difficult to trust. In our survey, **63.46%** of people said that paying in Govt accommodations is reliable, **25%** of people said maybe it's reliable, and **13.46%** people said no reliable.
3. **Location advantage-** Whether it's a pleasure tour or business travel, everyone seeks a better and convenient location. Most of the time tourists prefer staying at those properties where the natural view is prominent, like sea-facing or mountain-facing room simultaneously how easy to reach from railway station or airport. Besides, travellers who are the most frequent visitors seek easy to commute places. In our survey, **78%** of people said that location and the natural view from Tourist Lodges are the reason to choose. whereas, **22%** people said location and view was not their reason to choose.
4. **Monetary worth-** It is the human tendency to think wisely every time before investing a single penny. In the case of service, it is hard to concentrate on quantitative gains against your investment, so with Tourism, an intangible service. Therefore, justification of the paid amount is not always the same, sometimes it's up to the mark or in some other time is below the average, this count always

Figure 3. Location factor

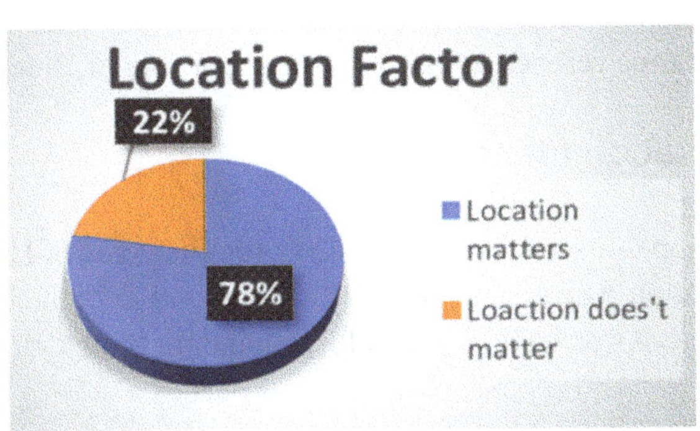

Figure 4. Monetary factor

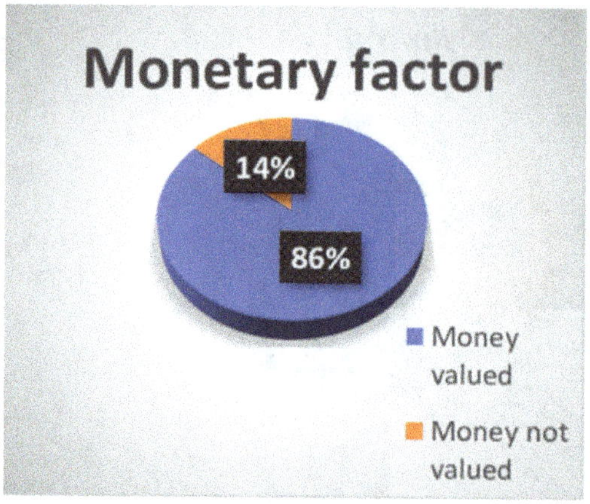

Figure 5. Service factor

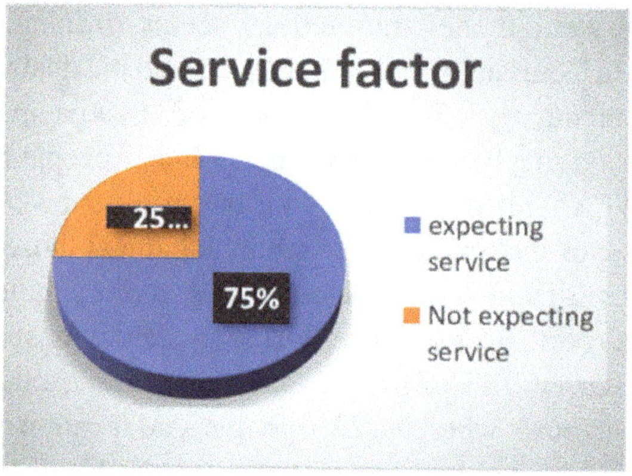

goes concurrently with the service we get vs the amount we paid. In our survey, **86%** of people said choosing govt. owned accommodation are worth for money, and **14%** said it's not valued.

5. **Expected Service** – In terms of service, intangible and irreplaceable object in the Hospitality industry. Tourists always expect an overwhelming and homely service. In our survey, **75%** of people said in Tourist Lodges they always expect to receive a standard service and homely ambiance. And **25%** of people said service was not the reason to choose.

Other than these, **acquittance reference** also plays a major role in choosing Govt. accommodation.

3.1 Suggestions and Recommendations

1. Since the accommodation units are situated in hilly / remote areas of Himalayan region of the West Bengal. So, the tourist should be provided with the facility of dedicated transportation option from nearest Railway (NJP) & Airport (Bagdogra) for enhancing their tourist experience.

2. The tourists should be provided with variety of cuisines including local cuisines in different accommodation units since the variety of foods are normally not available in remote regions.
3. To improve the satisfaction level of the tourist (Domestic /Foreigner) better internet facility, provision of 24hrs hot and cold water system, uninterrupted electricity supply, recreational indoor activities should be provided in those units.
4. Provision of elevator should be installed for facilitating the tourists mostly for the elder/ physically challenged tourists.
5. Trained driver cum guide facilities may be offered to the tourists for their local sight seen for better and hazardless tourist experience.
6. Since the activity options at evenings or leisure time is limited at these destinations, as a result the showcasing the cultural programmes and promoting local handicrafts will enhance the satisfaction level of the tourists.
7. Provision of bar and permit rooms for the tourists with flexible timings will enhance the tourist experience in these destinations.
8. For achieving better occupancy ratio in these accommodation units, promotion and marketing in the form of participating in international or national tourism related festivals as well as strong social media campaign will serve its purpose.

4. CONCLUSION

Government owned accommodation is having its own merits and demerits in providing proper satisfaction to the tourists especially in the hilly regions of West Bengal. The study conclude that the tourists are mainly comfortable in Govt. based accommodation in Himalayan region because they feel it is easy to rely on Govt. than on any unknown organization, it's secure in terms of local hazards, nicely located and ultimately they get value for money in terms of service and regular amenities including better service. Study also have proved that psychological security plays a vital role to choose hotels which they feel 'home-away home'. However, in this competitive world government accommodation should come up with better service facilities, variety of food facilities including maintenance of cleanliness and hygiene which are normally seen in the private hotels.

The accommodation unit owned by the government in Darjeeling, Kurseong and Kalimpong are found to be attractive for the tourists because of their locational advantage, historical significance and tourist attractions. Effort should be made to improve the infrastructural facilities, improving service standards, upgrading the existing facilities, improving the skills of the employees through imparting training on a regular basis will not only improve the business sustainability but also enhance the tourist experience to attract the repeat visitors and thereby generating profitability. Therefore, the concern department of the government of West Bengal who looks into the day to day function and activities of these above properties consider these above suggestions and recommendations so that more and more number of tourists will be interested in availing these government owned accommodations in this Himalayan region of West Bengal.

REFERENCES

Baruca, P. Z., & Civre, Z. (2012). *How do guests choose a hotel.* Academica Turistica- Tourism and Innovation Journal.

Callan, Roger & Bowman, Lisa. (2000). Selecting a hotel and determining salient quality attributes: A preliminary study of mature British travelers. *International Journal of Tourism Research.*

Chu, R., & Choi, T. (2000). *An importance-performance analysis of hotel selection factors in the Hong Kong hotel industry: A comparison of business and leisure travellers.* Tourism Management - Elsevier BV.

Clow, K. E., Garretson, J. A., & Kurtz, D. L. (1994). An exploratory study into the purchase decision process used by leisure travellers in hotel selection. *Journal of Hospitality & Leisure Marketing.*

Gunasekaran, N., & Anandkumar, V. (2012). Factors of Influence in Choosing Alternative Accommodation: A Study with Reference to Pondicherry, A Coastal Heritage Town. *Procedia: Social and Behavioral Sciences, 62*, 1127–1132. doi:10.1016/j.sbspro.2012.09.193

Hartline, M. D., Wooldridge, B. R., & Jones, K. C. (2003). Guest perceptions of hotel quality: Determining which employee groups count most. *The Cornell Hotel and Restaurant Administration Quarterly, 44*(1), 43–52. doi:10.1016/S0010-8804(03)90045-0

Pertiwi, P. & Sulistyawati, A. (2020). Factors of influence in choosing accommodation: a study with reference to Mas village, An Artistic Heritage Village In Ubud - Bali. *Journal of Business on Hospitality and Tourism.*

Swarbrooke, J., & Horner, S. (1999). *Consumer Behavior in Tourism.* Butterworth-Heinemann.

Tuan, L. (2019). *Study the factors affecting the choice of accommodation for new students: case in Da Nang City.* International Research Journal of Advanced Engineering and Science.

Wilkins, H. C. (2005). *A structural model of satisfaction and brand attitude in hotels. Griffith.* Griffith Business School, Griffith University.

Wuest, B. E. S., Tas, R. F., & Emenheiser, D. A. (1996). *What do mature travellers perceive as important hotel/motel customer service?* Hospitality Research Journal.

Yavas, U., & Babakus, E. (2005). Dimensions of hotel choice criteria: Congruence between business and leisure travelers. *International Journal of Hospitality Management, 24*(3), 359–367. doi:10.1016/j.ijhm.2004.09.003

Chapter 9
Fostering Sustainable Development Through Rural Tourism in the Himalayan Paradise of Sikkim

Anish Mondal
National Council of Hotel Management-Institute of Hospitality, Noida, India

ABSTRACT

The present review examines rural tourism in Sikkim as a potential driver of sustainable development. Sikkim's unique Himalayan landscape and rich culture make it an appealing destination. While global development often focuses on cities, rural tourism can economically and socially benefit local communities by promoting rural life, art, culture, and heritage. The study explores rural tourism's role in addressing poverty, employment, and community development in underdeveloped rural areas. It uses a descriptive approach to analyze factors contributing to sustainable rural growth and the promotion of Sikkim as a tourist destination. The findings highlight the potential for sustainable development, poverty reduction, community-based tourism, and the importance of pro-poor tourism, which prioritizes economic benefits for rural communities, environmental conservation, social empowerment, and cultural preservation in Sikkim.

1. INTRODUCTION

Rural tourism in Sikkim is a growing industry that holds significant potential for sustainable development and the empowerment of local communities. According to the Tourism and Civil Aviation Department of the Government of Sikkim, there has been a decline in the number of visitors from 2010 to 2021, with less than 25 visitors, mostly domestic, and less than 5 visitors on an annual basis (DEMKOVA et al., 2022). However, recognizing the importance of rural tourism in promoting economic and social well-being, the government has taken steps to promote community-based tourism in rural areas.

DOI: 10.4018/979-8-3693-0823-3.ch009

From an economic perspective, rural tourism in Sikkim has been regarded as an effective strategy for sustainable social and economic development (Epuran et al., 2020). Rural tourism provides an opportunity for rural villages to revitalize their communities by utilizing and commodifying existing local resources. Furthermore, it contributes to the development of infrastructure, improvement in the quality of life, increase in incomes for the local population, job creation, and the preservation of local traditions and culture. The development of rural tourism also plays a crucial role in the preservation of local identity, tradition, and customs (Vidović, 2018). By engaging tourists in local activities, such as traditional crafts, agricultural practices, and cultural events, rural tourism helps to safeguard and promote the unique heritage of Sikkim.

Moreover, rural tourism in Sikkim has a positive impact on the environment by promoting sustainable practices and enhancing the conservation of natural resources. It encourages the adoption of organic farming methods, which minimize the use of chemicals and reduce the ecological footprint. Furthermore, rural tourism in Sikkim contributes to the overall goals of sustainable development by diversifying the economy and reducing dependency on traditional agriculture. Rural tourism in Sikkim also contributes to the overall development of the national economy. The development of rural tourism in Sikkim not only benefits the local communities but also has a positive impact on the national economy. The revenue generated from rural tourism in Sikkim contributes to the overall GDP and helps to strengthen various sectors such as agriculture, construction, transportation, and hospitality. Additionally, rural tourism in Sikkim can contribute to the reduction of income disparities between urban and rural areas. The government's promotion of community-based rural tourism in Sikkim is based on recognizing the importance of achieving sustainable development in economic, ecological, and social aspects. In the context of Sikkim, rural tourism has been recognized as an effective strategy for achieving sustainable social and economic development.

Thus, rural tourism in Sikkim has emerged as a vital tool for rural development by leveraging local resources, preserving cultural heritage, promoting sustainable practices, and contributing to the overall economy of the region. Rural tourism in Sikkim has become an integral element of rural economies, generating additional employment and income for farmers and small communities. Particularly in marginal areas, rural tourism creates entrepreneurship opportunities for small and medium local rural enterprises and offers new forms of travel experiences (Khartishvili et al., 2019). Rural tourism in Sikkim not only generates additional employment and income for farmers and small communities, but it also creates entrepreneurship opportunities for small and medium local rural enterprises. It fosters economic diversification and reduces dependence on traditional agricultural activities, thus contributing to the overall development of the national economy. Rural tourism in Sikkim plays a significant role in promoting sustainable development, economic diversification, and reducing income disparities between urban and rural areas. Rural tourism in Sikkim contributes to the overall development of the national economy by generating additional employment and income for farmers and small communities. Moreover, it creates entrepreneurship opportunities for small and medium local rural enterprises, thereby fostering economic diversification.

1.1 Exploring the Beauty of Sikkim's Countryside

To enhance rural tourism in Sikkim, the government has identified and recognized several eco-treks in the state. These include the Yoksom-Dzongri-Goechela eco-treks in West Sikkim, the Tholung-Kishong in Dzongu Valley, North Sikkim, and the Golitar-Tinjurey and Pastanga-Khedi eco trails in Rongni and

Taktsom Chu valleys, East Sikkim (DEMKOVA et al., 2022). Rural tourism in Sikkim offers tourists the opportunity to explore the scenic beauty and cultural heritage of the state's countryside. Moreover, rural tourism in Sikkim contributes to the sustainable development of rural areas by improving infrastructure, enhancing the quality of life, increasing local incomes, creating job opportunities, supporting, and preserving local traditions and culture, and promoting environmental preservation. Rural tourism in Sikkim is not only a means of economic development but also an avenue for educating tourists on sustainable lifestyles and the preservation of natural resources. Rural tourism in Sikkim holds great promise for sustainable development and the well-being of local communities. Rural tourism in Sikkim is a form of sustainable economic development in tourism that aims to achieve certain economic benefits while effectively improving resource management for the benefit of local communities and future generations. The growth of rural tourism in Sikkim should be in harmony with the local ecological status, biodiversity, and cultural heritage. This approach ensures the integrity of ecological resources, maintains a good ecological system in the countryside, promotes the construction of rural landscapes, and improves the rural living environment. Rural tourism developers and local relevant departments and governments in Sikkim need to adopt a holistic view of rural tourism development (Mei & Han, 2022). They should consider the economic, social, and environmental aspects to ensure its long-term success.

The growth of rural tourism in Sikkim is closely tied to the overall economic and social reforms that have taken place in recent years. The number of visitors to Sikkim has declined sharply in recent years, with less than 25 visitors, mostly domestic, and less than 5 international tourists per year. However, rural tourism in Sikkim has the potential to revitalize the tourism industry and attract a larger number of visitors. Sikkim has provided valuable insights into the potential of rural tourism in Sikkim. According to the report "Our Common Journey: A Transition toward Sustainability," sustainable tourism is characterized by three interconnected aspects: environmental, socio-cultural, and economic. In the context of rural tourism in Sikkim, sustainability refers to the integration of these three aspects - environmental, socio-cultural, and economic - to ensure long-term benefits for the local communities and the preservation of natural resources. Rural tourism in Sikkim is an activity that seeks to achieve sustainability by simultaneously promoting economic development, preserving natural resources, and fostering socio-cultural interactions within the local communities. This approach to rural tourism in Sikkim recognizes the importance of balancing economic development with ecological and cultural preservation. In order to achieve sustainable development in rural tourism in Sikkim, it is crucial for stakeholders such as rural tourism developers, local departments, and governments to collaborate closely and adopt an integrated approach that takes into account the economic, social, and environmental aspects of tourism development. Rural tourism in Sikkim has the potential to contribute significantly to the economic, social, and environmental development of the region.

The sustainable development of rural tourism in Sikkim can play a vital role in improving the overall economic conditions and demographic trends. By diversifying the rural economy towards ecological, economic, and social sustainable development, rural tourism in Sikkim can contribute to improved quality of life for the local people, reduction of poverty, and mitigation of social and environmental degradation (Li et al., 2020). Furthermore, rural tourism in Sikkim can help in the sustainability and regeneration of rural areas. Rural tourism in Sikkim has the potential to foster economic growth by attracting tourists and generating income for local communities. Additionally, rural tourism offers a unique opportunity for urban tourists to escape the stresses of city life and experience an authentic rural environment (Chroneos-Krasavac et al., 2018). Moreover, rural tourism in Sikkim can play a crucial role in preserving the local culture and environment. In order to achieve sustainable rural tourism development in Sikkim, it is important to implement appropriate policies and strategies that ensure the benefits of tourism are maximum.

1.2 Understanding the Impact of Tourism on Sikkim's Economy

Tourism in general has been recognized as a tool for revitalizing rural areas and driving economic growth. Rural tourism in Sikkim plays a significant role in this aspect by offering economic and social benefits that can boost rural development.

Several studies have highlighted the positive impact of rural tourism on improving the rural economy (Shin et al., 2017). For instance, tourism in rural areas can diversify economies that are heavily reliant on agriculture, which is susceptible to weather and market fluctuations. This diversification can help rural communities mitigate the risks associated with agricultural dependency and enhance their economic stability. Moreover, Sikkim has the potential to accelerate economic transformation by tapping into its tourism potential. By improving infrastructure, promoting local entrepreneurship, and creating job opportunities in sectors such as hospitality, transportation, and handicrafts, rural tourism can stimulate the growth of other economic sectors and contribute to overall rural development. Additionally, Sikkim has the ability to enhance the quality of life for local residents. Improved infrastructure in rural areas, such as better roads and transportation systems, not only benefit tourists but also improve the accessibility of basic services for the local community, such as healthcare and education. Furthermore, the income generated from rural tourism can uplift the living standards of local residents and reduce poverty levels. While rural tourism in Sikkim can bring about significant economic benefits, it is crucial to approach its development with a holistic view that takes into account the sustainable development of the environment and the preservation of the local culture and heritage.

Tourism has become a key driver of Sikkim's economy (Rizal & Asokan, 2016). The influx of tourists has led to the creation of jobs in various sectors such as hospitality, transportation, and handicrafts. The tourism sector contributes significantly to the state's GDP (Bhat, Z. A., 2013). In fact, Sikkim's economy heavily depends on tourism, making it one of the primary sources of revenue for the state government (Sattar, S., 2013). Tourism also has several socioeconomic benefits for the local population. It has helped in poverty alleviation by providing employment opportunities (Muganda et al., 2010). Additionally, tourism has spurred infrastructure development, such as road construction and the establishment of healthcare facilities, which benefit both tourists and local residents. While tourism has brought economic and social benefits, it also raises environmental concerns. The fragile Himalayan ecosystem in Sikkim is vulnerable to the negative impacts of tourism, including deforestation and pollution (Mandal, J., & Sengupta, P., 2015). Sustainable tourism practices are essential to mitigate these adverse effects and ensure the long-term viability of the industry (Uddin et al., 2023). Tourists' expenditure plays a crucial role in stimulating local businesses. Many local entrepreneurs have benefited from the tourism industry, particularly in the form of homestays and small-scale handicraft businesses (Deyshappriya & Nawarathna, 1970). The spending by tourists not only supports the local economy but also helps in the preservation of indigenous cultures and traditions (KC et al., 2021).

Despite its positive impact, the tourism sector in Sikkim faces challenges, including seasonality and overreliance on a single industry (Sood, J., Lynch, P., & Anastasiadou, C., 2017). The COVID-19 pandemic further highlighted the vulnerability of the tourism-dependent economy. However, the state has an opportunity to diversify its tourism offerings by promoting adventure tourism, eco-tourism, and cultural tourism (Burman et al., 2007; Dam, S. 2013).

1.3 Cultural Experiences in Rural Sikkim

Sikkim have garnered increasing scholarly attention in recent years due to their significance in preserving and understanding the rich tapestry of Sikkimese culture. Sikkim, nestled in the Himalayan region of India, boasts a unique cultural amalgamation of indigenous traditions, Tibetan Buddhism, and Nepalese influences, all of which manifest vividly in its rural communities. Scholars like (Bhutia, K. D., 2018), have explored the profound impact of Buddhism on rural Sikkimese life, emphasizing its role in shaping rituals, festivals, and everyday practices. Additionally, studies by (Pradhan, P., 2016), have illuminated the vibrant ethnic diversity of Sikkim's rural areas, with a focus on the preservation of traditional languages, dances, and cuisines among various indigenous groups. Furthermore, the work of (DEMKOVA et al., 2022), has underscored the importance of sustainable tourism in rural Sikkim, highlighting its potential to promote and protect local cultural heritage. These studies collectively underscore the complex interplay between tradition and modernity in rural Sikkim and emphasize the need for holistic approaches to safeguard its cultural richness for future generations.

Researchers have emphasized the significance of religious festivals, such as Losar and Dashain, as pivotal moments for communal bonding and the preservation of traditional practices (Chettri, P., 2022). Furthermore, Sikkim's rural communities are characterized by their resilience in preserving indigenous knowledge systems, agricultural practices, and handicraft traditions, which are critical to their cultural identity (Lepcha, O, 2022). As tourism continues to grow in the region, there is a pressing need to investigate the impact of modernization on these rural cultures and the strategies adopted by local communities to negotiate change while safeguarding their heritage. This literature review highlights the emerging trends in research on cultural experiences in rural Sikkim, shedding light on the evolving dynamics of tradition and modernity in this captivating Himalayan landscape.

Arora, V. (2006) have emphasized the pivotal role of indigenous festivals, such as Losar and Saga Dawa, in preserving and transmitting the cultural heritage of the region. Additionally, works by Chettri, L., & Sharma, B. K. (2021), have delved into the intricacies of traditional dance forms like Chham and Maruni, highlighting their role in fostering communal identity and cohesion. Furthermore, studies like Lama, R., & Tyagi, N. (2019), have underscored the symbiotic relationship between cultural tourism and rural livelihoods, shedding light on the economic dimensions of cultural experiences in Sikkim's hinterlands. These collective efforts contribute to a comprehensive understanding of the multifaceted cultural tapestry that characterizes rural Sikkim and underscore the importance of preserving and celebrating its unique heritage.

Research on cultural experiences in rural Sikkim has gained increasing attention in recent years due to the region's rich and diverse cultural heritage. Sikkim is home to a multitude of ethnic communities, including the Bhutias, Lepchas, and Nepalis, each with their unique traditions, languages, and rituals (Shrestha, B. G., 2015). Scholars have explored various aspects of these cultural experiences, from traditional festivals like Losar and Bumchu to the preservation of indigenous knowledge systems (Maharana et al., 2009). Additionally, the impact of modernization and tourism on rural Sikkim's cultural dynamics has been a subject of research interest, shedding light on the challenges and opportunities for preserving traditional practices and fostering sustainable development (Ueda, A., 2003;Lama, R., & Tyagi, N., 2019) As this body of research continues to expand, it contributes to our understanding of the intricate interplay between culture, identity, and socio-economic changes in the context of rural Sikkim.

1.4 Eco-Tourism Initiatives to promote the rural tourist destinations.

Sikkim has gained prominence due to its potential for promoting conservation efforts while providing livelihood opportunities to local communities. The state's government has recognized the importance of eco-tourism and has initiated several projects to develop eco-friendly tourism infrastructure in rural areas (Government of Sikkim, 2018). Sikkim's commitment to preserving its fragile ecosystem while promoting tourism has led to the development of eco-tourism. The state government has taken several measures to encourage eco-friendly practices in the tourism sector. Initiatives such as creating nature-based tourism circuits and promoting homestays in rural areas aim to balance tourism growth with environmental preservation (Envis Hub: Sikkim status of environment and related issues). The Government of Sikkim has played a pivotal role in fostering eco-tourism. The Sikkim Tourism Policy (2018) highlights the importance of sustainable tourism development. The policy emphasizes community participation, local empowerment, and the protection of natural resources. In addition, the state government has actively collaborated with various organizations to implement eco-friendly practices. One of the most significant eco-tourism initiatives in Sikkim is the promotion of community-based tourism (CBT). CBT encourages rural communities to actively participate in and benefit from tourism activities. Initiatives like the Khangchendzonga Conservation Committee (KCC) have been successful in engaging local communities in the conservation of their natural surroundings and promoting traditional cultural experiences for tourists (Sikkim. KCC. n.d). Sikkim's rich biodiversity is a prime attraction for eco-tourists. Conservation initiatives, such as the declaration of Khangchendzonga National Park as a UNESCO World Heritage Site, demonstrate the state's commitment to preserving its unique flora and fauna. The park's protection has resulted in increased eco-tourism activities, contributing to both local economies and conservation efforts (UNESCO, 2021).

While Sikkim's eco-tourism initiatives have shown promise, they are not without challenges. Infrastructure development in remote rural areas remains a challenge, hindering accessibility. Moreover, the delicate balance between tourism development and environmental preservation is often difficult to maintain. Overcrowding, pollution, and waste management issues are concerns that require continuous monitoring and mitigation (Hardoy, J. E., & Satterthwaite, D., 1991).

Sikkim's eco-tourism initiatives have significantly contributed to the promotion of rural tourism while fostering sustainable practices and community involvement. The government's policies, community-based approaches, and biodiversity conservation efforts have set positive examples for eco-tourism development. However, addressing infrastructure challenges and maintaining the ecological balance remain ongoing concerns. Continuous collaboration between stakeholders and adaptive strategies will be crucial to ensuring the long-term success of eco-tourism in Sikkim.

1.5 Significance of Local Participation in Rural Tourism

The significance of local participation in rural tourism has garnered substantial attention in academic discourse, as it lies at the intersection of sustainable tourism development and community empowerment. Local participation is considered a critical determinant of the success and sustainability of rural tourism initiatives (Komppula, 2014). It fosters a sense of ownership and responsibility among local residents, leading to the preservation of cultural heritage, natural resources, and the overall enhancement of the visitor experience (El Moslem Badr, 2022). Additionally, local participation in planning and decision-making processes empowers rural communities economically by creating opportunities for income generation, employment, and the promotion of local businesses (Chirenje et al., 2013). Furthermore, it contributes to the dispersal of economic benefits beyond urban centers, addressing issues of rural development and alleviating poverty (Andereck & Nyaupane, 2011). As such, the significance of local participation in rural tourism is not only pivotal for the sustainability of tourism ventures but also for the holistic development of rural areas. This literature review explores the multifaceted dimensions and outcomes of local participation in rural tourism, shedding light on its vital role in achieving sustainable and inclusive tourism practices.

Local participation is increasingly recognized as a crucial factor in the sustainable development of rural tourism destinations (Komppula, 2014) The active involvement of local communities in planning, decision-making, and resource management not only fosters a sense of ownership and pride but also ensures that the benefits of tourism are distributed more equitably (Park & Kim, 2015). Furthermore, local participation can lead to the preservation of cultural heritage, promotion of authentic experiences, and the enhancement of the overall visitor experience (Mowforth & Munt, 2015; Richards & Hall, 2000). This study underscores the multifaceted significance of local participation in rural tourism, shedding light on its potential to foster sustainability, economic growth, and cultural preservation, thereby making it a pivotal area of investigation in contemporary tourism research.

Local participation empowers local residents to actively engage in tourism-related activities such as homestays, handicraft production, and guided tours (Romario et al., 2023). By actively involving the local population, rural tourism not only contributes to community development and economic diversification but also fosters cultural preservation and environmental conservation (Quaranta et al., 2016). Furthermore, it is important to recognize that the success of rural tourism in Sikkim hinges on striking a balance between the economic benefits and the preservation of the region's unique cultural and environmental assets (Wearing, S., & Neil, J., 2009). As Sikkim continues to attract tourists seeking authentic and sustainable experiences, understanding the significance of local participation is vital for policymakers, tourism practitioners, and scholars in ensuring the long-term success of rural tourism in the region.

1.6 Promoting Sustainable Tourism in Sikkim

Sustainable tourism, as defined by the United Nations World Tourism Organization (UNWTO), refers to tourism that meets the needs of present tourists and host regions while protecting and enhancing opportunities for the future. It involves the responsible management of all resources involved in tourism, including environmental, socio-cultural, and economic aspects (UNWTO, 2005). In the context of Sikkim, sustainable tourism implies preserving the state's fragile ecosystems, promoting local culture, and ensuring equitable distribution of economic benefits. Sikkim's pristine environment is one of its main attractions, but it is susceptible to degradation due to uncontrolled tourism activities. Issues such as deforestation, littering, and excessive water usage pose significant challenges (Onuoha, C. A., 2022). Rapid tourism growth demands infrastructure development, which can harm fragile ecosystems. Balancing infrastructure needs with environmental conservation is crucial.

Sikkim is rich in indigenous culture and traditions. The influx of tourists can lead to the commodification of these cultural elements (Macleod, D., 2006). It is vital to protect and promote the local way of life while engaging tourists. Involving local communities in decision-making and the tourism value chain is essential for sustainable tourism.

Encouraging eco-friendly tourism practices, such as waste management, energy conservation, and wildlife protection, can mitigate environmental impacts (Kim, A., Kim, K. P., & Nguyen, T. H. D.,

2021). Implementing strict regulations and regular monitoring of tourism activities can help control environmental degradation (Farrukh, B., Younis, I., & Longsheng, C., 2023). Developing and promoting cultural tourism experiences that are respectful of local traditions can help preserve Sikkim's rich heritage (Newpaney, R., & Lee, S., 2016). Supporting community-based tourism initiatives allows local residents to benefit economically from tourism and fosters a sense of ownership (Lapeyre, R., 2010). Providing training and education to both tourists and local communities on sustainable tourism practices can create awareness and change behaviour (Kiper, 2013).

1.7 Key Challenges and Opportunities in Sikkim's Rural Tourism

Sikkim's mountainous terrain presents significant challenges in terms of infrastructure development and accessibility to rural areas (Tambe, S., Arrawatia, M. L., & Ganeriwala, A. K., 2012). Limited road connectivity and the absence of robust transportation networks hinder the flow of tourists into remote regions. The fragile Himalayan ecosystem is under constant threat due to unchecked tourism activities (Singh, R. B., Mal, S., & Kala, C. P., 2009). Maintaining a balance between promoting tourism and preserving the environment remains a persistent challenge. The influx of tourists can lead to the commodification of indigenous cultures (Dorsey, E. R., Steeves, H. L., & Porras, L. E, 2004).). Striking a balance between showcasing traditions and avoiding cultural appropriation is an ongoing challenge. Rural communities often lack the necessary skills and resources to cater to tourists effectively (Liu, A., 2006). Training programs and capacity building initiatives are required to enhance local participation in tourism.

Engaging local communities in tourism initiatives can empower them economically and socially (Scheyvens, R., 2003). Community-based tourism models provide an opportunity for sustainable growth. Sikkim's rich cultural diversity can be leveraged to create unique and authentic experiences for tourists (Sattar, S., 2020). Traditional festivals, cuisine, and handicrafts can be highlighted. The state's biodiversity and pristine landscapes offer opportunities for ecotourism (Phil, T. M., 2022). Emphasizing responsible and low-impact tourism can protect the environment. The government of Sikkim has launched several initiatives to promote rural tourism, including incentives and policy support (Sikkim Tourism, n.d.). These initiatives create a favourable environment for investment.

Sikkim's rural tourism sector holds significant promise for the state's economic development and the empowerment of rural communities. However, addressing the challenges such as infrastructure, environmental sustainability, cultural preservation, and capacity building is imperative for the long-term success of rural tourism. Leveraging opportunities like community-based tourism, cultural experiences, ecotourism, and government support can help overcome these challenges and ensure a sustainable and enriching rural tourism experience in Sikkim.

1.8 Recommendations for Future Development of Sikkim's Rural Tourism

One key challenge in the development of rural tourism in Sikkim is the need for sustainable tourism practices that prioritize environmental conservation. Implementing eco-tourism initiatives and promoting responsible tourism practices can help preserve the natural beauty and biodiversity of rural areas in Sikkim. These initiatives include promoting environmentally friendly modes of transportation, encouraging waste management, and recycling practices, and educating tourists about the importance of preserving the local environment. Another challenge is ensuring the active participation of local communities in the development and management of rural tourism. This can be achieved by promoting community-based tourism initiatives, where local residents are actively involved in decision-making processes and benefit directly from tourism activities. Additionally, it is crucial to preserve and showcase the unique cultural heritage of rural Sikkim. This can be done through initiatives that promote cultural experiences and interactions between tourists and local communities. For example, organizing cultural festivals, handicraft workshops, or traditional food experiences can not only provide visitors with an authentic cultural experience but also contribute to the socio-economic development of local communities. Moreover, it is important to address the issue of rural poverty and provide better prospects for the future of rural youth.

One way to address this challenge is by creating opportunities for skill development and entrepreneurship in the rural tourism sector. This can include training programs for local residents in hospitality, guiding, and other tourism-related skills, as well as providing support for the establishment of small businesses in the tourism sector. Thus, the development of rural tourism in Sikkim should prioritize sustainable practices that ensure environmental conservation, preserve cultural heritage, and promote the active participation of local communities. By implementing eco-tourism initiatives and encouraging responsible tourism practices, Sikkim can protect its natural resources and foster sustainable development in rural areas. Promoting responsible tourism practices can contribute to the preservation of Sikkim's natural beauty and biodiversity. Additionally, integrating environmental concerns into the planning, management, and monitoring of rural tourism actions is crucial for sustainable development (Apostolopoulos et al., 2020). This can include implementing measures such as waste management systems, promoting energy-efficient practices, and raising awareness among tourists about the importance of preserving the local environment. Incorporating these measures will not only contribute to the long-term sustainability of rural tourism in Sikkim but also provide local communities with economic opportunities and improve their quality of life. Moreover, it is important that the local governments and NGOs work in collaboration with the communities to ensure their ownership and participation in the planning and management of tourism initiatives. Additionally, it is essential to consider the carrying capacity of rural areas in Sikkim to prevent over-tourism and avoid irreversible damage to the environment and the local communities. Rural tourism in Sikkim has the potential to play a significant role in the economic development of local communities. Through the implementation of rural homestays and community-based ecotourism initiatives, rural tourism can provide alternative livelihood options for disadvantaged indigenous communities in Sikkim. This can help to reduce dependence on traditional livelihoods and contribute to poverty alleviation.

2. CONCLUSION: THE FUTURE OF RURAL TOURISM IN SIKKIM

In conclusion, rural tourism in Sikkim has the potential to contribute significantly to the overall development of the region. By leveraging the natural and cultural resources of rural areas, rural tourism can create employment opportunities, support local businesses, and preserve the unique heritage of host communities (An & Alarcón, 2020). However, it is imperative that the development of rural tourism in Sikkim prioritizes sustainable practices to ensure long-term environmental conservation and cultural preservation. These practices should include waste management, energy efficiency, and raising awareness among tourists about the importance of preserving the local environment. Furthermore, active involvement and ownership of local communities in the planning and management of tourism initiatives is essential to ensure their well-being and economic empowerment. Overall, the future of rural tourism in Sikkim relies on sustainable development practices that balance economic growth with environmental protection and cultural preservation.

REFERENCES

An, W., & Alarcón, S. (2020, September 19). How Can Rural Tourism Be Sustainable? *A Systematic Review*. https://scite.ai/reports/10.3390/su12187758

Andereck, K. L., & Nyaupane, G. P. (2011). Exploring the nature of tourism and quality of life perceptions among residents. *Journal of Travel Research*, *50*(3), 248–260. doi:10.1177/0047287510362918

Apostolopoulos, N., Liargovas, P., Stavroyiannis, S., Makris, I. A., Apostolopoulos, S., Petropoulos, D. P., & Anastasopoulou, E. (2020, September 17). *Sustaining Rural Areas, Rural Tourism Enterprises and EU Development Policies: A Multi-Layer Conceptualisation of the Obstacles in Greece*. ScitePress. https://scite.ai/reports/10.3390/su12187687

Arora, V. (2006). The forest of symbols embodied in the Tholung sacred landscape of North Sikkim, India. *Conservation & Society*, 55–83.

Bhat, Z. A. (2013). Tourism industry and pilgrimage tourism in Jammu and Kashmir: Prospects and challenges. *International Journal of Research in Management & Technology*, *2*, 105–113.

Bhutia, K. D. (2018). Foxes, Yetis, and Bulls as Lamas: Human-Animal Interactions as a Resource for Exploring Buddhist Ethics in Sikkim. *Journal of Buddhist Ethics*, *25*, 45–70.

Burman, P. D., Cajee, L., & Laloo, D. D. (2007). Potential for cultural and eco-tourism in Northeast India: A community-based approach. *WIT Transactions on Ecology and the Environment*, 102.

Chettri, L., & Sharma, B. K. (2021). *Some Traditional Folk Songs and Dances of Sikkim Himalayas*.

Chettri, P. (2022). *Exploring the Potentiality of Cultural Tourism in Sikkim: A Study on Lepcha*.

Chirenje, L. I., Giliba, R. A., & Musamba, E. B. (2013). Local communities' participation in decision-making processes through planning and budgeting in African countries. *Zhongguo Renkou Ziyuan Yu Huanjing*, *11*(1), 10–16. doi:10.1080/10042857.2013.777198

Chroneos-Krasavac, B., Radosavljević, K., & Bradić-Martinović, A. (2018, January 1). *SWOT analysis of the rural tourism as a channel of marketing for agricultural products in Serbia*. ScitePress. https://scite.ai/reports/10.5937/ekopolj1804573k

Dam, S. (2013). Issues of sustainable ecotourism development in Sikkim: An analysis. *South Asian Journal of Tourism and Heritage*, *6*(2), 32–47.

Deyshappriya, N. P. R., & Nawarathna, A. M. D. B. (1970, January 1). Tourism and SME Development: Performance of tourism smes in coastal tourist destinations in southern Sri Lanka. *EconStor*. https://www.econstor.eu/handle/10419/238521

Dorsey, E. R., Steeves, H. L., & Porras, L. E. (2004). Advertising ecotourism on the internet: Commodifying environment and culture. *New Media & Society*, *6*(6), 753–779. doi:10.1177/146144804044328

El Moslem Badr, D. M. (2022). Challenges and future of the development of sustainable ecotourism. *International Journal of Modern Agriculture and Environment*, *2*(2), 54–72. doi:10.21608/ijmae.2023.214937.1006

Farrukh, B., Younis, I., & Longsheng, C. (2023). The impact of natural resource management, innovation, and tourism development on environmental sustainability in low-income countries. *Resources Policy*, *86*, 104088. doi:10.1016/j.resourpol.2023.104088

Government of Sikkim. (2018). *Sikkim Tourism Policy 2018-2023*. Retrieved from. [URL]

Hardoy, J. E., & Satterthwaite, D. (1991). Environmental problems of Third World cities: A global issue ignored? *Public Administration and Development*, *11*(4), 341–361. doi:10.1002/pad.4230110405

K C, A., Ghimire, S., & Dhakal, A.KC. (2021). Ecotourism and its impact on indigenous people and their local environment: Case of Ghalegaun and Golaghat of Nepal. *GeoJournal*, *86*(6), 2747–2765. doi:10.1007/s10708-020-10222-3

Khartishvili, L., Muhar, A., Dax, T., & Khelashvili, I. (2019, January 15). *Rural Tourism in Georgia in Transition: Challenges for Regional Sustainability*. ScitePress.https://scite.ai/reports/10.3390/su1102041

Kim, A., Kim, K. P., & Nguyen, T. H. D. (2021). The green accommodation management practices: The role of environmentally responsible tourist markets in understanding tourists' pro-environmental behaviour. Sustainability, 13(4), 2326.

Kiper, T. (2013). *Role of ecotourism in Sustainable Development*. Advances in Landscape Architecture. doi:10.5772/55749

Komppula, R. (2014). The role of individual entrepreneurs in the development of competitiveness for a rural tourism destination – A case study. *Tourism Management*, *40*, 361–371. doi:10.1016/j.tourman.2013.07.007

Lama, R., & Tyagi, N. (2019). Community-Based Tourism: A Tool for Sustainable Development (Case Study of Yuksom, Sikkim). *Emerging Trends in Indian Tourism and Hospitality: Transformation and Innovation*, 185.

Lapeyre, R. (2010). Community-based tourism as a sustainable solution to maximise impacts locally? The Tsiseb Conservancy case, Namibia. *Development Southern Africa*, *27*(5), 757–772. doi:10.1080/0376835X.2010.522837

Lepcha, O. (2022). *Youth Perception on Indigenous Knowledge and Ethnomedicinal Plants: A Study of Dzongu*.

Li, H., Nijkamp, P., Xie, X., & Liu, J. (2020, April 14). *A New Livelihood Sustainability Index for Rural Revitalization Assessment—A Modelling Study on Smart Tourism Specialization in China*. ScitePress. https://scite.ai/reports/10.3390/su12083148

Liu, A. (2006). Tourism in rural areas: Kedah, Malaysia. *Tourism Management*, *27*(5), 878–889. doi:10.1016/j.tourman.2005.05.007

Macleod, D. (2006). Cultural commodification and tourism: A very special relationship. *Tourism, Culture & Communication*, *6*(2), 71–84. doi:10.3727/109830406777410580

Maharana Uprety. (2009). *Iyatta & Sharma, Ghanashyam*. Cultural Attributes, Economic Valuation and Community Conservation in Holy Khecheopalri Lake of Sikkim in the Eastern Himalaya.

Mandal, J., & Sengupta, P. (2015). The impact of tourism on livelihood and environment in West Sikkim: A case study of Pelling. In ILEE (Vol. 38, Issue 1).

Mei, R., & Han, L. (2022, September 1). *Evaluation System of Rural Sustainable Tourism Land Based on Ecosystem Service Value*. ScitePress. https://scite.ai/reports/10.2478/eces-2022-0025

Mowforth, M., & Munt, I. (2015). *Tourism and Sustainability: New Tourism in the Third World*. Routledge. doi:10.4324/9781315795348

Muganda, M., Sahli, M., & A Smith, K. (2010). Tourism's contribution to poverty alleviation: A community perspective from Tanzania. *Development Southern Africa*, *27*(5), 629–646. doi:10.1080/0376835X.2010.522826

Newpaney, R., & Lee, S. (2016). Mountain Tourism Development and Impact of Tourism: A case study of Himalayan State Sikkim. *호텔경영학연구*, *25*(7), 329-349.

Onuoha, C. A. (2022). ENVIRONMENTAL CHALLENGES AWARENESS IN NIGERIA. *RE:view*.

Park, E., & Kim, S. (2015). The potential of cittaslow for sustainable tourism development: Enhancing Local Community's empowerment. Tourism Planning & amp. *Tourism Planning & Development*, *13*(3), 351–369. doi:10.1080/21568316.2015.1114015

Phil, T. M. (2022). *Ecotourism: A Sustainable Development Connect to Nature and A Strategy for Balancing Economic Growth*. Socio-Cultural Development and Conservation.

Pradhan, P. (2016). *Culture and Tourism: A Study of Sikkim* [Doctoral dissertation, Sikkim University].

Quaranta, G., Citro, E., & Salvia, R. (2016). Economic and social sustainable synergies to promote innovations in rural tourism and Local Development. *Sustainability (Basel)*, *8*(7), 668. doi:10.3390/su8070668

Richards, G., & Hall, D. (2000). *Tourism and Sustainable Community Development*. Routledge.

Rizal, P., & Asokan, R. (2016, September 16). *Measuring the impact of tourism industry on regional economy of Sikkim State, India*. SSRN. https://ssrn.com/abstract=2839694

Romario De Fretes, D., Caroline, C., Amin, A.-., Sirait, S., & Rizani, A. (2023). Exploring the economic impacts of sustainable tourism on locals: a comparative analysis between bali and north sumatera. *International journal of humanities, social sciences and business (INJOSS)*, *2*(3), 387–417. https://doi.org/. doi:10.54443/injoss.v2i3.88

Sattar, S. (2013). Potential of Tourism in Poverty Alleviation and Development of Sikkim. *IASSI-Quarterly*, *32*(3), 74–87.

Sattar, S. (2020). Culture, Creativity and Tourism: A Study of Sikkim. In The Cultural Heritage of Sikkim (pp. 289-302). Routledge.

Scheyvens, R. (2003). Local involvement in managing tourism. In *Tourism in destination communities* (pp. 229–252). CABI Publishing. doi:10.1079/9780851996110.0229

Shin, H., Kim, H. N., & Son, J. (2017, April 18). *Measuring the Economic Impact of Rural Tourism Membership on Local Economy: A Korean Case Study*. ScitePress. https://scite.ai/reports/10.3390/su9040639

Shrestha, B. G. (2015). *The Newars of Sikkim: reinventing language, culture and identity in the diaspora*. Vajra Books.

Singh, R. B., Mal, S., & Kala, C. P. (2009). Community responses to mountain tourism: A case in Bhyundar Valley, Indian Himalaya. *Journal of Mountain Science*, *6*(4), 394–404. doi:10.1007/s11629-009-1054-y

Sood, J., Lynch, P., & Anastasiadou, C. (2017). Community non-participation in homestays in Kullu, Himachal Pradesh, India. *Tourism Management*, *60*, 332–347. doi:10.1016/j.tourman.2016.12.007

Tambe, S., Arrawatia, M. L., & Ganeriwala, A. K. (2012). Managing rural development in the mountain state of Sikkim, India. *Mountain Research and Development*, *32*(2), 242–252. doi:10.1659/MRD-JOURNAL-D-12-00005.1

Uddin, H., Ahammed, S., Rana, M. M., & Majumder, S. C. (2023). Investigating the relationship between environmental quality and tourism industry in Thailand. *Environment, Development and Sustainability*. Advance online publication. doi:10.1007/s10668-023-03801-0

Ueda, A. (2003). *Culture and modernisation: From the perspectives of young people in Bhutan.*

UNESCO. (2021). *Khangchendzonga National Park*. UNESCO. https://whc.unesco.org/en/list/1513

Vidović, B. (2018, January 1). *Ecotourism management for development countries*. ScitePress. https://scite.ai/reports/10.5937/tehnika1802293v

Wearing, S., & Neil, J. (2009). *Ecotourism: Impacts, potentials, and possibilities?* Routledge. doi:10.4324/9780080940182

Chapter 10
Impact of Climate Change on Himalayan Tourism Destinations

Sanjeev Kumar Saxena
https://orcid.org/0000-0002-2139-9255
Assam University, India

Vandana Gupta
LM College of Science and Technology, Jodhpur, India

Wahengbam Priyalakshmi
Assam University, India

Mehzabin Ferdaus
Assam University, India

ABSTRACT

Climate change poses significant challenges to Himalayan tourism destinations, affecting their landscapes, ecosystems, and local communities. This article investigates the various impacts of climate change on these destinations and explores the implications for tourism activities. It also explores the consequences of altered weather patterns on adventure tourism activities and addresses the impact on local flora and fauna. Additionally, the socio-economic repercussions for local communities and the challenges faced by indigenous populations and cultural heritage preservation are discussed. The article emphasizes the importance of sustainable tourism practices to mitigate these impacts and foster resilience. Collaboration between local communities, governments, and tourism stakeholders is crucial, as is the development of effective policies at national and international levels. The study concludes by highlighting the need for further research to address the complexities of climate change impacts on Himalayan tourism destinations.

DOI: 10.4018/979-8-3693-0823-3.ch010

1. INTRODUCTION

Stretching across various countries in South Asia, the majestic mountain range of The Himalayas is known as the "Roof of the World". The Himalayas is a beauty to behold with its towering peaks, breathtaking landscapes and diverse ecosystems. Some of the highest peak of the world like the Mount Everest and Mount Kanchenjunga are in this range where mountaineers, trekkers and adventure enthusiasts are attracted from around the globe. Apart from the beauty, the Himalayas also has great significance of culture and spiritual. For centuries ancient traditions, religions and indigenous culture has been flourishing in this mountain range. The Himalayan range is a popular destination among many tourists for cultural tours, spiritual retreats and even eco-tourism experiences because of its blend in cultural heritage which is beautifies by its stunning natural beauty. But regardless of how mighty the Himalayan range is, it is not immune to the various challenges like the retreating of glaciers, altered weather patterns and ecological disruption which occurs due to climate change. For the future generations to cherish and appreciate the Himalayan ranges it is important for the present generation to ensure longevity and vitality of the mountain range by putting in efforts to preserve the eco system and promote sustainable tourism practices.

Though the Himalayas offer a wide range of attraction, the climate change has a huge impact on the destination. The mountain range with its majestic landscape and fragile ecosystem is very much vulnerable to the effects of global warming. The region's iconic attractions ad various tourism activities like the adventure tourism activities has been directly affected by the melting of glaciers, reduced snowfall, and also the altered weather pattern which is due to global warming. The Glaciers Lake Outburst Floods (GLOFs) and landslides which occurs due to the retreat of the glaciers is a threat to the human settlement and infrastructure. Also the change of the pattern of precipitation and extreme weather conditions is also a very big issue for the adventure tourism which impact the safety of the visitor and various activities like trekking and mountaineering which have been feasible to the tourists. Climate change is also adversely affecting the balance in the ecosystem of the Himalayan region which results in the loss of biodiversity and even allowing the eco-tourism experience to lose their authenticity. These results in the need and requirement for implementing sustainable tourism practices and various adaptation strategies. For the Himalayan tourism destination to have resilience and long term sustainability, it is high time that the local communities, government and tourism stakeholders to collaborate and work towards the same goal. Taking into consideration of global warming and climate change while tourism planning, promoting various sustainable practices and supporting the conservation of natural and cultural heritage in mitigating the harsh effects of global warming and climate change and preserve the unique destination for the future generations to enjoy.

1.1 Objectives of this Study

The research objectives of this study on the impact of climate change on Himalayan tourism destinations are as follows:

- To investigate the specific impacts of climate change on iconic attractions in the Himalayas,
- To identify sustainable tourism practices and adaptation strategies to mitigate the adverse effects of climate change on Himalayan tourism destinations.
- To provide policy recommendations for national and international stakeholders to integrate climate change mitigation and adaptation strategies into tourism planning and management.

2. IMPACTS OF CLIMATE CHANGE ON HIMALAYAN TOURISM DESTINATIONS

2.1 Vulnerability of Iconic Attractions

The Himalayan region is home to iconic attractions that draw tourists from around the world, such as the Everest region, Leh-Ladakh, and Bhutan. However, these attractions are particularly vulnerable to the impacts of climate change. One of the primary concerns is the accelerated melting of glaciers, which has direct consequences for the availability of water resources and affects the scenic beauty of the landscape. The retreat of glaciers not only alters the visual appeal of the destinations but also poses risks related to landslides and glacial lake outburst floods (GLOFs), endangering human settlements and infrastructure.

Another aspect of climate change that adversely affect the tourism activities is the reduced in snowfall. Recreational activities like the ski resorts and other snow dependent activities has to face a huge challenge as they got shorter and less reliable snow seasons. Thus affecting the overall tourism experience and economic viability of these destinations.

2.2 Altered Weather Patterns and Adventure Tourism

In the Himalayas region, the change in climate has also bought alteration in the weather patterns of the region that includes the change in precipitation and variability of the temperature. The changes in these pattern has a direct impact on the activities of adventure tourism like trekking, mountaineering and outdoor expeditions. Extreme weather events might also increase with the shifts in precipitation patterns like heavy rainfall or even intense storms which are a risks to the safety of the tourists and it also limits the window of opportunity for adventure tourism.

Additionally, changes in temperature and weather patterns affect the timing and feasibility of treks and expeditions. The availability of suitable weather windows for summit attempts may become more unpredictable, challenging the planning and logistics of adventure tourism operators and impacting visitor experiences.

2.3 Impact on Local Flora and Fauna

The rich biodiversity and the unique ecosystems which supports a wide range of plant and animal species has made the Himalayan region famous for decades. But in recent times the change in climatic condition of the region has become significant threat to the balance of these ecosystems. The natural habitats and reproductive cycles of the plants and animal species has been disrupting by the rise in temperature and altered precipitation.

The aesthetic appeal of the landscape can also be affected by the change in the vegetation patterns and phenology like the time of flowering and fruiting which in turn have a huge impact on the ecotourism activities. The traditional medicinal practices and cultural significance associated with certain plant species can also be a result from the loss of the species or shifts in their distribution.

Moreover, the delicate interactions between species, leading to shifts in animal populations and potential extinctions can also be disrupted by the change in climate. The attraction and authenticity of ecotourism experiences in the region can also be undermine by the disappearance of iconic wildlife like the snow leopards and Himalayan Monal pheasants.

The impacts of climate change on Himalayan tourism destinations are multifaceted and interconnected. The vulnerability of iconic attractions to melting glaciers and reduced snowfall, the implications of altered weather patterns on adventure tourism activities, and the impact on local flora and fauna all contribute to the need for effective climate change adaptation and sustainable tourism practices. Addressing these impacts requires collaborative efforts among governments, tourism stakeholders, and local communities to ensure the long-term sustainability and resilience of Himalayan tourism destinations.

2.4 Socio-Economic Consequences

In the Himalayan region, climate change plays a significant role in repercussions of socio-economy for the local communities and economies that derive from tourism. Due to the change in the environmental condition of the Himalayan region the economy which were dependent on the tourism industry are huge and there is even potentiality for loss of livelihoods in the region. Also the indigenous people faces challenges regarding the preservation of cultural heritage with the change of climate.

2.4.1 Economic Dependence on Tourism

Tourism plays a crucial role in the economy of many Himalayan communities. The influx of tourists contributes to employment generation, income generation, and local business development. Small-scale enterprises, such as hotels, guesthouses, restaurants, and tour operators, heavily rely on tourist arrivals for their sustenance. Additionally, local artisans, guides, and porters are directly involved in providing services to tourists.

But, there are impacts that greatly disrupt the tourism industry in the region due to the change in climate like the melting of glaciers and altered weather patterns. For example, in winter tourism, ski resorts can have a huge impact if there is less snowfall which will lead to the reduce in visitor and revenue from that particular tourism. With the change in weather patterns and increase in the occurrence of extreme events, the tourists are also deterring to engage in adventure tourism activities and this affect the income of the local operators and service providers.

2.4.2 Loss of Livelihoods

The local communities can lose their livelihood due to the changing environmental conditions which are associated with climate change. For instance, comuunities which were depending on agriculture might face challenges because of the altered patterns in precipitation which lead to the scarcity of water and in turn reduces the crop yields. This can in turn can become a driving force to migrate from rural areas to the urban areas to search for alternate livelihood opportunities, thus disrupting the traditional ways of life.

Additionally, the tourism sector can have a huge negative impact from the loss of iconic attractions which results from the melting of glaciers and disappearing of wildlife due to climate change. Thus aggravating the economic vulnerability of the local communities.

2.4.3 Challenges Faced by Indigenous Populations and Cultural Heritage Preservation

The indigenous people of the Himalayan region has their own unique cultural traditions, knowledge systems and lifestyles which are closely intertwined with the natural environment. Climate change is indeed a nuisance as it directly affects the livelihoods, cultural practices and traditional knowledge of the region.

For indigenous populations dependent on subsistence agriculture, changes in temperature and precipitation patterns can disrupt agricultural cycles and food security. The loss of traditional crops and medicinal plants due to changing environmental conditions undermines their cultural and economic resilience.

Additionally, climate change poses a threat to the preservation of cultural heritage sites and structures in the Himalayas. Increased temperatures, extreme weather events, and natural disasters can lead to the degradation and loss of historical sites, monasteries, and traditional settlements, eroding the cultural identity and heritage of the region.

To support the indigenous communities, preserve their cultural heritage and incorporate the traditional knowledge into climate change adaptation strategies efforts must be made. In the face of climate change it has become important to engage the local communities in various sustainable tourism activities and initiatives and empower them to actively participate in decision making process which will in fact enhance the resilience and ensure the preservation of their cultural heritage.

The socio economic condition of the Himalayan region is very unintended with climate change where the community is depending on the tourism sector. In the face of climate change, it has become very important to highlight the need for adaptive strategies and support mechanism so as to be sure of the well-being and sustainability of the local communities where they face potential loss of livelihoods and various other challenges faced by the indigenous populations and cultural heritage preservation.

3.STRATEGIES TO MITIGATE THE IMPACTS OF CLIMATE CHANGE ON HIMALAYAN TOURISM DESTINATIONS

3.1Sustainable Tourism Practices and Adaptation

To lessen the impacts of climate change, it has become crucial to adopt sustainable tourism practices in the Himalayan tourism destinations. The section will be dealt with various adaptive measures that can be implemented to promote sustainability and resilience. The section also emphasizes on the importance of collaborating among tourism stakeholders like the local communities, governments and tour operators that plays a significant role in developing effective strategies and contingency plans.

3.2 Promoting Energy-Efficient Infrastructure

The promotion of energy-efficient infrastructure is one of the key aspect for sustainable tourism. Alternate measures like the use of renewable energy sources, such as solar and hydroelectric power for hotels, guesthouses and many other tourism facilities can be encourage to reduce carbon emissions significantly. Also implementation of energy-efficient building designs and energy saving practices like the efficient lighting systems and insulation can also further helps in minimizing the environmental impact of tourism infrastructure.

3.3 Waste Management Systems

In the Himalayas region, proper waste management system is an essential part for developing sustainable tourism. With the increase in the number of tourist visiting the region, it also puts pressure on waste generation and disposal. The environmental footprint of tourism activities can be minimizing by the implementation of proper waste reduction, recycling and composting programs. Non the less, educating the tourists and providing easily accessible recycling and waste disposal facilities in tourist's areas can also help in contributing to more responsible waste management practices.

3.4 Sustainable Transportation Options

In the tourism sector, transportation plays a significant role in contributing to the green house emissions. so it has become crucial to promote sustainable transportation options to help in reducing the environmental impact of tourism activities. In many popular tourist destinations, encouraging the use of public transportations, introduction of the electric or hybrid vehicles for the transportation of tourists and also supporting the development of cycling and pedestrian infrastructure can be a great boon for minimizing carbon emissions and alleviate traffic congestion.

To address the challenges posed by climate change, it requires the collaboration between local communities, governments and tourism stakeholders. It is also essential to develop resilience strategies and contingency plans for adapting to the changing environmental conditions. To enhance the resilience of tourism destination it is important to conduct vulnerability assessments, identify risks and implement measures.

Engaging local communities in decision-making processes and involving them in sustainable tourism initiatives is crucial for building resilience. This can include providing training and capacity-building opportunities, supporting community-based tourism enterprises, and ensuring the equitable distribution of benefits from tourism.

Moreover, in the development and enforcement of policies and regulations for promoting sustainable tourism practices, it is essential for the tourism stakeholders and governments to collaborate. The collaborations can include setting of environmental standards for tourism operations, promoting eco certification programs and motivate sustainable practices through taxation or financial support.

3.6 Awareness and Education

It is essential to raise awareness among the tourists about the various impacts of climate change and the importance of sustainable tourism practices. Various tourism stakeholders can play a vital role in educating the tourist about their responsible behavior towards the environment like conservation of water, reducing waste, respecting local culture and traditions and also in supporting the local economy. All this can be done through campaigns, interpretive signage and educational programs that highlight the fragile nature of the Himalayan ecosystems and the need for their conservation.

On the Himalayan tourism destinations, it is very crucial to adopt sustainable tourism practices and implementation of adaptation strategies to mitigate the impacts of the climate change. Environmental footprint of tourism can be reduce by promoting energy efficient infrastructure, waste management systems and sustainable transportation systems. For developing a resilience strategies and contingency plans, collaboration between all the tourism stakeholder is a must. For the benefit of both the present and the future generations, it has become very crucial to raise awareness and educate the tourists about responsible behavior, preservation and sustainability of the Himalayan tourism destinations.

4.POLICY IMPLICATIONS AND FUTURE DIRECTIONS

4.1 Policy Implications at National and International Levels

At the national level, governments in Himalayan countries have a crucial role to play in addressing the challenges of climate change in tourism destinations. They need to develop comprehensive policies and regulations that integrate climate change mitigation and adaptation strategies into tourism planning and management. This can involve setting emissions reduction targets, promoting renewable energy initiatives, and implementing sustainable tourism standards and certifications.

To encourage the adoption of sustainable practices by the tourism industries, the governments can support capacity building initiatives and provide financial incentives. Also the government should prioritize the conservation and preservation of natural and cultural heriage sites through effective governance, enforcement of regulations and collaborations with local communities and relevant stakeholders.

Also at the international level, it is essential to collaborate among the government, NGOs, and international organizations. This helps in developing the best practices. Promoting knowledge exchange and providing technical and financial support for sustainable tourism development in the Himalayan region. In facilitation partnerships, providing funding opportunities and advocating for the inclusion of climate change considerations in tourism policies and strategies, International organizations play a very crucial role.

4.2 Supporting Sustainable Tourism Development and Resilience

To support sustainable tourism development and foster resilience in the face of climate change, governments, NGOs, and international organizations can focus on several key areas:

a) Capacity Building: It is crucial for the local communities' local communities, tourism operators and government officials to invest in capacity building initiatives. It might include training programs on sustainable tourism practices, climate change adaptation and disaster risk management. This helps in enhancing the knowledge and skills of the stakeholders and empower them to actively participate in the decision making processes and contribute to sustainable tourism practices.

b) Financial Support: To encourage the adoption of environmental friendly practices it is important to provide financial incentives and support for sustainable tourism initiatives. Grants, loans, and subsidies can be includes for tourism business so they can invest in renewable energy systems, energy efficient infrastructure, waste management solutions and sustainable transportation options.

c) Stakeholder Collaboration: It is also vital to encourage collaboration and partnerships between local communities, tourism operators, governments, NGOs and international organizations. Enhancing the resilience of tourism destinations and ensure the equitable distribution of benefits can be achieving by engaging the stakeholders in dialogue, fostering participatory decision making processes and promoting community based tourism initiatives.

d) Research and Monitoring: To understand the complexities of climate change impacts on the Himalayan tourism destinations, continued research is very much essential. For valuable insights into the ecological, socio-economic and cultural dimensions of type impacts, interdisciplinary research approaches is a must. The effectiveness of adaptation measures and guide evidence based decision making can be achieved by long term monitoring and data collection.

5. FUTURE RESEARCH DIRECTIONS

Further research is needed to deepen our understanding of the impacts of climate change on Himalayan tourism destinations and inform effective mitigation and adaptation strategies. Some potential avenues for future research include:

a) Assessing the vulnerability and resilience of different tourism destinations in the Himalayas to climate change impacts, considering factors such as elevation, geographic location, and infrastructure.
b) Investigating the socio-economic dynamics of climate change impacts on local communities, including the potential for alternative livelihood opportunities and community-based tourism models.
c) Examining the role of traditional knowledge and indigenous practices in climate change adaptation and the integration of such knowledge into sustainable tourism development.
d) Studying visitor perceptions, behaviors, and preferences regarding sustainable tourism practices and their willingness to support environmentally responsible tourism.
e) Exploring innovative approaches, such as nature-based solutions and sustainable tourism financing mechanisms, for enhancing climate resilience in Himalayan tourism destinations.

By addressing these research gaps, we can improve our understanding of the complexities of climate change impacts on Himalayan tourism destinations and develop evidence-based strategies to ensure their long-term sustainability and resilience.

6. CONCLUSION

In the Himalayan region, climate change poses a great risk by impacting their landscapes, ecosystems and local communities. Some of the key concerns are the melting of glaciers, altered weather patterns and biodiversity loss. Also the issues need to be address urgently due to its socio-economic consequences, including the economic dependence on tourism and potential loss of livelihoods etc.

So now it is crucial to adopt sustainable tourism practices to mitigate the impacts. Some of the steps to mitigate are promoting energy efficient infrastructure, implementing effective waste management systems and encouraging sustainable transportation options. It is also important to collaborate among all the stakeholders of the tourism industry whether government or NGOs to develop resilience strategies and ensuring equitable benefits.

Considering at the policy level, the key factor is to integrate climate change considerations into tourism planning and management and foster the collaborations at national and international levels. Future research should focus on vulnerability assessments, socio-economic dynamics, and innovative approaches to enhance climate resilience.

Thus we can ensure the long-term sustainability and resilience of Himalayan tourism destinations in the face of climate change by implementing sustainable tourism practices, supporting adaptation strategies and conducting further research.

Chapter 11
Mountain Tourism Development:
Evaluating Problems and Prospects of Nature-Based Tourism in the Himalayan Region of Darjeeling, India

Puspanjali Mohapatra
Utkal University of Culture, India

Soumendra Nath Biswas
Assam University, India

ABSTRACT

This study focuses on nature based sustainable tourism development in the Himalayan region of Darjeeling and other surrounding areas of West Bengal, India that affects tourist movement, protection of natural resources, and improves socio-economic conditions of local people. The purpose of nature-based tourism, its significance, and its effects on the tourism industry, local communities, and the environment of the tourism zone in the study area are discussed in detail here. Understanding the varied effects of nature-based tourism on the local economy and society is the paper's main concern. In some cases, it can contribute to conservation, protection of wildlife and economic benefit to the local people. In other cases, lack of proper planning can be detrimental to the environment, flora, fauna, and local people. Nature education, communication, awareness, impact analysis, measuring carrying capacity, visitor management, etc. are essential to reduce negative impact on the environment that affects socio-economic life of the local people, are focused on in this chapter.

1. INTRODUCTION

Any form of tourism that has a direct connection to the natural world is known as nature-based tourism. It includes ecotourism, adventure tourism, wildlife tourism and nature retreats. A prudent tour of

DOI: 10.4018/979-8-3693-0823-3.ch011

a natural region protects the environment and promotes community welfare. Nature based tourism is closely related with the concept of eco-tourism.

This paper explores many forms of natural attractions and how they affect the Darjeeling tourism industry, in turn influencing traveller behaviour and the socioeconomic well-being of the surrounding areas. This article explains the significance of nature-based tourism, its value to the industry, how it affects local communities, and how it affects Darjeeling's natural tourism resources. Understanding the varied effects of nature-based tourism on the local economy is the paper's main concern. In some cases it can contribute to conservation, protection of wildlife and economic benefit to the local people. In other cases lack of proper planning can be detrimental to the environment, flora, fauna and local people. Nature education, communication, awareness, impact analysis and survey, measuring carrying capacity, visitor management etc. are essential to reduce negative impact on the socio-economic life of the local people. Here, it is also highlighted how the government, local communities, and tourism service providers all have responsibilities.

2. OBJECTIVES

The main objective of this study is to evaluate problems and prospects of nature based Tourism in Himalayan region in the study area. However, the objectives in specific are:

1. To understand the concept and importance of nature based Tourism in the Mountain area.
2. To find out the problems and prospects of nature based Tourism in the study area
3. To evaluate the issues and challenges of the destinations in mountain region.

3. METHODOLOGY

Extensive interviews with residents, business owners, and visitors were conducted in order to learn more about the overall makeup of the local community as well as some of the problems with nature based Tourism. These experiences were included because they show a variety of problems in popular tourist areas. Utilizing self-administered questionnaires, this study was carried out. In order to collect information from the primary sources, two sets of structured questionnaires were developed for both tourists and locals, as well as local service providers. In addition, observation, and interview techniques were used for the aforementioned goal.

4. NATURE BASED TOURISM: MEANING AND CONCEPT

Responsible nature-based travel involves visiting natural regions while protecting the environment and promoting the well-being of the local populace. Both the destination and the local indigenous population receive financial incentives for conservation. Tourism related to nature helps the local economy and promotes the preservation of natural resources. Travel to natural area is meant by 'nature-based tourism'. 'Ecotourism', on the other hand, is described as ethical travel to natural regions that preserves the flora, fauna, and local culture while also enhancing the welfare of the local populace. Nature-based tourism promotes local communities and tourist destinations on an economic, cultural, and environmental level.

Sustainable development "meets the needs of the present without compromising the ability of the future generations to meet their own needs" (Bruntland Commission, 1987), where the ideas of sustainable tourism and responsible tourism originated. Agenda 21 for the Travel and Tourism Industry, published in 1992, described sustainable tourism as travel that "meets the needs of present tourists and host regions while protecting and enhancing opportunities for the future."

Excursions to wilderness areas and national parks, where a significant amount of the biodiversity is concentrated, are included in nature-based tourism (Olson et al. 2001). 'Adventure tourism' activities could potentially be a part of it. Urban dwellers and those who enjoy the luxuries and conveniences of globalization are becoming more and more disconnected from nature:

Boo (1990) defines nature tourism as "travelling to relatively undisturbed or uncontaminated natural areas with the specific objective of studying, admiring, and enjoying the scenery and its wild plants and animals, as well as any existing cultural manifestations,". Nature tourism, as defined by Lucas (1984), is "tourism that is based on the enjoyment of natural areas and the observation of nature," and it "has a low impact on the environment, is labour intensive, and contributes socially and economically to the nation," among other characteristics.

Tourism that is "responsible" or "ethical" (Kutay, 1989), "environmentally friendly" (Borst, 1990), "green," or "sustainable" (Lane, 1990) are other names that have been used to describe ecotourism and nature tourism. Other viewpoints have emerged in response to worries about the effects of tourism on the environment, such as the entire area of nature tourism ethics (Anon, 1989a, 1989b; Graham, 1979).

Nature based Tourism is a major business now. The preservation of natural systems and animals may result in financial gain and foreign exchange. However, nature-based tourism also poses a threat to the resources it depends on. The sustainability of natural systems is threatened by this maniacal activity. Sometimes it seems like we are killing the natural world (Berle, 1990). Aside from the well-known connection between most tourism and the environment, recent writings on the subject (Ingram and Durst, 1987; Kutay, 1989; Boo, 1990) show that "nature-based tourism" has been receiving more attention as a niche form of travel. Both local people and conservation agencies have indicated this interest (Young, 1986).

Natural surroundings are necessary for anyone who wants to view animals in the wild to appreciate their experience. Such an experience is undoubtedly dependent on nature, and successful tour firms are built on this dependency. Camping is frequently made better by the natural environment; most people like to camp in forests rather than quarries. These preferences do depend on nature; however some users may be able to complete the activity without a totally natural setting with the same level of happiness.

5. NATURE BASED TOURISM IN HIMALAYAN REGION OF DARJEELING

The Eastern Himalayan region's Darjeeling, known as the "Queen of Hills" in West Bengal, draws tourists because of its distinctiveness. The tea industry, breathtaking views of Kangchenjunga, and the Darjeeling Himalayan Railway, a UNESCO World Heritage site, are just a few of the specialties that make the Queen of the Hills unique. Apart from these, other attractions include the Mall Road, Himalayan Mountaineering Institute, Tiger hill, tea gardens etc.

Three 'T's like: Tea, Tourism, and Timber are three industries that are significant in Darjeeling. Darjeeling now offers tea, but the tourists find the term "organic tea" more fascinating than regular tea.

The third is tourism, which is intriguing but must now be referred to as eco-tourism due to the Darjeeling ecosystem's fragility. Ecotourism is now crucial for maintaining the ecosystem's attractiveness because the first two "T"s placed threats on it.

According to Banerjee et al. (1984), the sub-alpine forest is primarily made up of evergreen conifers with some mixing of species with broad leaves. The hub of the Darjeeling Himalaya's attractions, soil, climate, culture, community, and environment are remarkable. The current subject of ecotourism has been established in order to rediscover the picturesque view of Mt. Kanchenjungha as well as nature and natural resources.

The hilly regions of the Darjeeling District depend heavily on tourism as a source of income and employment for the local populace. Everyone benefits from tourism, from the Tea vendor to the hotel owner. In addition to raising income and boosting demand for regional goods, tourism has a multiplier effect. The multiplier is defined as the relationship between a change in one of the aforementioned factors and the change in visitor spending that caused it.

6. FINDINGS AND SUGGESTIONS

The findings of this study show that legislation, government programmes, and protection of protected areas all have a significant impact on the growth of environmentally responsible tourism in the studied

Figure 1. Issues and challenges of nature based tourism
(Author)

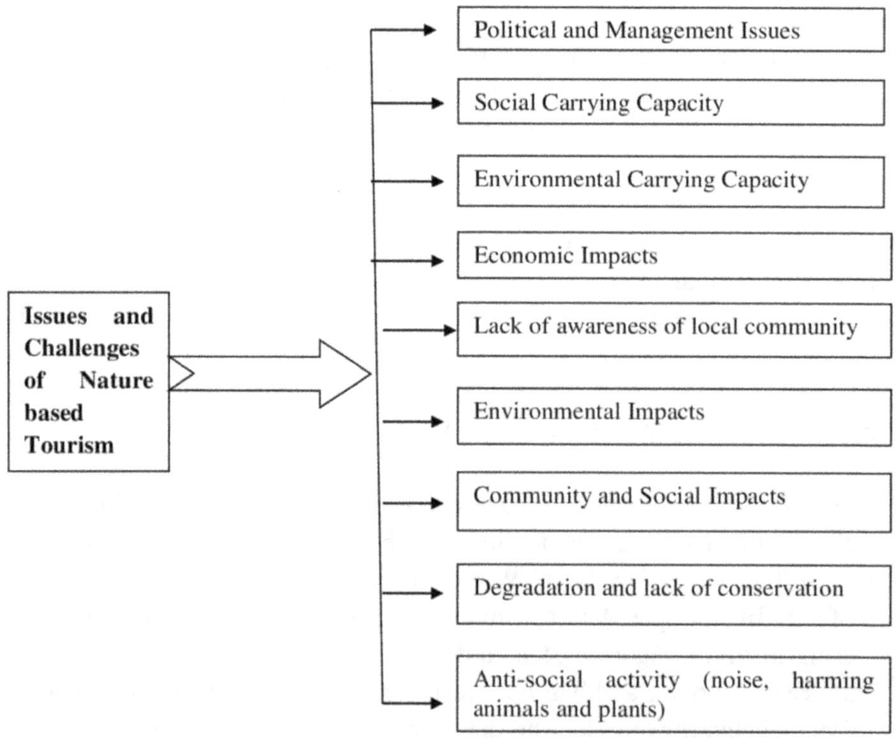

region. Active participation of local people, Government and NGOs are necessary for optimum utilization of natural resources. Some of the major findings and suggestions are:

1. Nature based tourism destinations in mountain region attract huge numbers of tourists every year. These eco friendly destinations with unique flora and fauna catch the attention of tourists of every demographic profile.
2. Himalayan destinations have limited access to natural resources, cultural heritage, and infrastructure, making it crucial to manage destination carrying capacity (getting the volume, timing, and location of visitors just right, as well as reducing traffic and ensuring socio-environmental compliance) in order to carry out tourism development.
3. Darjeeling is situated in the eastern Himalayan region and a famous Tourism destination in India. A land of spectacular beauty crowned by the mighty Himalayas, Darjeeling evokes images of snow-capped mountains, tranquilly of brilliant green hills steeped in grandeur. One of the world's most stunning hill resorts is Darjeeling.
4. Kangchenjunga, the third highest peak is clearly visible from here. Spectacular scenic beauty with snow peaks, the breathtaking colonial architecture, which includes homes and churches, the impact of several cultures on the magnificent crafts and cuisine make Darjeeling a famous Tourism destination.
5. The acreage and productivity of tea gardens should both be increased, but not at the expense of forests.
6. Deforestation needs to be stopped in order to prevent the region's rainfall from being disrupted. The need for and significance of soil conservation are highly significant in both agricultural and other regions.
7. Infrastructure planning will need to be refocused on the developments of urban settlements, taking into account the economic contributions made by these settlements relative to their feeder regions.
8. Efforts should be made to teach people about environmental awareness and incorporate instruction on environmental conservation.
9. Establishing agro business based on the use of natural resources will help local ethnic communities to earn money. Local government should provide facilities to the communities in terms of loan and training for pro poor development.
10. Raising public awareness of ecological degradation's wide-ranging effects, teaching people about them, and obtaining their support for eco-development initiatives are essential for maintaining sustainability.
11. By reducing the use of chemical fertilisers and pesticides on tea gardens in accordance with good agricultural practises, biodiversity has also been preserved to a large extent; however, the conversion of estates from conventional to organic production has contributed significantly more.
12. Tourist attractions need to be protected from strange factors like overcrowding, which immediately reduce carrying capacity. The site must be defined by the varied actions that are brought about by various visitors for distinct purposes. Illegal activities and damaging practises might affect a location's threshold, which obviously reduces tourism attraction and ought to be stopped.
13. Need-based training for service providers is crucial for reducing unfavourable interactions. In nature based tourism, it is crucial for providing better service to the tourists.
14. Various Tourism organisations must be actively persuaded to support efforts for neighbourhood development and local infrastructure like signage, hoardings, beautification etc. Creating awareness

about the effects of climate change along with providing facilities of education, training, health, and sanitation are essential.

15. A strategy for motivating businesses to reduce, reuse, and recycle solid waste is required in destinations. The 'polluter pays' principle must be upheld, and any leftover solid waste that cannot be recycled or reused must be disposed of safely and sustainably.
16. States must make sure that existing laws and regulations governing the purchase of real estate for tourist attractions are upheld, in compliance with community and indigenous rights, and with requirements for public participation.
17. Tourist security and safety should be prioritised, and cooperation amongst local search and rescue organisations should be a top concern.
18. In forested areas, physical infrastructure such as lodging facilities, dining establishments, and road and trail systems must be environmentally responsible. While there isn't a specific environmental legislation that regulates the tourism industry, integrating tourism-related activities, mitigating air, water, noise, and environmental pollution can help to some extent reduce that pollution.

7. CONCLUTION

The three key conclusions are nature protection, business concept development, and nature education. All Tourism stakeholders have responsibilities to protect nature. Tourist attractions need to be protected from strange factors like overcrowding, which immediately reduce carrying capacity. The site must be defined by the varied actions that are brought about by various visitors for distinct purposes. Illegal activities and damaging practises might affect a location's threshold, which obviously reduces tourism attraction and ought to be stopped. To lessen the occasional unfavourable interaction, both hosts and visitors must receive need-based training. In eco-tourism, it is crucial to clearly define the site boundaries and include areas for orientation.

To minimise pollution and ensure proper land use, government agencies like the Forest Department and the Pollution Control Board must take particular measures. It is crucial to promote research and development (RD) through the media in order to examine ecotourism operations. For the start of successful integrations during the visit, locating global heritage sites, monuments, stunning woodland views, locations with wild flora and wildlife, and native people is important.

The need to educate the stakeholders about the future of tourism, ecosystem damage, and the connection between ecotourism and eco-degradation is urgent. To manage the delicate ecosystem and create an environment that is both economically viable and ecologically sustainable, it is crucial to apply local as well as indigenous techniques.

During the busiest times, the amenities are unable to keep up with the massive demand, which causes the facilities and services to rapidly deteriorate. Water shortages, power outages, deteriorating municipal services, traffic jams on public transport, and a housing shortage all contribute to the chaotic situation. Even the locals become irritated when guests are inconvenienced and put through needless stress.

REFERENCES

Banerjee, S. K., Pal, D. K., & Banerjee, S. P. (1984). Soil characteristic and Floristic composition of some sub-alpine forests of Darjeeling (West Bengal). *Proc. Indian. Nat. Sci. Acad., 50A*(3), 269–277.

Basu, S., & Starkel, L. (2000). Rains, Landslides and Floods in the Darjeeling Himalaya. Indian National Science Academy, New Delhi.

Berle, P. A. A. (1990). Two faces of eco-tourism. *Audubon, 92*(2), 6.

Bhattacharya, D., & Mitra, A. (2013).. . *Paripex-Indian Journal of Research, 2*(9), 222–224.

Bhutia, S. (2014, April). Human Response on Socio-Economic Development in Darjeeling Himalaya of West Bengal. *Asian Academic Research Journal of Social Sciences and Humanities, 1*(22), 166–173.

Biswas, S. K. (1990). Strategy of Development in the Himalaya: A Profile of Socio-Economic Change. Institute of Social Research and Applied Anthropology, Calcutta.

Boo, E. (1990). *Ecotourism: the potentials and pitfalls* (Vols. 1–2). World Wildlife Fund.

Boo, E. (1990). *Ecotourism: the potentials and pitfalls* (Vol. 2). World Wildlife Fund.

Borst, B. 1990, Ecotourism: environment friendly international travel. *International Herald Tribune.*

Chakraborty, S. (1976, September). The Role of Transportation in the Economy of the Hill Areas of the Darjeeling District. *Indian Journal of Hill Economy, Darjeeling, 1*, 15–34.

Chakraborty, S. (1980, September). A Geographical Sketch of Darjeeling District. *Indian Journal of Hill Economy, Darjeeling, 5*, 1–7.

Dash, M. C., & Dash, S. P. (2010). Fundamentals of Ecology. McGrew-Hill Companies, Tata McGrew-Hill Education Private Limited, New Delhi.

Ingram, C. D., & Durst, P. B. (1987). *Marketing nature-oriented tourism for rural development and wildlands management in developing countries: a bibliography, USDA Forest Service general technical report SE44*. US Department of Agriculture, Forest Service. doi:10.2737/SE-GTR-44

Kutay, K. (1989). The new ethic in adventure travel. *Buzzworm: The Environmental Journal, 1*(4), 31–36.

Kutay, K. (1989). The new ethic in adventure travel. *Buzzworm: The Environmental Journal, 1*(4), 31–36.

Lucas, P. H. C. 1984, How protected areas can help meet society's evolving needs. In J.A. McNeely, and K.R. Miller, (eds.), National parks, conservation, and development. Smithsonian Institution Press, Washington D.C.

Olson, D., Dinerstein, E., Wikramanayake, E. D., Burgess, N. D., Powell, G. V. N., Underwood, E. C., D'amico, J. A., Itoua, I., Strand, H. E., Morrison, J. C., Loucks, C. J., Allnutt, T. F., Ricketts, T. H., Kura, Y., Lamoreux, J. F., Wettengel, W. W., Hedao, P., & Kassem, K. R. (2001). Terrestrial eco regions of the world: A new map of life on earth. *Bioscience, 51*(11), 933–938. doi:10.1641/0006-3568(2001)051[0933:TEOTWA]2.0.CO;2

Rao, R. R., & Sharma, B. D. (1990). A Manual for Herbarium Collections, BSI. *Brabourne Road, Kolkata, 1*, 1990.

Tisdell, C. 1984, Tourism, the environment, international trade and public economics. Asian-Australia Joint Research Project, Australian National University, Canberra.

Young, A. M. (1986). Eco-enterprises: Eco-tourism and farming of exotics in the tropics. *Ambio*, *15*, 361–363.

Chapter 12
Overtourism in the Himalayan Region:
Envisioning Sustainable Visitors Management Practices

Shreeansh Mishra
https://orcid.org/0009-0004-8847-6587
Central University of Tamil Nadu, India

Jitendra Mohan Mishra
https://orcid.org/0000-0002-5840-9701
Central Tribal University of Andhra Pradesh, India

ABSTRACT

Due to inadequate management practices historically associated with tourism development in the Himalayan region, significant repercussions have arisen. While studies have explored the phenomenon of over-tourism in the region, there is a need for a comprehensive understanding of the impact of over-tourism in the Himalayan region, and to revisit the existing tourism management practices and effective mitigation strategies concerning the ecological, socio-cultural, and economic impacts caused by excessive tourist influx to be evolved. This study explores sustainable visitor management practices as a means to address the challenges posed by over-tourism in the Himalayas. Existing contextual literature has been reviewed to understand the pros and cons of over-tourism in the Himalayan region. Accordingly, the study suggested measures to overcome the impacts of over-tourism through sustainable visitor management practices and offers valuable insights for policymakers, destination managers, and local communities in fostering sustainable tourism practices in the Himalayan region

1. INTRODUCTION

The Himalayan region, spanning across multiple countries, including India, Nepal, Bhutan, Tibet, and Pakistan, has long been a hotspot for tourists seeking solace in its majestic mountains, serene valleys,

DOI: 10.4018/979-8-3693-0823-3.ch012

and rich cultural heritage. Moreover, with its unique geological features and rich natural and cultural heritage, has attracted tourists worldwide, leading to significant tourism-driven economic growth and development (Chauhan et al., 2021; Nepal, 2000). The region is predominantly inhabited by indigenous societies that rely heavily on the rich biodiversity of the area to meet their livelihood needs (Chauhan et al., 2021, Samant et al., 2011). However, the increasing footfall of tourists in recent years has raised concerns about the sustainability of tourism activities in the region. Tourism in the Himalayas faces adverse effects including youth outmigration, disruptive road construction, economic vulnerability, and potential long-term impacts from climate change.(Bennike & Nielsen, 2023). Overtourism, characterized by an excessive influx of visitors beyond the carrying capacity of the destination, has become a significant challenge.

The Himalayas have always been a captivating destination, for tourists due to their landscapes diverse wildlife, and vibrant mix of cultures (Nyaupane & Chhetri, 2009). From the revered peaks of Mount Kailash to the tranquil valleys of Bhutan this region offers a combination of natural wonders and cultural treasures. However, the recent surge in tourist numbers has put a strain on the balance of the area ecosystems (Chandra & Kumar, 2021) The rapid growth of cities, construction projects, and unregulated tourist activities have left lasting effects on these mountain environments leading to deforestation, soil erosion and depletion of water resources (Rawat et al., 2023). What's more the influx of visitors has also worsened issues concerning waste management and pollution posing a threat to the pristine beauty of this region. The lack of coordination between agencies is causing inconsistencies between tourism development activities and regional sustainability (Chandra & Kumar, 2021). Exploitative practices that prioritize profit and uncontrolled growth neglecting the enduring repercussions of accelerated resource utilization cause damage to the social and natural environment and are linked to people's deteriorating well-being (Bhalla & Chowdhary, 2022)

At risk is the rich cultural heritage held by Himalayan communities. As traditional societies become increasingly integrated into the tourism market there is a growing concern that their customs and way of life are being commodified. This runs the risk of diluting their authenticity and diminishing their significance. Although numerous factors contributing to the disaster have been the subject of research and media discourse, there has been a noticeable lack of extensive examination and debate surrounding the historical evolution and continued expansion of tourism, as well as the socioeconomic dependencies and vulnerabilities it has engendered (Lorenz & Dittmer, 2021). While there are studies and frameworks available that discuss sustainability in tourism, there remains a distinct gap in research that holistically examines the design, implementation, and impact of sustainable visitor management practices tailored specifically to Himalayan regions, including natural and cultural heritage sites, urban areas, and remote regions.

This study aims to explore trends and practices in tourism management within the Himalayan region with an eye, toward sustainability. It seeks to identify strategies that can help minimize the impacts caused by overtourism. Additionally, the study seeks to make a contribution, to discussions surrounding tourism practices and policies. It aims to provide insights that are beneficial for policymakers, local communities, and stakeholders, within the tourism industry.

2. LITERATURE REVIEW: CAUSES AND CONSEQUENCES OF OVERTOURISM IN THE HIMALAYAN REGION

Bennike & Nielsen (2023) emphasized the need for a critical examination of tourism as a rural development strategy, highlighting the transformative potential and potential destructiveness of frontier tourism development when local realities are not thoroughly studied. The Himalayan ecosystem is susceptible to various factors, including natural changes, human activities leading to climate change, and the modern way of life, all contributing to the growing disparity between resource consumption and replenishment in affluent regions, primarily driven by increased per capita consumption (Chauhan et al., 2021). The unregulated and extensive influx of tourists, coupled with insufficient coordination among coordinating agencies, poses a significant threat to the ecological sustainability and tourism sector potentially increasing human disruptions and conflicts between regional sustainability and tourism growth (Chandra & Kumar, 2021). Some of the major factors responsible for over-tourism are:

a. ***Promotion and Accessibility:*** The advent of low-cost airlines, improved road networks, and improved technologies have made the Himalayan region more accessible to tourists, resulting in an unprecedented surge in arrivals causing a threat to the environment (Kapoor and Wangdus 2022; Ziegler et al., 2023). Such factors fuel overtourism by generating high demand and attracting an unsustainable number of visitors.
b. ***Social Media and Digital Influence:*** The proliferation of social media platforms has led to the rapid dissemination of images and stories, enticing more travelers to explore the region. Social media platforms enable brands to engage with customers, ultimately helping them shape a distinctive online presence (Harrigan et al., 2017). The travel and tourism sector has recognized the significance of social media as a primary marketing tool, with numerous countries adopting it as a key platform for promoting their tourism industry (Jamal et al., 2019). Taking into account all the benefits, it cannot be denied that social media can create significant hype for a destination, leading to a surge in tourism at that destination for a particular period. It is important to recognize that this increased influx of tourists, often driven by viral trends and social media influence, can contribute to overtourism in certain areas, straining local resources and infrastructure.
c. ***Lack of Regulation:*** Insufficient regulation, inadequate municipal planning strategies, and a dearth of rigorous law enforcement contribute to the issue of overtourism (Badola et al., 2018). Inadequate enforcement of tourism policies has allowed unchecked growth in the number of accommodations and tour operators. It, in turn, fails to manage and harmonize the demand for tourism with the destination's sustainable capacity, thereby significantly contributing to the phenomenon of over-tourism.
d. ***Commercialization:*** The recent commercialization of tourism activities/services in the region has primarily benefited the government and a select group of tour operators, where the intention was to make tourism accessible to individuals with limited financial resources and capabilities at an affordable cost, it has led to a significant increase in tourist arrivals (Ziegler et al., 2023). Tourism services, such as accommodations, restaurants, and attractions, are heavily commercialized to cater to a large number of tourists. In such scenarios, the primary focus is often on maximizing profits, which leads to overtourism in destination. Manali and Shimla serve as prime illustrations of commercialization and overtourism.

2.1 Consequences of Overtourism

The effects of over-tourism have broad-reaching implications, affecting numerous facets of both the environment and the communities in tourist destinations, which are briefly elucidated below.

a. ***Environmental Degradation:*** Uncontrolled tourism activities have resulted in deforestation, habitat destruction, waste accumulation, and pollution, putting immense pressure on fragile ecosystems due to factors such as unplanned random development and inadequate waste disposal (Joshi & Dhyani, 2009; Nepal, 2000; Rawat et al., 2023). In the Himalayas, additional tourism-related challenges encompass widespread littering and insufficient sanitation infrastructure, with the placement of many toilets near streams leading to local water pollution (Chauhan et al., 2021; Nepal, 2000). The forest resources have faced significant pressure from unregulated tourism activities for several reasons, encompassing the conversion of a substantial forested area into built-up land to accommodate the growing number of tourists and the extensive harvesting of fuelwood from local forests for food preparation (Singh et al., 2009). Hence, uncontrolled tourism and its associated consequences are placing immense pressure on the fragile Himalayan ecosystems.

b. ***Cultural Erosion:*** Uncontrolled tourism growth in the Himalayas can lead to the degradation of local resources and cultures, with local communities often not benefiting from the influx of tourists, and it may also result in conflicts between hosts and visitors (Joshi & Dhyani, 2009) The influx of tourists often leads to the commercialization of local cultures, eroding traditions, and values. One of the detrimental consequences of unregulated tourism in the Himalayas is the erosion of local culture, which includes the diminishing preservation of traditional ecological knowledge (Nyaupane and Thapa 2004).

c. ***Strained Resources:*** As destinations attract a surge in visitors, their natural and cultural resources often become overburdened, leading to environmental degradation, ecological problems, and increased pressure on local communities (Singh et al., 2009). Scarce resources such as water, energy, and waste management facilities are stretched beyond their capacity, negatively impacting both locals and tourists (Kala & Maikhuri, 2011). The surge in demand for foodgrains/groceries due to tourist growth, and seasonal variations may necessitate reliance on external sources, resulting in increased living costs and price rise for related commodities and perishables (Joshi & Dhyani, 2009).

3. DISCUSSION AND RECOMMENDATION

In light of these challenges, it is imperative for Himalayan destinations to adopt sustainable tourism practices. The absence of strategic planning, the incongruity of tourism practices with existing policy frameworks, unplanned developmental activities, and the overwhelming surge of visitors are all detrimental to the ecological integrity, stability, and long-term viability of the tourism business in the Indian Himalayan Region, highlighting the urgent need for sustainable visitor management practices. Visitor management in protected areas necessitates a comprehensive approach that encompasses a range of management actions, extending beyond solely restricting the number of visitors (Stankey and Cole 1998; Takahashi and Cegana 2005; Manning 2007). The Sustainable Visitors Management Practices in the context to Indian Himalayan Region are discussed below:

a. ***Carrying Capacity Assessment:*** Carrying Capacity of a destination determines the maximum number of visitors a destination can support without compromising its ecological and cultural integrity. According to Santos & Brilha, (2023), the concept of Tourism Carrying Capacity involves establishing acceptable limits for ecological degradation by defining a maximum visitor capacity, with the ecological dimension focused on impacts to the natural environment, and the physical dimension centered on elements of the built environment, including infrastructure systems like supply services, transportation, health services, and electricity, in order to manage sustainable tourism practices. The Himalayan tourism destination has tourism carrying capacity, especially in season time (Singh et al., 2009). The capacity for disaster risk reduction in the Himalayas has lagged behind the rising number of tourists and their growing vulnerability, indicating a lack of preparedness to respond to disasters, despite the long-standing awareness of potential risks within the region's tourism sector (Ziegler et al., 2023). Therefore tourism carrying capacity assessment is vital for visitor management at destinations, as it safeguards resources, maintains visitor quality, and sustains the destination's long-term viability, striking a balance between economic benefits and environmental preservation.

Bhutan's policy of limiting the number of tourists through high daily tariffs and emphasizing cultural preservation has been a successful example of sustainable tourism management.

b. ***Regulations and control mechanism:*** Tourism seasonality encompasses diverse manifestations, including changes in tourist arrivals, international visitors, and average length of stay, all of which result from fluctuations in tourism activity levels. The detrimental outcomes of seasonality are associated with impacts that occur not only during peak tourist periods but also during periods of minimal attendance. Ensuring a steady flow of tourists to the destination is essential to provide a reliable foundation for economic growth, job opportunities, and responsible resource utilization (Martín Martín et al., 2020). Local institutions have the capability to hinder the transition of nature-based tourism into mass tourism by controlling the number of tourists and the expansion of physical infrastructure (Badola et al., 2018). Implementing measures such as restricting visitor numbers during peak seasons and limiting the duration of visits to sensitive areas is imperative because it serves as essential strategies for managing the adverse impacts of over-tourism during tourist season time (Ziegler et al., 2023). Jamin & Rahmafitria (2022) in their study suggested three potential approaches to implement visitor limitations: raising ticket prices during peak seasons, enforcing visitor quotas for the crater area, and diverting visitors to alternative zones to alleviate pressure on the primary core zone. Case study 1 on Bhutan demonstrates how regulating visitor numbers by implementing a 'sustainable development fee' is a strategic visitor management activity that can contribute to the sustainable development of a tourism destination, particularly in the context of managing seasonality and ensuring a high-quality experience for tourists.

c. ***Community Involvement:*** Communities not only experience economic benefits or the effects of infrastructure development due to tourism but also adapt and respond to the changes it brings. To assess the impact and sustainability of tourism development, it is imperative for stakeholders to incorporate local perspectives and engage local communities in decision-making processes and revenue-sharing mechanisms, thus ensuring that they derive benefits from tourism. (Chan & Bhatta, 2023; Sahani, 2021). The economic benefits of tourism, including job opportunities, revenue, infrastructure improvements, and foreign exchange earnings, frequently motivate local communi-

Figure 1.

Case Study 1: Bhutan's High-Value, Low-Impact Tourism Model

In September 2022, the Kingdom of Bhutan reopened its borders to international tourists after the COVID-19 pandemic. Bhutan introduced a new tourism strategy focusing on sustainable development policies, infrastructure upgrades, and enhancing the guest experience. The country aims to reinforce its policy of "High Value, Low Volume" tourism, emphasizing the importance of preserving its values and natural beauty. Bhutan raised its Sustainable Development Fee (SDF) to fund projects related to economic, social, environmental, and cultural development, including sustainability initiatives and free healthcare and education for all. The SDF also contributes to offsetting the carbon footprint of visitors and supporting Bhutan's status as a carbon-negative nation. The government invested in infrastructure improvements and upskilling tourism industry workers during the pandemic. Bhutan seeks to provide high-value experiences for tourists, maintain a low sustainable footprint, and create well-paying jobs for its citizens. The new tourism strategy includes limiting the number of visitors at sacred sites and upgrading itineraries to showcase the best of Bhutan. The transformation project in Bhutan extends beyond tourism, aiming to develop human capital and equip the population with advanced skills and knowledge. The unveiling of 'Brand Bhutan' signifies the country's optimism and renewed ambition as it reopens to the world, focusing on its young citizens' future.

Source: Tourism Council of Bhutan (2022, September 2023)

ties to actively support tourism development, which is crucial for their inclusive participation in infrastructure planning and tourism for promoting equitable and sustainable development (Chan & Bhatta, 2023). Hence community involvement in tourism is essential for effective visitor management as it leverages local knowledge, preserves culture and the environment, promotes sustainability, educates visitors, ensures equitable economic benefits, resolves conflicts, and contributes to continuous improvements in the tourism industry.

d. ***Eco-friendly Infrastructure:*** Develop sustainable infrastructure like eco-lodges, waste management systems, and renewable energy sources to minimize the environmental footprint. The significance of eco-friendly technologies lies in their substantial reduction of carbon emissions and their capacity to mitigate the adverse environmental effects of international tourism (Erdoğan et al., 2022). Such infrastructure includes energy-efficient buildings, waste management systems, sustainable transportation options, and eco-sensitive designs. By incorporating these elements, destinations can reduce their ecological footprint, enhance the visitor experience, and ensure that tourism activities

Figure 2.

Case Study 2: Annapurna Conservation Area, Nepal

The Annapurna Conservation Area (ACA) serves as a compelling case study exemplifying best practices in sustainable tourism management. In this pristine mountain region, nature-based tourism plays a pivotal role in conservation and rural development. Over 1,000 local establishments, including teashops, lodges, and hotels, directly benefit from tourism. However, challenges emerge as the aspirations of local communities for infrastructure, particularly roads, may significantly transform the nature of tourism in the area. Key components of these best practices include meaningful participation of local communities through institutions such as Tourism Management sub-Committees (TMsCs) and mothers' groups. TMsCs develop local tourism policies, including initiatives to reduce firewood use, improve service quality, standardize rates, and ensure safety for tourists. Capacity enhancement programs regularly train local lodge operators and tourism entrepreneurs to elevate service quality, address sanitation and hygiene, manage waste, and enhance hospitality. Awareness workshops educate communities on the impacts of tourism, garbage management, nature conservation, and clean energy usage. Exchange visits facilitate knowledge sharing among community groups.Responsible travel practices are promoted among visitors, emphasizing the minimization of negative impacts and support for local economic development and heritage conservation. Infrastructure upgrades, environmental management, and exploration of new destinations within ACA aim to enhance visitor experiences while preserving the environment and culture.

However, potential threats loom, such as the desire for extensive road connectivity and large-scale projects. Despite challenges, ACA showcases how comprehensive planning, community involvement, environmental conservation, and responsible tourism practices can harmonize economic benefits with the preservation of natural and cultural heritage. These best practices serve as a model for sustainable tourism management, ensuring that nature-based tourism thrives while safeguarding the very essence of the destinations it touches.

Source: UNDP Nepal. (n.d.). Biodiversity Conservation and Impacts of Nature-Based Tourism in Annapurna Conservation Area.

are carried out in a manner that respects and preserves the environment, contributing to long-term sustainability in tourism management.

e. ***Education and Awareness:*** Initiatives must be taken by tourism stakeholders to create awareness among tourists about responsible and sustainable travel practices through signage, guidebooks, and online platforms. It should involve efforts to inform and enlighten tourists about responsible and sustainable travel behaviors. Strategies include the use of informative signage at key locations, the creation of guidebooks or brochures dedicated to responsible tourism, and the development of content for online platforms for disseminating information and engaging with tourists. Visitor centers can serve as hubs for educating tourists, while involving local communities in sharing their knowledge and culture fostering mutual respect. Collaboration among various stakeholders, such as government bodies, tour operators, and NGOs, is crucial to amplifying the message and ensuring its widespread adoption. By implementing these education and awareness initiatives, destinations empower tourists to make environmentally and culturally responsible choices, contributing to sustainable visitor management and the preservation of the destination's resources.

f. ***Capacity enhancement programs:*** Such programs empower local residents by providing training and skills development enhancing the quality of services offered by them to tourists, promoting sustainable environmental practices, preserving indigenous cultures, and fostering a sense of community ownership overtourism assets. Additionally, capacity building empowers communities to engage in decision-making processes related to tourism development, ensuring that their interests are represented and aligned with responsible and sustainable tourism practices (Singh et al., 2009). To enhance community involvement in small-scale tourism business startups and protect their common interests, it is imperative to implement capacity-building programs that encompass initiatives like hospitality training and facilitating financial access for communities (Kebete & Wondirad, 2019). These programs empower the community to provide better services, handle the influx of tourists, and mitigate the negative impacts of tourism, ultimately contributing to sustainable visitor management.

g. ***Visitor Quotas:*** Visitor quotas, a vital practice in visitor management at tourism destinations, involve setting limits on the number of tourists allowed into specific areas or sites to mitigate issues like overcrowding, and environmental degradation, and to maintain a high-quality visitor experience, thereby ensuring sustainable tourism and the preservation of natural and cultural assets. This approach aligns the economic benefits of tourism with the long-term viability of the destination, guided by carrying capacity considerations. Several destinations worldwide, such as the Galapagos Islands and Machu Picchu, have successfully implemented visitor quotas to maintain ecological balance. Adverse environmental effects can be controlled and limited through monitoring and controlling the influx of visitors and their activities in the destination (Jamin & Rahmafitria, 2022).

h. ***Certification:*** The promotion and adoption of ecotourism practices are essential for fostering sustainable development in various regions. To effectively implement ecotourism, region-specific planning strategies must be devised, considering the unique physical, socio-cultural, economic, and ecological characteristics and challenges of different parts of the Himalayas. Programs like the Global Sustainable Tourism Council's (GSTC) certification provide guidelines and standards for sustainable tourism practices. In a similar vein, organizations such as the Responsible Tourism Society of India (RTSOI), formerly known as the Ecotourism Society of India, and Green Globe, a prominent global leader in sustainable tourism certification, should actively endorse travel agencies, tour operators, or hotels that adhere to responsible tourism practices.

i. ***Coordination and Cooperation:*** The lack of coordination among planning, management, and scientific agencies, along with inadequate strategy and marketing expertise, as well as the mismanagement of waste and unregulated activities at multiple sites, hinders sustainability and sustainable business prospects in the sector (Chandra & Kumar, 2021). Sustainable visitor management in the Himalayas necessitates effective coordination and cooperation among tourism stakeholders and local communities, underpinned by responsible and ethical attitudes, along with a shared commitment to agreed-upon regulations, to ensure proper implementation.

j. ***Diversified Tourism:*** Diversifying tourism destination attractions by shifting the focus from religious tourism to the development of ecotourism, rural tourism, and other niche forms of tourism is a vital strategy for sustainable visitor management in the Himalayas (Chandra & Kumar, 2021).

This approach not only disperses tourist traffic more evenly across various attractions, reducing the strain on specific sites but also ensures a more balanced distribution of economic benefits to a wider range of communities. By catering to diverse interests and preferences, these niche tourism segments promote responsible and immersive travel experiences, aligning with the principles of sustainability and benefiting both the environment and local communities while safeguarding the region's delicate ecosystems and cultural heritage.

4. CONCLUSION

Overtourism in the Himalayan region is a complex issue that requires immediate attention and proactive measures. The substantial influx of tourists in the Himalayan region has the potential to evolve into sustainable business and developmental opportunities if taken care of properly and responsibly. Key strategies involve assessing and adhering to carrying capacity limits, implementing regulations and control mechanisms to manage seasonality and visitor numbers, actively involving local communities in decision-making processes, developing eco-friendly infrastructure, promoting education and awareness among tourists, initiating capacity enhancement programs for communities, imposing visitor quotas where necessary, encouraging certification for responsible tourism practices, fostering coordination and cooperation among stakeholders, and diversifying tourism offerings. These practices collectively aim to strike a balance between economic benefits and environmental preservation, ensuring the long-term viability of tourism destinations while minimizing negative impacts on the environment and local communities. By adopting these practices, the Indian Himalayan Region can aspire to achieve sustainable and responsible tourism management, safeguarding its unique natural and cultural assets for future generations. The study is a comprehensive guide for addressing the challenges of tourism growth in the Indian Himalayan Region while preserving its ecological and cultural integrity. By prioritizing the preservation of the region's unique natural and cultural assets, we can ensure that future generations can continue to enjoy the beauty and wonders of the Himalayas.

REFERENCES

Badola, R., Hussain, S. A., Dobriyal, P., Manral, U., Barthwal, S., Rastogi, A., & Gill, A. K. (2018). Institutional arrangements for managing tourism in the Indian Himalayan protected areas. *Tourism Management, 66*, 1–12. doi:10.1016/j.tourman.2017.10.020

Basu, S., Singh, J., Shrivastava, S.K., Paliwal, A., & Chakravartty, A. (2016). *Remembering 2013 Uttarakhand floods.*

Bennike, R. B., & Nielsen, M. R. (2023). Frontier tourism development and inequality in the Nepal Himalaya. *Journal of Sustainable Tourism, 0*(0), 1–22. doi:10.1080/09669582.2023.2174129

Bhalla, R., & Chowdhary, N. (2022). Green workers of Himalayas: Evidence of transformation induced regeneration. *Journal of Tourism Futures, 8*(3), 380–392. doi:10.1108/JTF-12-2021-0273

Chan, R. C. K., & Bhatta, K. D. (2023). Trans-Himalayan connectivity and sustainable tourism development in Nepal: A study of community perceptions of tourism impacts along the Nepal–China Friendship Highway. *Asian Geographer*, *40*(1), 59–80. doi:10.1080/10225706.2021.1943470

Chandra, P., & Kumar, J. (2021). Strategies for developing sustainable tourism business in the Indian Himalayan Region: Insights from Uttarakhand, the Northern Himalayan State of India. *Journal of Destination Marketing and Management, 19*, 100546. doi:10.1016/j.jdmm.2020.100546

Chauhan, N., Upadhyay, S. K., & Singh, R. (2021). The Himalayan natural resources: Challenges and conservation for sustainable development. *Article in Journal of Pharmacognosy and Phytochemistry, 10*(1), 1643–1648. www.phytojournal.com

Erdoğan, S., Gedikli, A., Cevik, E. I., & Erdoğan, F. (2022). Eco-friendly technologies, international tourism and carbon emissions: Evidence from the most visited countries. *Technological Forecasting and Social Change*, *180*, 121705. doi:10.1016/j.techfore.2022.121705

Harrigan, P., Evers, U., Miles, M., & Daly, T. (2017). Customer engagement with tourism social media brands. *Tourism Management*, *59*, 597–609. doi:10.1016/j.tourman.2016.09.015

Jamal, M. T., Sujood, M. A., & Arwab, M. (2019). Influence of Social Media on Travel and Tourism Industry: A Case Study of Pauri Garhwal District of Uttarakhand. *Amity Global Business Review.* https://www.researchgate.net/profile/Brighton-Nyagadza/publication/348297189_Embracing_new_banking_innovations_through_ICT_Firms_pursuing_growth_in_emerging_economies/links/5ff6d13245851553a026dad0/Embracing-new-banking-innovations-through-ICT-Firms-pursu

Jamin, A., & Rahmafitria, F. (2022). Visitor Management Concept Through Carrying Capacity Analysis In Forest Recreation. *Journal of Indonesian Tourism. Hospitality and Recreation*, *5*(1), 1–12. doi:10.17509/jithor.v5i1.46821

Joshi, R., & Dhyani, P. P. (2009). Environmental sustainability and tourism - Implications of trend synergies of tourism in Sikkim Himalaya. *Current Science*, *97*(1), 33–41.

Kala, C. P., & Maikhuri, R. K. (2011). Mitigating people-park conflicts on resource use through ecotourism: A case of the Nanda Devi Biosphere Reserve, Indian Himalaya. *Journal of Mountain Science*, *8*(1), 87–95. doi:10.1007/s11629-011-1010-5

Kapoor, S., & Wangdus, J. (2022). Impact of overtourism on sustainable development and local community wellbeing in the Himalayan region. *International Journal of Leisure and Tourism Marketing*, *7*(3), 199–214. doi:10.1504/IJLTM.2022.122454

Kebete, Y., & Wondirad, A. (2019). Visitor management and sustainable destination management nexus in Zegie Peninsula, Northern Ethiopia. *Journal of Destination Marketing & Management*, *13*(March), 83–98. doi:10.1016/j.jdmm.2019.03.006

Lorenz, D. F., & Dittmer, C. (2021). Disasters in the 'abode of gods'—Vulnerabilities and tourism in the Indian Himalaya. *International Journal of Disaster Risk Reduction, 55*, 102054. doi:10.1016/j.ijdrr.2021.102054

Manning, R. E. (2007). *Parks and carrying capacity: commons without tragedy*. Island Press.

Martín Martín, J. M., Salinas Fernández, J. A., Rodríguez Martín, J. A., & Ostos Rey, M. (2020). Analysis of Tourism Seasonality as a Factor Limiting the Sustainable Development of Rural Areas. *Journal of Hospitality & Tourism Research (Washington, D.C.)*, *44*(1), 45–75. doi:10.1177/1096348019876688

Nepal, S. K. (2000). Tourism in protected areas. *Annals of Tourism Research*, *27*(3), 661–681. doi:10.1016/S0160-7383(99)00105-X

Nyaupane, G. P., & Chhetri, N. (2009). Vulnerability to climate change of nature-based tourism in the Nepalese Himalayas. *Tourism Geographies*, *11*(1), 95–119. doi:10.1080/14616680802643359

Nyaupane, G. P., & Thapa, B. (2004). Evaluation of Ecotourism: A Comparative Assessment in the Annapurna Conservation Area Project. *Journal of Ecotourism*, *6*(1), 50–58. doi:10.1080/14724040408668148

Rawat, A., Joshi, S., & Rai, S. K. (2023). Evaluating the issue of sustainable tourism with a system dynamic approach: Evidence from Uttarakhand, India. *Environment, Development and Sustainability*, *3*(0123456789). doi:10.1007/s10668-023-03711-1

Sahani, N. (2021). Application of hybrid SWOT-AHP-FuzzyAHP model for formulation and prioritization of ecotourism strategies in Western Himalaya, India. *International Journal of Geoheritage and Parks*, *9*(3), 349–362. doi:10.1016/j.ijgeop.2021.08.001

Samant, S. S., Vidyarthi, S., Pant, S., Sharma, P., Marpa, S., & Sharma, P. (2011). Diversity, distribution, indigenous uses and conservation of the medicinal plants of Indian Himalayan region used in cancer. *Journal of Biodiversity*, *2*(2), 117–125. doi:10.1080/09766901.2011.11884732

Singh, R. B., Mal, S., & Kala, C. P. (2009). Community responses to mountain tourism: A case in Bhyundar Valley, Indian Himalaya. *Journal of Mountain Science*, *6*(4), 394–404. doi:10.1007/s11629-009-1054-y

Stankey, G. H., & Cole, D. N. (1998). Historical development of limits of acceptable change: conceptual clarifications and possible extensions. In *Limits of acceptable change and related planning processes: progress and future directions. Ogden* (pp. 5–9). EUA.

Takahashi, L. Y., & Cegana, A. C. (2005). Como monitorar o impacto dos visitants utilizando o sistema LAC-Limite Aceitável de Câmbio. *Revista Turismo Em Análise*, *16*(2), 206–222. doi:10.11606/issn.1984-4867.v16i2p206-222

Tourism Council of Bhutan. (2022). *Final Tourism Rules and Regulations.* Tourism Council of Bhutan. https://www.tourism.gov.bt/uploads/attachment_files/tcb_7uivTQtN_23.09.22%20Final%20Tourism%20Rules%20and%20Regulations%202022.pdf

U. N. D. P. (n.d.). *Biodiversity Conservation and Impacts of Nature-Based Tourism in Annapurna Conservation Area.* United Nations Development Programme. https://www.undp.org/nepal/biodiversity-conservation-and-impacts-nature-based-tourism-annapurna-conservation-area

Ziegler, A. D., Wasson, R. J., Sundriyal, Y., Srivastava, P., Sasges, G., Ramchunder, S. J., Ong, C. E., Nepal, S. K., McAdoo, B. G., Gillen, J., Bishwokarma, D., Bhardwaj, A., & Apollo, M. (2023). A call for reducing tourism risk to environmental hazards in the Himalaya. *Environmental Hazards*, *22*(1), 1–28. doi:10.1080/17477891.2021.1984196

Chapter 13
Promotion of Homestay for the Sustainable Livelihood of the Local Community in the North Eastern Himalayan Region of India

Hawaibam Jhalari Devi
Assam University, Silchar, India

Soumendra Nath Biswas
Assam University, Silchar, India

ABSTRACT

Northeastern India is a region with unique culture and traditions and a varied landscape offering a greater potential to promote homestays. It is a locally owned and operated family business that provides visitors with a more enriched experience of nature as well as amenities like local cuisine, traditional culture, lifestyle, and local resources. This idea is ideal for the Northeastern Himalayan region of India. This paper will concentrate on understanding the importance of homestay in developing Community-based tourism and comprehending the potential of sustainable Community-based tourism in the Northeastern Himalayan region of India. A review was undertaken and the paper is designed to identify the key areas for the promotion of homestays and their contribution to the growth of sustainable community-based tourism and the benefits it offers in the Northeastern Himalayan Region of India.

INTRODUCTION

Tourism, being the world's largest and fastest expanding industry (Scheyvens & Momsen, 2008), is seen as a key industry and a significant source of socioeconomic factors. Community-Based Tourism (CBT) is a type of sustainable tourism in which the local community engages and is involved in vari-

DOI: 10.4018/979-8-3693-0823-3.ch013

ous tourism-related activities for the growth of tourism as well as the conservation of the cultural and natural treasures of their places, which benefits them (Mann, 2000). Northeast India, with its rich natural resources, history, culture, traditions, festivals, delectable cuisines, handloom and handicrafts, and art forms, is a region with magical beauty and bewildering diversities where people of all ages, interests, and levels of experience are drawn to by the breathtaking alpine scenery with its visions of snow-capped mountain peaks and the serenity of its lush, green hills (Ministry of Culture, Government of India). The development of Community-based tourism in the Northeastern Himalayan Region can boost job creation and economic empowerment in local communities by providing sustainable livelihoods (Roy, 2022).

Northeast India has an area of approximately 2, 62,379 km2 and is located between latitudes 22°N and 29°5'N and longitudes 88oE and 97°30'E, sharing international borders with Bangladesh, Bhutan, China, and Myanmar (Rodgers and Panwar, 1988). Based on biogeographic zones, the region is separated into Eastern Himalaya (Arunachal Pradesh and Sikkim) and Northeast India (Assam, Manipur, Meghalaya, Mizoram, and Nagaland) (Rao, 1994). According to Conservation International, the Eastern Himalayas encompass the entire state of Northeast India (Myers, 1988).

Northeastern India has a distinct culture and heritage, as well as a diverse geography, making it an ideal location for promoting homestays. The growing tourism trend of community-based homestays would make it possible to provide a sustainable source of income for the residents. Homestay is one of CBT's activities; it is a locally owned and maintained family business that offers guests an enhanced experience of nature as well as amenities like local cuisine, traditional culture, lifestyle, and local resources. A homestay is when tourists stay in someone's home for a short length of time, providing accommodation and several other services to the tourists, ensuring that they are satisfied with the services offered by the homestay owner and the local community (Timlasana, 2012). It gives tourists a wonderful opportunity to learn about the local people, interact with the community, and spend quality time with the homestay owners, as well as giving them a royal experience during their stay (Magar, 2021). It is one of the most important CBT activities that can be promoted in India's Northeastern Himalayan Region to reduce rural poverty (Leksakundilok, 2004).

OBJECTIVES

1. To find out the potential of homestay in the Northeastern Himalayan region.
2. To understand how homestay programs can contribute to the sustainable livelihood of local communities.
3. To identify the benefits of homestay for the local communities.
4. To suggest measures to promote homestay as a tool for Community based tourism development in the Northeastern Himachal region.

COMMUNITY-BASED TOURISM IN NORTH EASTERN HIMALAYAN REGION OF INDIA

Community-based tourism (CBT) is regarded as a responsible and sustainable tool for the livelihood of rural communities by utilising tourism resources with a high potential for income generation, preserving local culture, and environmental conservation that will help provide long-term benefits to the local people while also offering satisfying products and services to tourists and all visitors to the destinations (Dutta, & Dutta, 2019). The importance of community-based tourism lies in the importance of local people working together for tourism development in their community, which results in developing and generating income for a sustainable livelihood that will raise their living standards through the use of available resources (Blackstock, 2005).

Given the growing importance of tourism as an economic instrument. CBT will be a formal development tool rather than an alternative development model (Hamzah & Khalifah, 2009). CBT generates a sustainable life for the local community, lowers poverty in society, and provides people with several chances for employment, as service providers, and the sale of diverse local items such as handloom and handicrafts, local cuisines, and so on (Lee, 2013). Because of its incredible biodiversity, Northeast India has been highlighted as a tourism hotspot. The development of CBT should be well established by the local community, which facilitates tourism with their involvement at every step of tourism development in India's Northeastern Himalayan Region.

Sustainable Tourism, Eco-Tourism, and Community-Based Tourism are three subdivisions of the notion of Sustainable Development, according to the United Nations World Tourism Organisation (UN-WTO 2004). Considering the characteristics of rural and community-based tourism, the Government of India's Tourism Policy (2002) created the concept of rural tourism to highlight rural life, art forms, the country's rich culture and traditions, and handloom and handicraft products. In 2002-2003, the Ministry of Tourism launched a number of rural tourism initiatives, resulting in increased interaction between locals and tourists for a more enriching experience, as well as social and economic benefits to the locals.

Various state governments in India's Northeastern Himalayan region, as well as the national government, are eager to encourage rural tourism and homestays through Community-Based Tourism. This will contribute to the expansion of tourism in the remote Himalayan region and make it more accessible to all visitors and tourists. Darap village homestay in Darap village (Sikkim), Tenzin homestay in Tawang, Arunachal Pradesh, Maipakchao family homestay in Manipur, Mawlynnong village homestay in Meghalaya, and others are available across the Northeast.

Tourism in the Northeast has grown rapidly, and as one of the most lucrative industries, all eight states' governments are now focusing more on sustainable tourism that creates jobs, promotes local culture, traditions, and products, and conserves and protects the environment. It also plays a vital part in improving the region's economic position by creating job opportunities and generating revenue, so alleviating poverty and improving the livelihoods of the local community in Northeast India.

Community-Based Tourism will go a long way towards improving income-generation opportunities in rural communities. Conservation of community natural resources is critical for tourism to stay sustainable and beneficial to a local community (Kayat, 2014). Local communities' participation in tourism will support their livelihood needs and, as a result, generate a stake in the survival of local culture, ecology, and environment. Some Northeastern Himalayan regions, such as Sikkim and Arunachal Pradesh, have already implemented homestay programmes, and community-based tourism is being pushed successfully in many communities. With its rich unique culture and customs, fairs and festivals, sports, and the availability of diverse tourism resources, Northeast India has a stronger potential to promote homestay.

The Ministry of Tourism, Government of India, in May 2022, clearly emphasised the expansion of rural tourism through the rural homestays programme as part of the National Strategy for Promotion of Rural Homestays. The Ministry also developed a strategy for the development of sustainable tourism in the country, with the goal of recognising the role of homestays in boosting rural and sustainable tourism and developing the rural homestay support system.

THE POTENTIAL ASPECTS OF HOMESTAY IN THE NORTH EASTERN HIMALAYAN REGION

Homestay programmes can be implemented in Northeast India through community-based tourism, as the region possesses both natural spaces and community culture to encourage tourism (Barooah, 2018). The region is rich in potential and offers a variety of options to pioneer an ecotourism path that leads to the protection and preservation of nature and culture through active community engagement.

Homestay is something that entices travelers to leave their typical surroundings and spend time in the rural splendorous natural environment. Homestay in Northeast India has enormous potential to create various job opportunities for local communities by offering various activities and resources available in the region, such as its rich culture and heritage, ethnic cuisines, various sightseeing, handloom and handicrafts, arts and sports, and so on. At the same time, tourists can get a chance to experience the unique lifestyle, customs, and culture while the homestay owners provide them with clean, comfortable, and eco-friendly accommodations and food and provide an alternative source of income to the rural communities (Gangotia, 2013).

As a biodiversity hotspot in the world, Northeast India, with its pristine forests, hills, wildlife sanctuaries, and national parks, attracts a diverse range of tourists, resulting in the promotion of sustainable tourism. All of northeast India's potential can let travelers have a community-based homestay experience while also learning about its traditional culture and lifestyle.

Through homestays, Northeast India gives visitors a taste of the true Northeast. There is no better way to learn more about a destination than to stay among the people. Homestays might serve to introduce people to rural culture and nature. As a result, community-based tourism homestay is a fresh concept for India's Northeastern Himalayan region.

Arunachal Pradesh is ideal for homestay enthusiasts who want to be rejuvenated by snow-capped mountains, Himalayan ranges, and rivers. The Arunachal Pradesh homestay is the best entryway to a tranquil and peaceful vacation surrounded by lush flora, stunning landscapes, and Himalayan peaks. The Arunachal Pradesh State Tourism Department produced policy guidelines to involve the local villagers in administering all community-based tourism-related activities in the village, including homestays (Dutta, 2015).

Assam, India's picturesque northeastern state, is well known for its tea gardens, wildlife sanctuaries, religious sites, and cultural history. Assam homestays may be as exciting as traveling to this region of India since the state offers a colorful culture, delectable cuisine, and a wealth of sites to see. There are numerous excellent homestays in the state. The Assam State Tourism Department developed this new type of accommodation with the goal of offering new dimensions and thrust to the state's rural homestay services in tourist areas.

Manipur, the ideal vacation location, has a comfortable climate, a rich natural environment, wildlife, cultural legacy, distinctive flora and fauna, colorful festivals, and scrumptious indigenous cuisines. The homestay concept has been outlined in Manipur's State Tourism strategy in order to promote it in a structured manner, particularly in the interior areas where tourists will be able to explore and experience the pure beauty of the place. Manipur's State Tourism Department has developed such community-based tourism homestay ideas in order to attract tourists to the state. It has also issued a number of common rules and recommendations for the establishment of homestays throughout the State.

Meghalaya is a lovely state with majestic mountains, abundant flora, and cascading waterfalls. Homestay in Meghalaya provides a gateway to the state's heart and spirit, assuring a memorable and rewarding stay for all visitors. Meghalaya's State Tourism Department promotes community-based tourism in rural regions through the Meghalaya Village Development and Tourism Promotion Cooperative Society Ltd. Many homestays are being created by village groups throughout the state.

Mizoram is well-known for its natural beauty, verdant valleys, Blue Mountains, and exceptional Mizo culture. Since the concept of community-based tourism is relatively new in the state, the Tourism Department of Mizoram has introduced a new scheme for homestays under the former New Economic Development Policy (NEDP) to supplement the demand for accommodation at various tourist destinations throughout the State.

Nagaland, a picturesque state in Northeast India noted for its festivals, is a country of magical forests, majestic mountains, age-old folklore, and rich history, all of which have established the state's potential for rural tourism. There are already several homestays in the state where there is significant community participation, particularly among women.

Sikkim, a popular tourist destination, has led the way in developing an ecotourism route. Community-based tourism is well developed in the state, with homestay programs effectively implemented by local communities in the state's numerous ecotourism destinations. Sikkim's Tourism Department has launched several projects and established several laws and regulations to promote village tourism. Through the building of homestays in various sections of the state, the state's homestay programs have given a significant boost to the growth of rural tourism in the State.

Tripura is considered as the key trade gateway to the rest of the country. Tourists can explore and enjoy the state's rich history and archaeology, cultural heritage, traditional art, and music. The state also offers a lot of potential for developing community-based tourism. The concept of homestay is relatively new in the state, but it offers great potential for the development of community-based tourism homestays throughout the state.

WHY SHOULD ONE CONSIDER STAYING AT A HOMESTAY IN NORTH EAST INDIA?

Homestays offer an authentic experience by allowing guests to engage in residents' daily activities. So, when an individual wants to learn more about Northeast India, it is better to choose such type of accommodation sector in order to experience and learn about the local culture and languages, while also having the feeling of being at home, and such accommodation sector can give you a true experience of local life, thus contributing to the livelihood of the local community.

Northeast India has traditionally been noted for its different communities coexisting with their own distinct traditions, customs, and ways of life, as well as its simplicity and cleanliness.

Visitors or guests might consider staying at a homestay in Northeast India since the homestay owner treats them as family members rather than as strangers. The tourists will be treated well and will be able to enjoy village life with good authentic local food and live in a region's authentic culture.

Apart from this, it has raised the living standards of the locals and this makes the locals more actively participate in various tourism-related activities. So, the following are the reasons why one should consider staying at a homestay.

Table 1. Homestay SWOT analysis for the NE Himalayan region

Strengths	Weaknesses
• The natural resources available at the destination are considered one of the major strengths of the region. • The presence of diverse ethnic tribal cultures with unique customs and traditions is also another strength of the region. • Another strength is the existence of various flora and fauna. • The availability of forests, tea gardens, and organic farms is a significant asset for the region's development.	• Lack of active community participation. • Lack of awareness and training programs. • Lack of skilled laborers. • Lack of proper management. • Lack of transportation and communication systems. • Lack of availability of tourist amenities. • Language barriers. • Lack of financial support.
Opportunities	**Threats**
• The local communities are encouraged to preserve the local culture and provide knowledge of local customs and culture to enhance the tourist's experience. • Provide opportunities to showcase the local ethnic cuisines and also provide various activities to the tourists. • It also provides opportunities to promote unexplored areas. • It also enables to offer the rich culture of the NE Himalayan region. • The region has a great opportunity for the development of handloom and handicraft products. • Development of border trade is also another great opportunity for the region. • Provide employment opportunities to the local communities. • It helps to develop a bond between the guest and the owner during their stay in the homestay.	• Increase in competition from similar mountain destinations in the same segment. • Political instability. • Law and order condition. • Increased in insurgencies. • Derailing from the core concept of homestay tourism.

1. *To experience the real rural lifestyle or their way of living:* There is no better way to experience the beauty of nature and at the same time visitors get to know and experience the local culture, traditions, local food, and lifestyle.
2. *Provide specialized service:* Homestay owners provide visitors with high-class services. The guests receive plenty of attention and build a strong bond between the guests and the host family and spend most of the time giving attention to the guest's needs and wants during their stay.
3. *Safety and Security:* The homestay owners create a safe and productive environment and protect the guests from any harm and the guests will feel safer when they are staying with a family.
4. *Provide healthy authentic local cuisine:* The homestay owners provide meals which is a home cooked with different meal plans. The meals are simple and provide authentic local cuisines.
5. *Provide opportunities for a variety of activities:* Homestays provide tourists opportunities for a variety of activities depending on the geographic location and cultural ethos of the place. Such as taking a tour of the village to learn local history and insights, etc.
6. *Supporting Local Families:* Such community-based tourism programs provide a chance for the local people to earn and improve their sustainable livelihood and this will create a pathway to contribute to the local economy of the region.

BENEFITS OF SUSTAINABLE HOMESTAYS FOR LOCAL COMMUNITIES

Homestays can benefit both individual homestay owners and local rural communities on the whole. The local communities not only provide accommodation to the tourists; but also provide various other services such as tourists are given a chance to experience the authentic local culture, the rural lifestyle, and local food, they get to know about the handloom and handicrafts, arts and sports, and also get chance to have an interaction with the local communities. To understand the various benefits of homestay for the growth of sustainable livelihood of the local community and to suggest measures to promote homestay in the Northeastern Himalayan region (Thakur & Monga 2022).

MEASURES TO PROMOTE HOMESTAY AS A TOOL FOR THE CBT DEVELOPMENT IN THE NORTHEASTERN HIMALAYAN REGION

- Benefits should be provided to local residents through self-employment and enterprise development programmes. Such rural accommodation should be regarded eligible projects under schemes such as the PMEGP to promote self-employment, village industries, and other Central and State Government initiatives for the promotion and growth of MSME service firms. As a result, rural homestays should be eligible for incentives under these schemes.
- The State Government should also provide Schemes for financial support to Rural Homestays.
- The State Government should provide requisite support under various schemes for setting up common facilities and public infrastructure to enhance the engagement of visitors with the local community and visitor experience.
- The government should immediately focus on building good roads, transport connectivity, electricity, medicine and healthcare, and all other residential facilities.
- Proper marketing and promotion are very much needed to generate new business or to run the existing business.
- Proper training on leadership, and hospitality along with good communication and personality development for more enhancing skills for the local people are to be organized by the related authorities.

Table 2. The below shows the multiple benefits of homestay to the local communities and how it can be used for the growth of its sustainable livelihood

Economic Benefits	• Create employment opportunities for the local communities. • Led to the increase in additional source of income. • Generate foreign exchange. • Economic benefits to the locals through various products such as handloom and handicrafts, local food, etc. • Growth and development in various public amenities.
Social and Cultural Benefits	• Raise the living standard of the local communities. • It also rises in the cultural exchanges through fairs and festivals. • Embeds development of local products. • Improves in the preservation of local culture and values of the region. • Nurturing relationships with guests and the local community.
Environmental Benefits	• Homestays have helped the locals aware of the importance of conserving and preserving the rich culture, natural resources, and environment. • Homestays have helped the locals to develop a healthy environment with proper hygiene and sanitation for better living. • Helps in maintaining the cleanliness of the surroundings of the local. • Homestay has helped to eradicate poverty and contributes to economic empowerment. • Homestays have encouraged economic activities. • Sustainable use of natural resources.

- Thus, in order to promote homestays, various tourism stakeholders should take measurable steps and responsibility and also organize various campaigns for homestay tourism.

CONCLUSION

This paper investigates the role of homestay in community development and its impact on the sustainable livelihood of the local people. It can be concluded from the preceding discussion that the natural beauty of the region, the diverse communities, climatic conditions, agricultural business, national parks and wildlife sanctuary, handloom and handicrafts, adventure sports, flora and fauna are the main attractions to people visiting the Northeastern Himalayan region of India, and these may be considered the destination's strengths. It is evident from the study that homestay programs promote rural tourism and can benefit both the homestay owners and the local communities. So, there is a need for the development of homestay programs as an alternative source of income for the sustainable livelihood of the local people. Thus, homestays can act as a potential tool to promote and boost sustainable community based tourism in the Northeastern Himalayan Region of India.

REFERENCES

Barooah, S. (2018). Touristy feet in muddled Northeast India. Paper presented in *National Conference on Livelihoods through Rural Tourism in Northeast India: The Way Forward*. Research Gate.

Bhuyan, R., & Deka, S. M. (2019). Homestay for whom-lessons of ASEAN countries for North East India. *International Journal of Tourism and Travel, 12*(1&2), 21–26.

Blackstock, K. (2005). A critical look at community-based tourism. *Community Development Journal: An International Forum, 40*(1), 39–49. doi:10.1093/cdj/bsi005

Chakravarty, S., Suresh, C. P., Puri, A., & Shukla, G. (2012). North-East India, the geographical gateway of India's phytodiversity. *Indian Forester, 138*(8), 702.

Dutta, M., & Dutta, P. K. (2019). Community-based tourism model for conservation: A case study of Arunachal Pradesh, part of Himalaya Biodiversity Hotspot. SSRN *Electronic Journal*. doi:10.2139/ssrn.3376193

Dutta, P. K. (2015). *Guidelines for the promotion of homestays in Arunachal Pradesh*. WWF, GOI-UNDP Project.

Gangotia, A. (2013). Home stay scheme in Himachal Pradesh: A successful story of community based tourism initiatives (CBTIS). *Global Research Analysis, 2*(2), 206–207.

Gebreyesus, A., Tesfay, G., Tesfay, M., & Meheretu, Y. (2020). *Opportunities and challenges of community-based tourism development in Southern Tigray*.

Hamzah, A., & Khalifah, Z. (2009). *Handbook on community based tourism 'How to develop and sustain CBT*. APEC Secretariat.

Hussin, R., & Kunjuraman, V. (2014). Sustainable community based tourism (CBT) through homestay programme in Sabah, East Malaysia. *Proceeding of the social sciences research ICSSR*. Research Gate.

Kayat, K. (2014). Community-based rural tourism: A proposed sustainability framework. *SHS Web of Conferences, 12*. IEEE. 10.1051/shsconf/20141201010

Lama, R. (2014). *Community based tourism development (a case study of Sikkim)* [Doctoral Dissertation, Kurukshetra University].

Lee, T. H. (2013). Influence analysis of community resident support for sustainable tourism development. *Tourism Management, 34*, 37–46. doi:10.1016/j.tourman.2012.03.007

Leksakundilok, A. (2004). *Ecotourism and community based tourism in the Mekong region*, (working paper series no. 10). Australia Mekong Resource Center.

Magar, D. A. (2021). Economic contribution of homestay tourism: A case study of Magar homestay of Naruwal, Lamjung. *Marsyangdi Journal*, 129–139. doi:10.3126/mj.v2i1.39972

Mann, M. (2000). *The community tourism guide. Tourism concern*. Earthscan Publications.

Mendoza, M. M. R. (2017). *How they do it: Community based tourism; case study of the community participation of Kikil, Yucatan, Mexico.*

Myers, N. (1988). Threatened biotas: "hot spots" in tropical forests. *The Environmentalist, 8*(3), 187–208. doi:10.1007/BF02240252 PMID:12322582

Rao, R. R. (1994). *Biodiversity in India: Floristics Aspects*. Bishen Singh Mahendra Pal Singh.

Rodgers, W. A., & Panwar, H. S. (1988). *Planning a Wildlife Protected Area Network in India* (Vol. I and II). Wildlife Communications Trust of India.

Roy, S. (2022). *Rural homestay as a driver of community development and ecological sustainability, a case study of Buxa tiger reserve forest*, West Bengal. *International Journal of Mechanical Engineering, 7*(7).

Sarkar, S. (2022). Home stay tourism and its sustainable approach in Rural Development: An appraisal from eastern Himalaya. *International Research Journal of Modernization in Engineering Technology and Science.*

Scheyvens, R., & Momsen, J. H. (2008). Tourism and poverty reduction: Issues for Small Island states. *Tourism Geographies, 10*(1), 22–41. doi:10.1080/14616680701825115

Thakur, J., & Monga, N. (2022). Role of homestay in reviving tourism in Shimla. *PUSA Journal of Hospitality and Applied Sciences, 8*(2), 49–59. doi:10.48165/pjhas.2022.8.2.7

Timlasana, P. (2012). Homestay tourism boosts Ghale Gaon"s Economy". http://www.gorkhapatra.org.np/risin Retrieved 10/6/2016

UNWTO. (2004). *A guidebook on indicators of sustainable development for tourism destinations*. World Tourism Organization.

Chapter 14
Role and Importance of Nature-Based Health Tourism in the Himalayan Region:
With Special Reference to Different Natural Health Techniques

Arunesh Parashar
Dev Sanskriti Vishwavidyalaya, India

Prachi Agarwal
Dev Sanskriti Vishwavidyalaya, India

ABSTRACT

Himalayas are a multifaceted site of subjective, global, and metaphysical importance that are responsible in maintaining a form of tourism that satisfies the person's main motive for maintaining and strengthening the physical, spiritual, emotional, and psychological health. There are numerous opportunities that are availed in contrast with natural resources available in the Himalayan region that will help in boosting natural healing procedures and foster physical and psychic well-being of the tourists. In the circumstances of this fast-paced urbanisation, the Himalayas are the only option to get nature therapy and able to enjoy their close connection with it. Human-nature relationships have always been considered very important. So, theory of wellness in the Himalayas is supported by the research with the fact that nature has a great impact on different facets of person's physical, psychological, social, and spiritual life. This chapter seeks to shed light on the region's potential as a global wellness destination.

1. INTRODUCTION

1.1 Nature

Secrets of being healthy are hidden in the lap of nature. "Nature" can refer to the phenomena of the physical world, and also to life in general. Human health and well-being depend on the natural environ-

DOI: 10.4018/979-8-3693-0823-3.ch014

ment, which is the source of clean air, water, healthy soils and food. In ancient times, human beings lived close to nature and its resources and took advantages of nature's bounty. To create a balance between physical and mental health of a person, there is a need to improve one's own lifestyle and daily routine. For which nature is the ultimate source of improving a lifestyle that helps us in moving according to the laws of nature. Nature offers us not only the physical benefits of fresh air, exercise, and nourishing food but also serves as a sanctuary for our mental and emotional well-being. In the embrace of nature's beauty, stress subsides, and a sense of peace envelops the soul. The soothing rustle of leaves, the scent of blooming flowers, and the tranquil vistas of mountains or seascapes provide respite from the frenetic pace of modern life. Recent decades have also shown that how man's increasing alienation from his natural environment can have a diametric effect on health and well being. These high peaked majestic mountains that stands erect, gushing waterfalls, snow capped areas, blooming flowers, and lavishing greenery on the earth has created a ecosystem that sustains everyone's life around themselves. No one can live without their merciness.

1.2 Health

Health is the invaluable treasure of man, on which happiness, prosperity and activities of a man depends. According to W.H.O, "Health is defined as the state of Physical, Mental, Social and Spiritual well being of a person". Health is not a creative activity, it encourages and restores people. Health is a multifaceted concept that encompasses not only the absence of disease but also the physical, mental, and social well-being of an individual. It is a state of dynamic equilibrium where the body's systems function optimally, the mind is resilient and balanced, and social interactions contribute positively to one's overall quality of life. Achieving and maintaining good health is a lifelong journey that involves a combination of factors, including nutrition, exercise, rest, stress management, and access to healthcare. It is not merely the absence of illness but the presence of vitality, energy, and a sense of fulfillment in daily living. Health is a fundamental asset that empowers individuals to pursue their dreams, engage in meaningful activities, and enjoy the richness of life. It is a precious resource that deserves our attention, care, and investment, as it forms the foundation upon which we build our futures.

1.3 Nature and Health

"When we protect nature, we are protecting our lives. If there was never nature, there would never be life."

All life on earth is connected like a chain. The need for restoration through nature has a long history, originating from ancient Chinese healers and Greek philosophers, and the life belief that humans can improve their body and mind in natural environments repeatedly appears throughout human history. **(Selhub E.M.. Logan A.C. Your Brain on Nature: The Science of Nature's Influence on your Health, Happiness and Vitaluty. John Wiley & Sons; Hoboken, NJ,USA:2012.p.12).** Researchers have identified from over the last 20 years that the benefits to human health and wellness that attributed them to re-connect with physical activity in natural environment. Health and Science experts have determined that deepening our connection to nature is extremely important for boosting our overall health, mood and mental clarity as it gives time to recharge, soak up some sun, and restore one's energy. Human body which comprises of five elements of earth i.e. Earth (Prithvi), Water (Jal), Fire (Agni), Air (Vayu), Ether(Aakash)and therefore, man can survive only by the association of the elements of nature. In fact, the whole process of human development takes place in the lap of nature. Many studies performed and found that people got several improvements in their well-being while visiting such green and blue spaces.

1.4 Natural Beauty and Health

"Earth and sky, woods and fields, lakes and rivers, the mountains and the sea, are excellent schoolmasters, and teach some of us more than we can learn from books."

- *John Lubbock*

No one has meant to seclude themselves indoors and staring at white walls for most of their days. The best concept is to be escape from their various social and daily activities and connect with nature

1.5 Objectives of this Chapter

- The aim of the study is to focus on the opportunities of nature-based health tourism offered by Himalayas.
- To explore the potentialities of nature-based health tourism in Himalayas.
- To recommend the possible ways of developing nature-based health tourism in Himalayan region.
- To suggest the concept of natural healing techniques for being healthy.

1.6 Health Tourism

Health tourism is defined by the World Tourism Organization (*WHO/Europe | Urban Health - Protecting Nature Protects Health – Lessons for the Future from COVID-19*, n.d.) as a service that improves an individual's health and morale (via mineral water, weather, or medical procedures) that takes place in the person's home for more than 24 hours but less than a year. This type of tourism helps in boosting natural healing procedures and foster physical and psychic well-being of the tourists. They offer a peaceful, natural setting for unwinding away from the crowded daily commotion of modern cities. It is also important to keep in mind that the Himalayas are home to a variety of religious places, therapeutic rivers, minerals, and other natural attractions that are capable of luring visitors seeking wellness even without charging them more.

1.7 Classification and Types of Health Tourism

The concept of Health Tourism originated from the definition by International Union of Official Travel Organisation (1973), the predecessor of UNWTO, defined as "providing health facilities in rural areas with natural resources, especially hot spring areas and unique climate areas." Health tourism is a branch of general tourism foreseeing tourists travel with the aim of receiving specific healing treatments or enhancing their mental, physical or spiritual well-being (Pessot et al., 2021).

Dr. Arunesh Parashar in his study "Features of Health Tourism in Uttarakhand with special reference to Rishikesh – A Critical Study", 2012 has divided Health Tourism in three different parts – Nature – based Health Tourism in which treatment involves all the natural resources to stay healthy and for caring their holistic health, second type involves Wellness tourism, in which tourists uses both natural and medical techniques and also adopts lifestyle and spirituality in their life that directly or indirectly

Figure 1. Parts of health tourism
(Parashar, 2012)

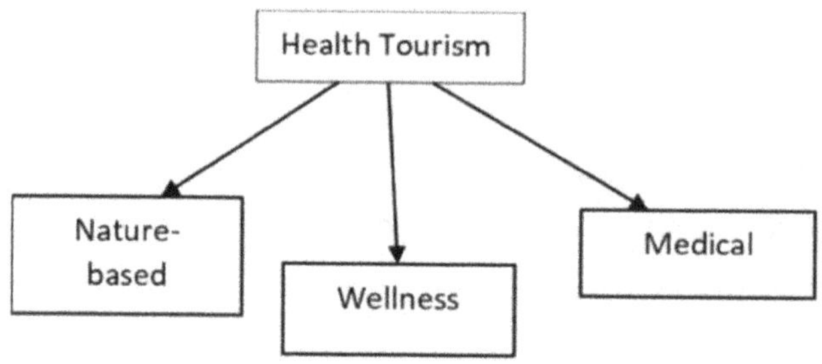

benefits to tourists mental health, third one is Medical tourism in which tourists travel from one place to another for surgeries and different medical science treatments to cure their health issues (Parashar, 2012).

1. Nature-based Health Tourism

Tourism, an act of movement from one place to another for the purpose of leisure, pleasure and recreation for not less than 24 hours, defined by UNWTO. This movement has started since the human civilisation has stepped to our earth with different needs and purposes. With the passage of time and development numerous changes and forms have arrived but travelling for health purpose is the common need of human beings which they fulfil either by searching for food, shelter etc. Invention of fire and wheel has paved more ways to travel for trading, spirituality and then good networks of communication and roads encourages people to move because of their curiosity or in search of finding some answers to their questions. To provide more services many kings ordered to construct rest houses for the tourists. This development and needs started building the new realms of a new and modern world with all basic amenities. Therefore, tourism acts as a bridge that joins two different phenomenon and in this chapter we are going to understand the importance of nature and it's resources available in Himalayan region in maintaining one's health with respect to travelling. As defined in the report of UNWTO, Health Tourism covers those aspects of tourism which have as a primary, motivation the contribution to physical, mental, and/or spiritual health through medical and wellness- based activities.

Nature-based health tourism in the Himalayan region is one of a burgeoning trend that combines the breathtaking natural beauty of the Himalayas with holistic wellness practices to offer a unique and rejuvenating experience for travelers seeking physical, mental, and spiritual well-being. This form of tourism capitalizes on the region's serene landscapes, fresh mountain air, and rich biodiversity to provide visitors with a range of wellness activities, outdoor adventures, and relaxation opportunities. Since ancient times, Himalayas have always been an attracting point and a place of immense peace for the Indian sages and priests.

India with its varied geographical features has a strong and vibrant system of traditional medicines as also mentioned in one of our ancient text book Charaksamhita written by Sage Charak. It can create a correspondingly large potential new market for commercial nature tourism, as a provider of nature therapies and therefore India would like to offer to the world, post Covid- 19 calm, peace and health through yoga, ayurveda and other forms of traditional treatments and spiritual surroundings. Even WHO has also published a set of prescriptions for a healthy, green recovery from Covid-19 – **"Protect and Preserve the source of Human Health: Nature".**

Ancient sages said that Himalayas are the rich source of herbals and ayurveda and therefore said:

औषधनिःपराभूमःहिमिवन शैल उत्तमः

As Himalayas are the richest source of such herbal and organic medicine. In the way sages also said that:

गरियस्य परवता हमिवंतो उरण्य ते पृथ्वी स्योनमस्तु

which means that Himalayan region can capitalize on the area's stunning landscapes and serene environment to offer visitors rejuvenating experiences.

Since, ancient times people started to travel towards the nature's asthetic element Himalayas for their happiness. The main purpose of these journeys is to understand the spiritual importance, entertainment and to visit new natural destinations. Due to which, people were able to face and attracts towards the richness, greatness, secrets and as well as the beauty of mountains. The Himalayas offer a unique and serene environment that is conducive to various wellness practices, making it an ideal destination for individuals seeking holistic well-being. The combination of breathtaking landscapes, clean mountain air, and a rich cultural heritage provides a perfect backdrop for nature-based health tourism.

1.8 Nature-Based Health Tourism Model

With the help of nature-based health tourism model, tourists can able to take care for their health, by utilizing the resources available in nature based health tourism destinations such as by travelling to natural places such as near mountains, rivers, forests etc., surrounding ourselves with its majestic beauty and surroundings, by performing different activities and in natural settings and using ancient health

Figure 2. Nature-based health tourism model
Parashar, A. & Agarwal, P. (2021)

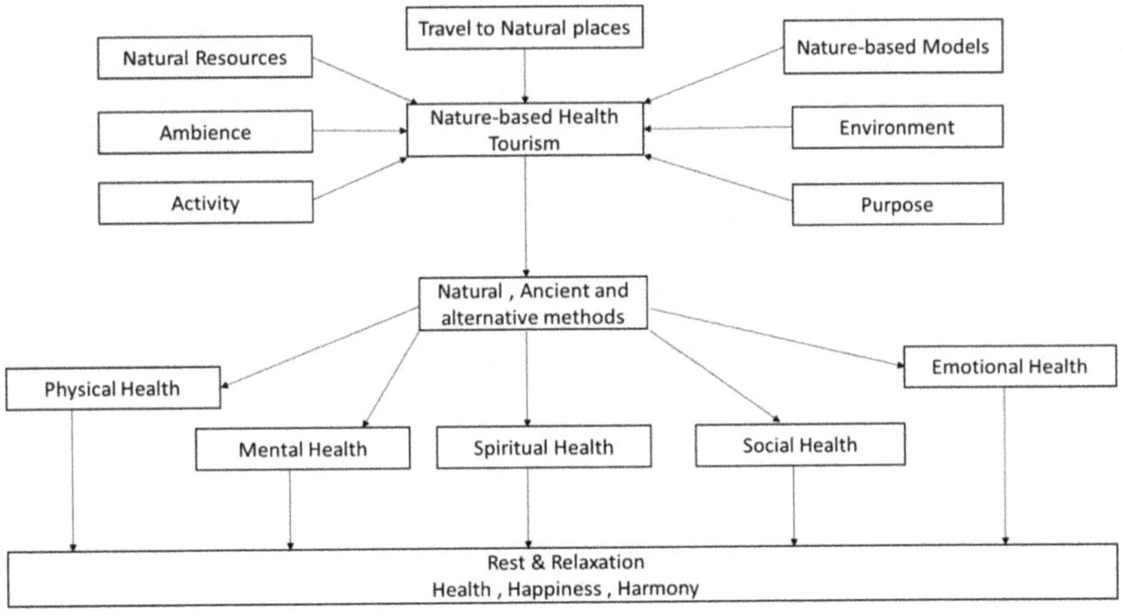

techniques such as yoga, ayurveda, meditation. All these different dimensions will open the new doors to new ways of attaining good physical health by building good strength and stamina, mental health by getting relaxation, spiritual health by feeling a sense on interconnectedness, social health by involving with the local communities and emotional health. This holistic health approach leads to achieve happiness and creates a balance in their life. Different parts of environment such as mountains, rivers, lakes, forests etc. plays a vital role in the life of nature-based health tourists that takes us to new heights, have new learning experiences and to ultimately help us be better humans.

2. HOLISTIC VIEW OF NATURE BASED HEALTH TOURISM IN HIMALAYAN REGION

Home to one of the most varied ecological landscapes in the world, the source of many a myth and stories spun out of travelling voices since millenia. The Himalayan region, with its majestic peaks, lush valleys, and serene landscapes, has long been revered for its spiritual significance and natural beauty. In recent years, this region has witnessed a transformative trend in the world of tourism – the emergence of nature-based health tourism, where people attracts towards the natural resources to take care for their holistic health by using different natural techniques.

Himalayas are the great mountain system of Asia, which stands with its majestic landscapes, towering peaks, rich biodiversity and cultural heritage. Over the centuries, Himalayan region has often emerged as a central motif in terms of health. As Rabindranath Tagore has said – "The Himalayas condense a well of endless spiritual knowledge and truth that sustains human values". Himalayas as a multifaceted site of subjective, global and meta physical importance that are responsible in maintaining a form of tourism that satisfies the person's main motive for maintaining and strengthening the physical, spiritual, emotional and psychological health. This chapter engages with health benefits that are hidden in the lap of Himalayas. There are numerous opportunities that is availed in contrast with natural resources available in the Himalayan region that will help in boosting natural healing procedures and foster physical and psychic well-being of the tourists. Himalayas have also been the centre for the spirituality since ages for Indian sages. As it provides a way to be with the natural environment for getting relax and be away from the congested day to day hustle and bustle of today's scenario. It is very much true that himalayas are the abode of all gods, unique natural features & varied attractions such as healing waters, scenic beauty, serene atmosphere that attracts everyone for their well-being through naturalistic approach towards wellness that are not properly defined in terms of Himalayas.

In ancient times people can take refuge, shelter and comfort in such natural settings during their moments of sadness and joy as a place where body and mind can both heal. Our historians have been telling us time and again that Himalayas are bestowed with the beauty of nature and also serves with the best retreat to many lifestyle problems. In the circumstances of this fast pace urbanisation, Himalayas are the only option to get the nature's therapy and able to enjoy their close connection with it. Human - nature relationship have always been considered very important. The five elements of earth out of which human body is made of and therefore, it is very much responsible in maintaining the person's health, according to ayurveda. So, theory of wellness in Himalayas is supported by the researchers with the fact that nature has a great impact on different facets of person's physical, psychological, social and spiritual life. Therefore, Himalayan region, with its awe-inspiring landscapes and spiritual heritage, has emerged as a significant destination for health tourism.

2.1 Role of Nature Based Wellness Techniques in Himalayan Region

Activities that encourage people to engage with nature in order to boost their physical, mental, and emotional wellbeing are known as nature-based wellness activities. With the help of these activities, stress can be reduced, moods can be lifted, and general health can be improved. Here are some instances of wellness activities based in nature:

- **Forest- Bathing (Shinrin Yoku):** One of the Japanese practice forest immerse themselves in a forest setting and pay attention to its sights, sounds, and fragrances. There is evidence that taking a forest bath can lower blood pressure, reduce stress, and improve mood. A nature-based forest, often known as a "natural forest" or "wild forest," is an ecosystem that has been left intact and undisturbed by considerable human intervention or disruption. Nature-based forests are protected to conserve their biological integrity and biodiversity, in contrast to managed forests that are prone to logging, farming, or other human activity.
- **Nature Walks & Hiking:** People can do exercise, fresh air, and the therapeutic effects of being in nature by taking leisurely strolls or challenging hikes in natural surroundings.
- **Wildlife Watching:** Getting close to and observing animals in their natural settings may be a source of delight and surprise. Taking part in activities like birdwatching, whale watching, or animal observation might help people develop awe and an appreciation for nature.
- **Accomodation in Natural Areas:** To have a more profound relation with nature, staying in the midst of such areas and just viewing the breathtaking views of peaceful scenery, will be more easier to keep nature on the brain.
- **Practicing Yoga and Meditation:** Performing and practicing yoga and meditation in natural setting, can enhance relaxation, can deepen the mind- body connection that provides a serene backdrop for yoga and meditation.
- **Art & Creativity:** Engaging in different creative activities such as painting, photography, journaling, writing encourages more feeling of self-expression while connecting with the beauty of outdoors.
- **Adventure Activities:** Camping, hiking, cycling etc. offers an opportunity to disconnect ourselves from technology, to reconnect with environment by playing and engaging themselves in different activities that keeps us close and helps to be the part of nature.

These nature-based wellness activities, if perform in Himalayan region cater and to diverse range of interests and offers opportunities to all people. It fosters a deeper appreciation for the environment and also creates awareness that makes realize everyone to felt the importance of its conservation.

2.2 Role of Eco-Activity

The emergence of sedentary lifestyle that created a huge gap between human and nature, which has created a havoc in human's lifestyle that leads to many physical, social and psychological problems. According to Pt. Shri Ram Sharma Acharya ji, founder of All World Gayatri Pariwar has said that if a person wants to live for hundred years then he has to move parallely with the nature. There are various ailments available in the states of himalayan region like Uttarakhand, Sikkim, Himachal Pradesh etc. Also Pretty et al. (2005), in his study that involves simulated green exercise (exercise on a treadmill while exposed to photographs of green spaces compared with other spaces), found that green exercise appears to have benefits both for cardiovascular health and mental health.

Nature brings us back to our roots, reminds us of a simple time when humans used to cherish nature by viewing, practicing various activities in nature and as well as living according to the laws of nature that automatically gives us a mental break and fills our soul with peace.

All the activities that a traveller, tourists or locals perform for the nature and in the nature are considered as an eco- activities such as:

- wildlife watching,
- camping, adventure,
- mindfulness practices,
- yoga and meditation,
- healing techniques,
- spa therapies,
- forest bathing,
- zero waste living,
- eco-friendly transportation,
- hotels with green and blue spaces, home stay tourism etc.

These are some of the effective ways to foster a sense of responsibility towards the planet and actively participate in the preservation of the environment and improving their one's good health by caring their physical, mental and social well-being to achieve good health and prosperity.

3. NATURE BASED HEALING PRACTICES

Nature-based healing practices focuses on the healing power of nature by getting interacted with flora, fauna and natural landscapes. Some studies have shown that hospital patients recovered faster when they view nature. The environment plays a crucial role in human beings overall wellbeing. Combining the appreciation and interaction with nature here are some of the elements to integrate in your life for a more balanced and fulfilling existence:-

- **Feel and View Nature:-** Nature immersion can rejuvenate your spirit and reduces stress. Spending time in natural setting and just observing the changing phenomena of nature, by looking towards the cloud, majestic beauty of plants and flowers that blooms our minds, experiencing sunrise and sunset, sitting near the flow of sacred rivers in peace can evokes a sense of wonder and presence. Feeling the sun on face and breathing in some much needed fresh air can help us feel alive, to pursue ourselves naturally.
- **Mantra Healing:-** Choose a mantra that resonates your goal and intentions and practice to recite it by just sitting between the natural environment. Explore the healing properties of sound by listening to soothing sounds of nature or mantras. As mantras are the sounds or vibrations that generates our own bodily vibrations which harmonizes with our external environment to create and restores the balance, harmony and comfort to us.

- **Affirmations with Nature:-** "Nature is fuel for the soul", wrote Dr. Richard Ryan in the study. Affirmations and manifestations are the sources that radiates good feelings and vibrations. Practicing affirmations with nature because it can be remarkable friend for practicing such things as it provides a peaceful ambience by reducing the distraction in natural settings, allows you to focus on your inner thoughts, nature's beauty inspires to amplify the positive emotions and keeps your mind relaxed to more positive affirmations.
- **Physical and Mental Practices:-** Practicing yoga, meditation, sports activities, adventure, etc. or many different practices that requires physical strength and stamina, natural is ultimate power booster to rejuvenate and recover to perform good physical activities. This will help to step out of comfort zone and challenge to try something new and exciting. Practicing deep breathing exercise in fresh air, which is crucial for any physical growth, can connect with the rhythms of environment and promote relaxation.
- **Natural Phenomena:-** Phenomena like- animal migrations, meteor showers, changing weather conditions, experiencing different seasons fill one's mind with various scenic beauties. It also helps the body to learn to acclimatize itself that strengthens the body physically and mentally. Different phenomena builds up a sense of awe and respect to natural world. It aligns your personal growth and took on healing journey with the cycles of nature.
- **Improving lifestyle:-** Introducing plant-based diet to align with lifestyle nature's cycle and promotes good health. Accepting suitable choices in your lifestyle such as reducing waste, conserving energy, spending time with nature can reduce stress, a great way to work connect with nature.
- **Biodiversity:-** Both flora and fauna are the part of biodiversity. Practicing different physical and bodily movement under or between some of the specific plants and trees such as Peepal, Neem, Deodar, Banyan etc. that are considered as the 24*7 source of oxygen. Similarly, spending time with different faunas like cows, birds, dogs etc. lowers stress and blood pressure, boosts mood and self-esteem. Caring them and showing kindness towards them leads to empathy and sense of responsibility in an individual.
- **5 Elements of Nature:-** Our body is made up of five elements of nature- Water (Jal), Earth (Prithvi), Fire (Agni), Air (Vayu), Space/Ether (Aakash). According to Ayurveda, there is a need to balance these five elements to attain good health as they are responsible in balancing Vatta, Pitta and Kapha in our body. If these are in balance, then it helps to detoxify the body naturally. Element **earth** helps to purify the blood which requires to eat greenery vegetables, walking bare foot on the earth to aids digestion. **Water** helps to hydrates the cells, there is 70% of water in our body that cleanse and removes waste from body. **Fire** requires exposure to sun, fire helps to break down the toxins in our body. **Air,** helps to clean lungs and remove impurities from the air we breathe and **Space** creates a sense of peace that helps us in the clarity of thoughts. So, daily exposure with these five elements balances our body from internally to achieve good health.

This holistic approach encourages a lifestyle that is not only suitable but also spiritually fulfilling and in tune with the natural world.

3.1 Supporting Factors and Resources for Nature-Based Health Tourism in Himalayas

The Himalayan region can position itself as a premier destination for nature-based health tourism, offering a unique blend of natural beauty, cultural richness, and holistic wellness experiences. The Himalayas, on the northern border of India, surrounded by the streams of the Brahmaputra in the east and the Indus River in the west, have been the area of intense contemplation and research for sages, inventors and researchers since time immemorial. At the foot of the Himalayas, the culture of the land of India, irrigated by many Ganges emanating from it, is Ganga-culture. No other geographical landscape has had more influence on the Indian way of life, culture and history than the Himalayas. Even though Indian sages choose this place to meditate and for experiencing immense peace and tranquility. Utilizing these resources in a sustainable creates a sense of awareness towards their environment. With its lofty peaks and green and serene mountains and hills, Himalayan region as Nepal, Sikkim, Himachal Pradesh, and Uttarakhand are the perfect destinations for yoga and meditation lovers. Introduction of yoga into the trekking experience makes up for a richer experience as the physical and mental landscape of the body and mind are opened through daily meditation, yoga and trekking.

4. NATURE BASED HEALTH TOURISM: NEW INSIGHTS

4.1 Wellness and Spa Retreats

Wellness and spa retreats evolved to offer new insights into holistic health and rejuvenation. Today most of the hotels, spas and resorts are offering such green and blue spaces to their guests to experience good health and harmony. So that we can see and better appreciate the role nature plays in our lives. Different health techniques like naturopathy, panchkarma, aroma therapy, yoga, meditation, organic and healthy diet are offering by different centres. Post Covid-19, again woke up everyone and alarms their mind to get back live their lives with nature. So different destinations are offering such practices to make tourists and travelers to experience naturally in one go. The new trend in wellness and spa retreats reflects a growing awareness of the importance of self-care, mental health, and personal development. As people seek more meaningful and transformative experiences, these retreats provide an opportunity to recharge, gain new insights, and foster a deeper connection to oneself and the world around them. This trend is characterized by various factors and elements in today's scenario: Digital detox, culinary wellness, fitness and movements, holistic therapies, educations workshops, explorational, community and connection.

4.2 Creating Infrastructure Service

The integration of green and blue spaces into hotels and travel destinations is running broadly that not only enhances the guest experience but also demonstrates a commitment to sustainability and responsible tourism. As G20 summit has also focused towards sustainability for which tourism plays a crucial role in implementing such new insights to its travel. Travelers seeking authentic and nature-focused experiences are increasingly drawn to such destinations and accommodations, making it a valuable trend in the hospitality industry. Increasing more number of eco-friendly transportation, naturals pools, organic food, green certification, zero waste are the aspects and services that are in-builting to many different hotels and restaurants that are providing a new, memorable, natural and asthetic experience to their visitors which either directly or indirectly but connecting people to their roots that is Nature. Moreover in Eco-psychology or nature-guided therapy the various aspects of the human-nature relationship have been considered very important. Under such circumstances of fast pace urbanization, Himalayas are the only option to get the nature's therapy and to enjoy the close relationship with it. Also the theory of wellness in Himalayas is supported with the fact that the indulgence with nature affects numerous facets of a person's physical, mental and social life such as reducing anger, accepting sense of belongingness etc.

4.3 Eco-lodges

Eco lodges are typically situated in pristine natural environment which are eco-friendly and sustainable in nature. These are the accomodations designed and operated with strong responsible tourism practices. According to the Center For Responsible Travel's Travel Trends & Statistics report, around one in five consumers (21%) say they would be willing to pay more for a trip with a company that has a better environmental and social record. A report by The Travel Foundation found that 66% of travelers surveyed would like to be able to identify a "greener" holiday more easily. Travelers enjoy such unique environment friendly experiences and gets a chance to be with nature.

4.4 Meditation Centers

Himalayan region has a serene and spiritually rich environment, which is also a home to numerous meditation centers. These centres offer a peaceful setting for spiritual seekers, relaxation and self assesment.

4.5 Nature-Based Home Stay

Travelers can immerse themselves in a sustainable way while experiencing a destination's local culture and natural beauty by staying in nature-based homestays. These lodgings provide visitors a chance to interact with nature while savoring genuine hospitality. They are frequently owned by local families and are frequently found in rural or natural locations.

5. IMPORTANCE AND IMPACT OF NATURE BASED HEALTH TOURISM IN HIMALAYAN REGION

Nature-based health tourism provides a holistic approach to health and wellbeing of a person, which combines various benefits with wellness rituals and cultural experiences. Its significance is in creating environmental awareness, supporting local communities, promoting physical and mental health, and fostering a more balanced and fulfilling life for people looking for a break from the stresses of modern living. This form of tourism combines the healing and rejuvenating aspects of nature with wellness practices, promoting physical, mental, and emotional well-being. Himalayan region which is loaded with abundance of therapeutic benefits and natural resources. It has its multi-facets benefits to various aspects of health, healing, development and conservation. Such as –

- Health and Wellness

- Alternative Healing Modalities
- Physical Fitness
- Cultural Exchange
- Environmental Conservation
- Vasudhaiv Kutumbakam
- Natural Health Techniques etc.

5.1 Socio-Cultural Impact

Nature-based Health Tourism causes socio-cultural impact to host and travelers in both positive and negative ways:

- **Positive Impacts**

Cultural Exchange: Nature-based health tourism often involves travelers from diverse cultural backgrounds interacting with local communities. This exchange can lead to a rich cultural experience for both tourists and hosts, fostering mutual understanding and appreciation.

Preservation of Traditions: Host communities may preserve or revive traditional healing practices, herbal medicine, and cultural rituals to cater to the needs of health tourists. This can help safeguard local traditions and knowledge.

Health and Wellness Awareness: The presence of health tourists can raise awareness about the importance of health and wellness within host communities, leading to healthier lifestyles and improved healthcare access.

- **Negative Impacts**

Cultural erosion: The influx of health tourists can sometimes lead to cultural erosion as locals adapt their customs and traditions to cater to the preferences of visitors. This can result in the loss of authentic cultural practices.

Social Disruption: Rapid tourism development can disrupt the social fabric of host communities, leading to increased land speculation, displacement, and conflicts over resource allocation.

Dependency: Over-reliance on nature-based health tourism can make host communities vulnerable to economic downturns in the tourism industry. A sudden decline in visitors can lead to financial instability.

5.2 Economic Impact

Nature-based health tourism have significant economic impacts on the host regions and countries:

Revenue Generation: Nature-based health tourism brings in revenue through various channels, including accommodation bookings, healthcare services, spa and wellness treatments, guided tours, and the sale of local products and services. This revenue directly contributes to the local economy.

Job Creation: The tourism industry, including nature-based health tourism, creates employment opportunities for local residents. Jobs are generated in hospitality, healthcare, transportation, tour guiding, and other related sectors, reducing unemployment rates and improving livelihoods.

Business Development: The demand for nature-based health tourism often leads to the establishment and growth of local businesses. These include wellness centres, eco-lodges, restaurants, souvenir shops, and transportation services, which stimulate entrepreneurial activities.

5.3 Environmental Impact

Development of Natural based Health Tourism in natural setting leads to various impacts on environment that needs to be carefully managed.

- **Positive Impact**

Environmental Conservation: To protect the natural environments that attract health tourists, communities may become more environmentally conscious. This can lead to conservation efforts, reduced pollution, and a greater emphasis on sustainable practices.

Health and Wellness Awareness: The presence of health tourists can raise awareness about the importance of health and wellness within host communities, leading to healthier lifestyles and improved healthcare access.

Moving towards Nature: Human beings, today started practicing such nature based health techniques to care for their health. They are trying to connect more and more with nature and it's resources. A sense of acceptance and feeling of moving back to our old traditions and solutions of being healthy that are suggested by Indian sages in different ancient texts.

- **Negative Impact**

Resource Consumption: Increased tourist activity in natural areas can lead to greater resource consumption, including water, energy, and food. Overuse of these resources can strain local ecosystems and put stress on water sources.

Loss of Biodiversity: Habitat destruction and disruption can lead to a loss of biodiversity in the area. This loss can affect both plant and animal species, potentially leading to extinction or endangerment.

Waste Generation: Tourism generates waste, including solid waste, wastewater, and pollution from transportation. If not properly managed, this waste can harm local environments and water bodies.

6. CHALLENGES AND RECOMMENDATION

Nature-based health tourism in Himalayan region can create some challenges that needs to be counter for reducing it's negative environment. Environmental sensitivity needs to be taken care of by reducing waste disposal, development of infrastructure services, Safety measure should be applied to overcome different season variations, Encouraging sensitive cultural practices of local people, Development should be restricted in disaster prone areas, Proper management of resources in eco-friendly accomodations. By balancing the growth of nature-based health tourism in Himalayan region that requires careful planning, active involvement and sustainable practices reduce the negative impact that can harness the regions development to safeguards its ecological and cultural heritage for future generations.

7. CONCLUSION

The chapter will conclude by summarizing the unique attributes of nature-based health tourism in the Himalayan region and its potential as a leading destination for wellness and rejuvenation. It will highlight the importance of preserving the region's cultural heritage, fostering sustainable practices, and embracing innovation to ensure the growth and long-term success of nature-based health tourism in the Himalayas. Himalayas that are endowed with numerous wellness resources, which has not yet been in focused. Such development of nature based health tourism will not only help in promoting the destinations and traditional aspects of wellness but also support in developing a harmony among nature and human being. It could also prove to be one of the sustainable approaches to tourism development especially in Himalayan regions. By adopting sustainable practices and actively addressing potential environmental impacts, nature-based health tourism can contribute to the preservation of natural ecosystems while offering visitors enriching and health-promoting experiences in harmony with the environment.

REFERENCES

Chang, L., & Beise-Zee, R. (2013). Consumer perception of healthfulness and appraisal of health-promoting tourist destinations. *Tourism Review, 68*(1), 34–47. doi:10.1108/16605371311310066

Fernandes, J. P., & Guiomar, N. (2018). Nature-based solutions: The need to increase the knowledge on their potentialities and limits. *Land Degradation & Development, 29*(6), 1925–1939. doi:10.1002/ldr.2935

Finnicum, P., & Zeiger, J. C. (1996). Tourism and wellness: A natural alliance in a natural state. *Parks & Recreation, 31*(9), 84–90.

Gössling, S., Scott, D., & Hall, C. M. (2020). *Pandemics, tourism and global change: a rapid assessment of COVID-19*. doi:10.1080/09669582.2020.1758708

Health and Wellness Tourism. (n.d.). Research Gate. https://www.researchgate.net/publication/302140629_Health_and_Wellness_Tourim

Health Promotion. (n.d.). WHO. https://www.who.int/teams/health-promotion/enhanced-wellbeing/first-global-conference

Health Tourism: An Overview By Kim Ross (HSMAI Marketing Review). (n.d.). Hospitality.net. https://www.hospitalitynet.org/news/4010521.html

Lee, C.-W., & Li, C. (2019). The Process of Constructing a Health Tourism Destination Index. *International Journal of Environmental Research and Public Health, 16*(22), 4579. Advance online publication. doi:10.3390/ijerph16224579 PMID:31752340

Lee, T. J., Han, J. S., & Ko, T. G. (2020). Health-oriented tourists and sustainable domestic tourism. *Sustainability (Basel), 12*(12), 20–22. doi:10.3390/su12124988

MacKenzie, B. (2004). Ecotourism: It's good for your health. *Sustainable Tourism, 9*, 161–170.

Parashar, A. (2012). *Feature of Health Tourism in Uttarakhand With Special Reference to Rishikesh- A Critical Study*. Dev Sanskriti Vishwavidhayalaya.

Sharma, P. S. R. (2008). *Swasthay Raksha – Prakarti ke Anusaran Se Sambhav*. Yug Nirman Yojna, Gayatri Tapobhumi.

The 10 Biggest Benefits of Nature Travel. (n.d.). Smarter Travel. https://www.smartertravel.com/nature-travel-benefits/

Van Den Bosch, M., & Sang, O. (2017). Urban natural environments as nature-based solutions for improved public health-A systematic review of reviews A R T I C L E I N F O. *Environmental Research*, *158*, 373–384. doi:10.1016/j.envres.2017.05.040 PMID:28686952

WHO/Europe. (n.d.). | *Urban health - Protecting nature protects health – lessons for the future from COVID-19*. WHOEuro. https://www.euro.who.int/en/health-topics/environment-and-health/urban-health/news/news/2020/6/protecting-nature-protects-health-lessons-for-the-future-from-covid-19

Chapter 15
Strengthening Livelihoods and Empowerment:
Craft-Based Capacity Building for Lahaul-Pangi Artisans in the Higher Himalayas

Shipra Sharma
https://orcid.org/0000-0002-1137-1793
National Institute of Fashion Technology, India

Paramita Sarkar
National Institute of Fashion Technology, India

ABSTRACT

The high-altitude Himalayan landscape of Lahaul-Pangi is marked by deforestation, land deterioration, diminished biodiversity, restricted agricultural productivity, and other environmental challenges. Although agriculture remains the primary occupation, handloom and handicrafts have emerged as potential revenue-generating resources, given the abundant availability of wool and the innate skills of the local artisans. NIFT conducted a field-based survey, involving 218 artisans across 09 panchayats, to grasp the region's needs and determine the most suitable skill development approaches. Following an analysis of the responses, NIFT developed and delivered training workshops to 09 panchayats in the Lahaul-Pangi region, ultimately empowering the artisans. The chapter presents key findings regarding the regional artisans, including information on their income and education levels, craft involvement and affiliations, assets, access to medical facilities, design, and technological proficiency, product pricing details, marketing and branding strategies, and training levels.

1. INTRODUCTION

The high-altitude Lahaul-Pangi region in the Himalayas exhibits deforestation, land deterioration, diminished biodiversity, and scarce livelihood prospects, among other challenges. This area is characterized by diverse climatic conditions, severe weather fluctuations, and significant climate variability.

DOI: 10.4018/979-8-3693-0823-3.ch015

Consequently, local communities face limited livelihood options and experience considerable marginalization due to their heavy reliance on natural resources and inadequate infrastructure. The Himalayan ecosystem is highly susceptible to climate-related risks that can impact both economic and biophysical conditions. The challenges faced by the Lahaul-Pangi region in the Himalayas are indeed formidable. The consequences of limited livelihood prospects are deeply intertwined with the unique ecological conditions of this high-altitude terrain. The Himalayas, known for their grandeur, are equally known for their vulnerability to climate-related risks, such as erratic weather patterns and glacial retreats. These risks, when coupled with the heavy reliance of local communities on natural resources and inadequate infrastructure, create a complex web of challenges that demand innovative solutions. Recognizing the pressing need for intervention, the National Institute of Fashion Technology (NIFT) undertook a mission that extended far beyond the realms of fashion and design. Their holistic approach encompassed community capacity building, design interventions, product functionality enhancements, and market-driven strategies. These strategies were tailor-made to cater to the unique needs and constraints of the Lahaul and Pangi regions, where livelihoods are intertwined with the harsh realities of high-altitude living. The National Institute of Fashion Technology (NIFT) embarked on the ambitious endeavor of discovering various avenues for community capacity building, design interventions, product functionality enhancements, and market-driven approaches to benefit the challenging terrains of the Lahaul and Pangi regions in the high Himalayas. The area presents difficulties in terms of terrain and livelihoods. Consistent efforts are needed to secure livelihoods and promote sustainable resource use to conserve these rugged landscapes. Appropriate interventions can aid in the creation of diverse livelihood opportunities and foster community capacity building. In the project landscape, handloom and handicrafts were recognized as potential resources for alternative livelihoods for the local community, owing to the easy availability of wool and the skills of the local population. The Lahaul-Pangi community members possess expertise in creating handicraft-based products, which they typically produce during the winter months. These local crafts, centered around artistic products, have evolved with the support of royal patronage. The choice to focus on handloom and handicrafts as potential resources for alternative livelihoods was a strategic one. The abundant availability of wool and the inherent skills of the local population made this an ideal avenue for economic growth. Traditionally, the people of Lahaul and Pangi have been adept at crafting handmade items, often during the long winter months when other livelihood options are limited. These crafts, steeped in history and tradition, have evolved with the support of royal patronage. While the community members possess intrinsic expertise in crafting handmade items, it was apparent that small, incremental developments were needed to harness their full potential. The key lies in introducing new designs, enhancing product functionality, refining production processes, and implementing market-level interventions. This approach aimed to rejuvenate wool-based handicrafts by making the best use of the readily available local resources. One of the remarkable aspects of this initiative is its ability to tap into the rich tapestry of indigenous knowledge. The craftspeople of Lahaul and Pangi, deeply connected to their land and culture, bring a wealth of traditional wisdom to their work. NIFT's interventions sought not to replace this heritage but to augment it with contemporary design sensibilities and market awareness. This harmonious blend of tradition and innovation is a testament to the adaptability and resilience of these communities.

Furthermore, the project's focus on sustainable resource use aligns perfectly with the fragile Himalayan ecosystem. By emphasizing responsible sourcing of materials and eco-friendly practices, it ensures that economic development is in harmony with environmental conservation. This approach safeguards the region's unique biodiversity, which is not just a source of pride but also essential for the well-being of its inhabitants. NIFT's ambitious venture in the Lahaul-Pangi region serves as a beacon of hope in the face of formidable challenges. By recognizing the potential of traditional handicrafts, harnessing local expertise, and infusing fresh design perspectives, it has paved the way for diversified livelihood opportunities. Through community capacity building and market-driven strategies, it has not only empowered the artisans but also offered a sustainable path forward for these rugged landscapes. This initiative is a testament to the resilience and creativity of the people of Lahaul and Pangi and their ability to adapt to changing circumstances while preserving their heritage and environment. Community members in both Lahaul and Pangi exhibit proficiency in crafting handmade items, although the methods and types of crafts differ significantly between the two regions. This implies that there is no requirement for extensive capacity development of community members but requires small incremental developments in terms of introducing new designs, products, process and functional upgradations, and market-level interventions. The suggested approach involves revitalizing wool-based handicrafts by utilizing resources readily available in the local area. The advancement of handicrafts in the project villages is vital from a livelihood standpoint, offering opportunities for capacity development.

To the project objectives, a comprehensive artisan database was required to gain insights into demographics, income opportunities, involvement and affiliation with crafts, design and technology skill levels, and product marketing, among other aspects. A total of 218 respondents across 09 panchayats in the Lahaul-Pangi region were surveyed to gather this information. The Lahaul region encompasses Tindi, Udaipur, and Tingret panchayats, while the Pangi region consists of Sural, Sach, Dharwas, Karyas, Killar, and Purthi panchayats. This database would aid in assessing the training needs of local community members.

2. METHODOLOGY

A survey was carried out within the specified region to gather general information about the artisans, such as contact information, education level, craft affiliation and involvement, income particulars, assets, medical facilities, design and technology proficiency, pricing details, marketing methods, and previously received training. Focus group discussions were also conducted with a diverse range of women practicing craft for both domestic and commercial purposes. This made it possible to explore the various forms of traditional craft practiced and the scope of craft-based livelihood in the region. In-depth interviews were held with key informants like village panchayat heads, skilled artisans, and traditional wool-based craft manufacturers/ suppliers of the region

3. CRAFT-BASED CAPACITY BUILDING FOR LAHAUL-PANGI ARTISANS IN THE HIGHER HIMALAYAS

The sample included mainly women artisans practicing wool-based handicrafts and handlooms in the Lahul & and Pangi regions of Himachal Pradesh. Data were collected from women from the age group of 10 to 80 years practicing wool-based crafts at both domestic and commercial levels from the 09 project panchayats that were selected as samples. Field visits and workshops were conducted across all 09 Panchayats in the Pangi and Lahaul regions, encompassing a total of 218 artisans. Training programs emphasized design input, product packaging, and digital marketing, with efforts made to bridge the gap between local artisans and modern products. Given the background and objectives, activities such as field explorations, hands-on work, artisan education, new product development, and demographic surveys were deemed appropriate for securing the livelihoods of the people in the region. The project team personally engaged with local community members to raise awareness about the training modules' content. Encouraging locals to attend and benefit from these programs was another significant undertaking carried out by the NIFT design team. The NIFT team extensively examined existing products made locally in the region, providing appropriate interventions that preserved traditional elements while introducing contemporary features. They worked with various raw materials and tools used in the practice of these crafts. Maintaining traditional aspects of existing crafts while designing contemporary pieces was a challenging task, yet the NIFT team skillfully addressed this by educating artisans on appropriate elements and motifs that incorporated modern elements, thus increasing their visibility and marketability. During the workshops, NIFT provided training on project awareness and its initiatives, design interventions, product display and packaging, and digital marketing. The team reached out to local community members through field visits and stakeholder consultations. Drawing from landscape-level studies and insights obtained, NIFT designed interventions that focused on community capacity building, harnessing abundant natural resources, introducing new and innovative designs, and implementing branding and marketing strategies.

3.1 Field Survey Results

Based on the field surveys conducted in the Lahaul region constituting panchayats like *Tindi, Udaipur,* and *Tingret* and the Pangi region including panchayats like *Sural, Sach, Dharwas, Karyas, Killar*, and *Purthi*, key findings were developed which are as follows:

1) **Distribution of respondents across different panchayats**: It was observed that 8.7% of the respondents were from Dharwas. 17.9% of the respondents were from Karyas. 19.3% of the respondents were from Killar. 5% of the respondents were from Purthi. 17.9% of the respondents were from Sach Panchayat. 6.0% of the respondents were from Sural. 5.5% of the respondents were from Tindi. 8.3% of the respondents were from Tingret. 11.5% of the respondents were from Udaipur.

The data reveals the distribution of respondents across various panchayats, indicating the geographic representation of the survey. The survey captures respondents from a range of panchayats, highlighting the diversity of experiences and perspectives within the Lahaul-Pangi region. This information is valuable NIFT to in creating inclusive, relevant, and effective educational programs. Understanding the local context and the needs of artisans from various family backgrounds allowed the NIFT team to make a meaningful impact on both education and community capacity building.

2) **Gender and age disparities of respondents:** It was observed that craft work is predominantly carried out by middle-aged women in the age range of (31-40 years) constituting the largest portion of artisans. There was a presence of younger artisans (18-30 years), crucial for the sustainability of artisan traditions, and the elder artisans (>50 years) possessing valuable traditional knowledge.

Figure 1. Data of panchayat-wise artisans surveyed

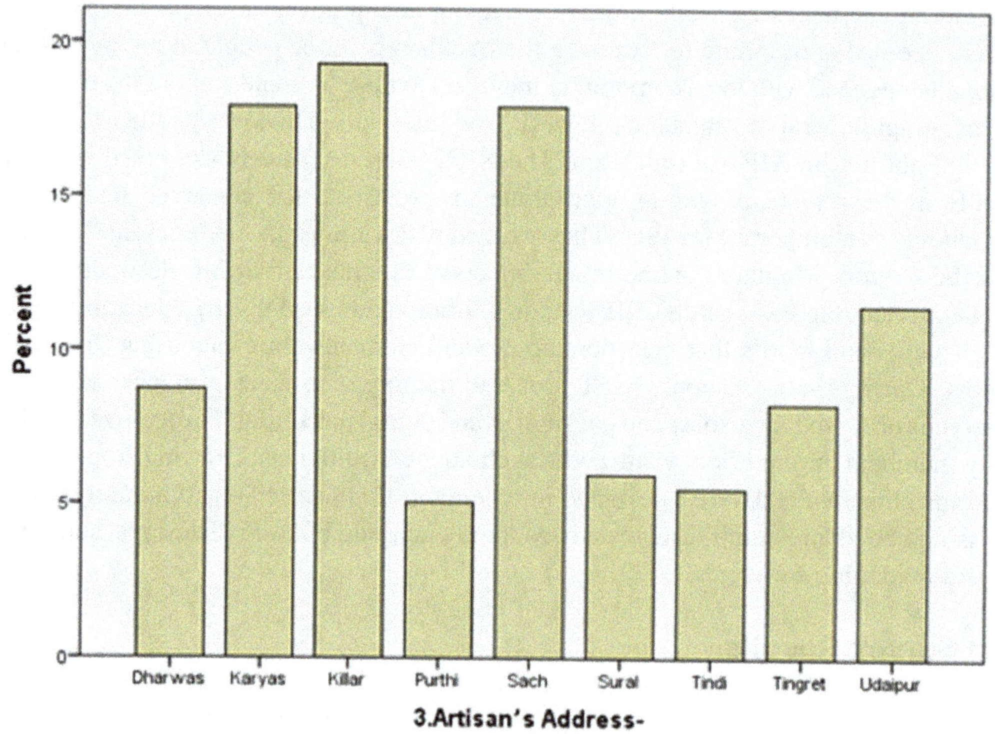

Figure 2. Age range of the artisans from the Lahaul- Pangi landscape

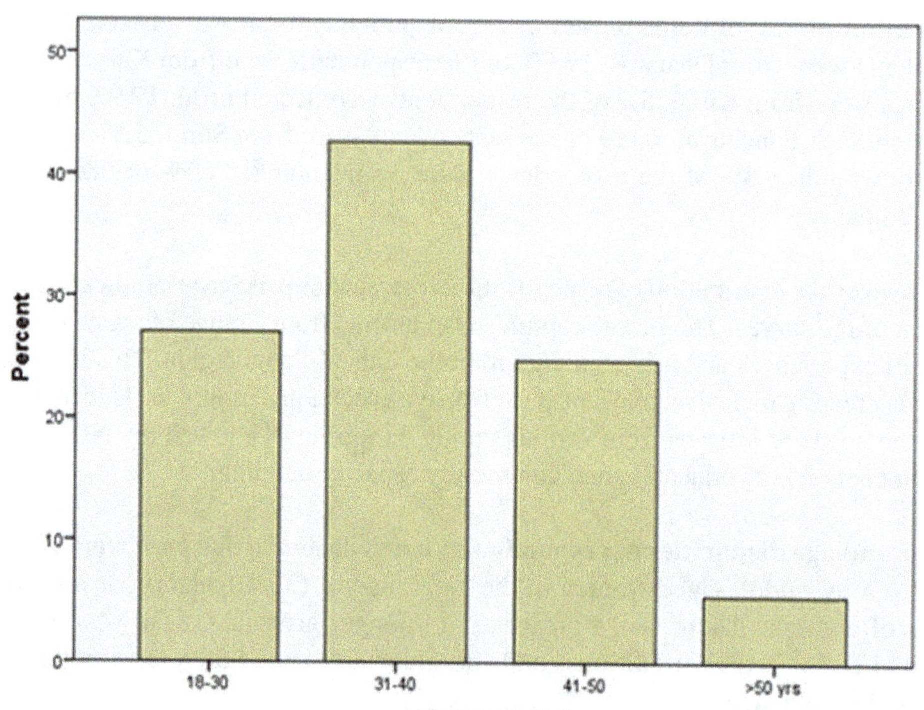

The age diversity within the artisan community is a positive aspect as it ensures a balance between tradition and innovation. Younger artisans can bring new perspectives and creativity, while older artisans can provide guidance and preserve heritage. Middle-aged artisans may need support in terms of training and access to markets to sustain their livelihoods. Younger artisans can potentially contribute to economic growth if their skills are harnessed effectively. The presence of both younger and older artisans is crucial for the preservation of cultural and traditional heritage. It allows for the continuity of craft practices that have been passed down through generations.

3) **Education level of the rural artisans:** As per the given data, 29.4% of the surveyed women are illiterate, indicating that they have not received any formal education. 13.3% of the women surveyed have received education only up to the primary level. A significant portion, 44.5%, have received education at the secondary level implying that they have completed at least up to the 10th grade or equivalent. 12.8% of the women surveyed have had the opportunity to attend college, indicating a higher level of education beyond secondary school.

This underscores the urgent need for initiatives that promote basic education, skill development, and gender equity in education. The data highlights significant disparities in education levels, with a notable portion of women being illiterate, emphasizing the importance of targeted efforts to improve literacy rates. The presence of women with secondary education suggests opportunities for skill development programs, while those who attended college indicate potential for further education and skill enhancement.

Figure 3. Education level of artisans from Lahaul- Pangi landscape

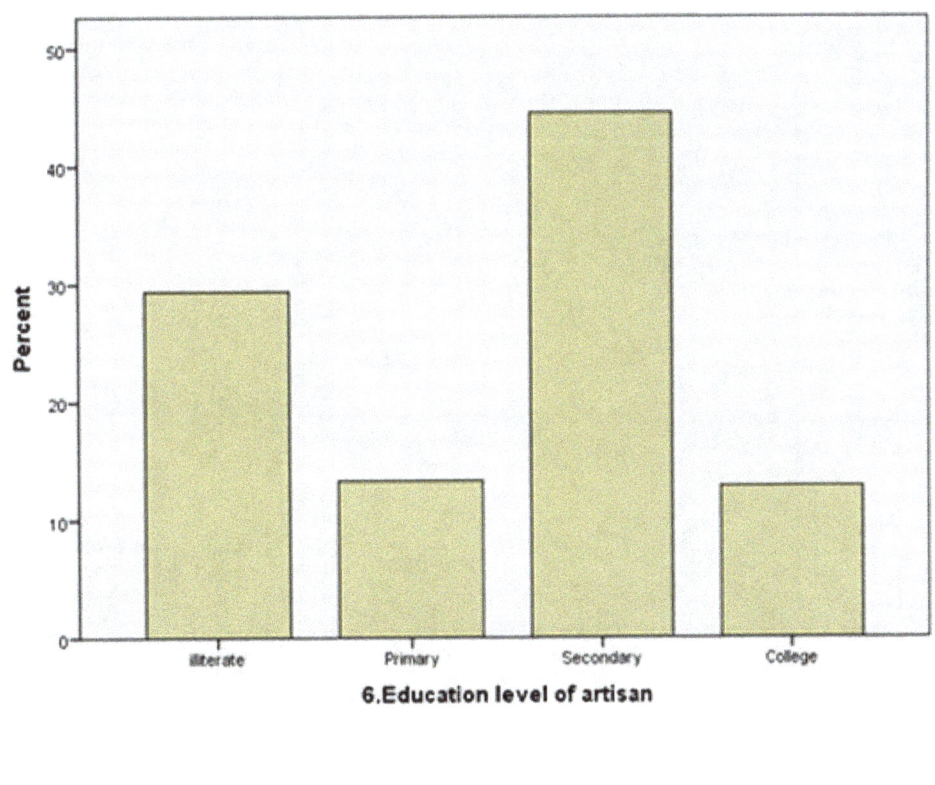

Promoting gender equity in education is vital to bridge these disparities and provide equal opportunities for women's empowerment.

4) **Linguistic skills of artisans:** As per the data collected, 54% of the surveyed artisans can speak Hindi, indicating that more than half of them have basic spoken proficiency in Hindi. Only a very small fraction, 0.5%, can write in English, suggesting that English literacy is extremely low among the surveyed artisans. 6.9% of the artisans can write in Hindi, which is higher than the number who can write in English but still relatively low. 30.3% of the artisans can both speak and write in Hindi, indicating a reasonable level of proficiency in the Hindi language. 8.3% of the artisans can read and write both Hindi and English, suggesting a higher level of language proficiency.

The survey data highlights a significant lack of English literacy among the surveyed artisans, which could limit their access to specific educational and economic opportunities. The majority of artisans are proficient in Hindi, making it a practical language for local communication. While a small percentage can write in Hindi, this suggests a basic level of literacy that can serve as a foundation for further skill development and education. Notably, about 8.3% of artisans possess the ability to read and write both Hindi and English, indicating bilingual proficiency. This proficiency can be advantageous for roles requiring communication in both languages.

Figure 4. Linguistic skills of artisans at Lahaul-Pangi

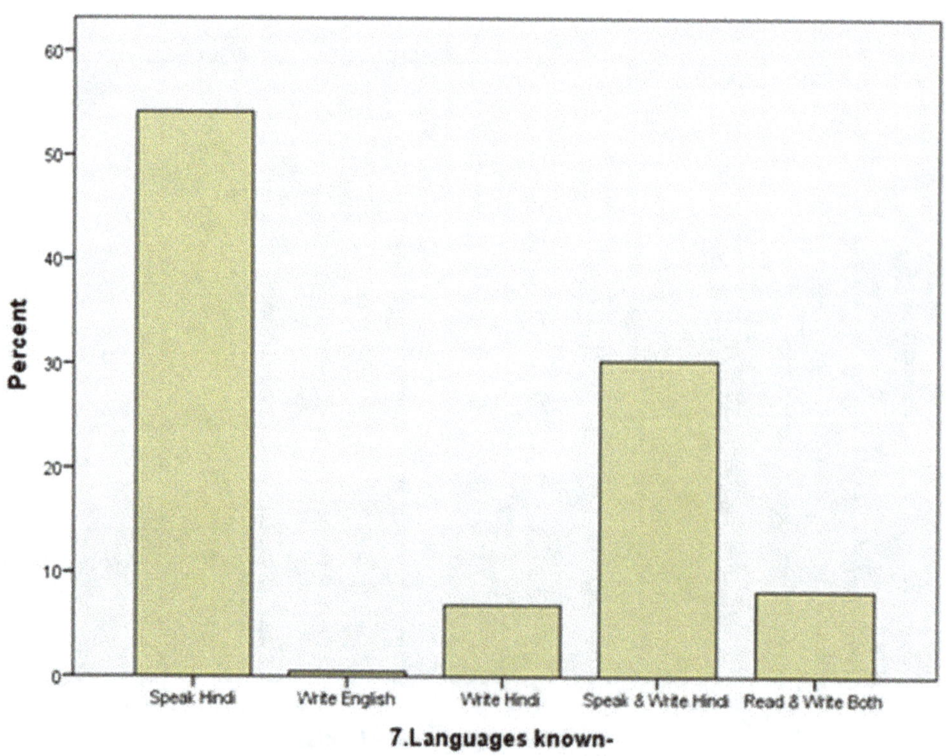

5) **Family size units of artisans**: The data collected from the field survey in the Lahaul-Pangi region provides insights into the size of family units among the respondents. 45% of the surveyed respondents reported having family units consisting of 1-4 members. 30.3% of the respondents mentioned that their family units contain 4-6 members. 24.8% of the respondents indicated that their family units have more than 6 members.

As per the data, smaller family units (1-4 members) might have different household dynamics and resource allocation compared to larger family units, which can impact decision-making and resource management. These different family sizes have implications for household dynamics, economic considerations, social support networks, and access to essential services like education and healthcare. Understanding the diversity in family sizes is essential for designing targeted interventions and policies that address the unique needs of families in the region at a broader level.

6) **Artisan's experience in practicing their crafts:** The given data indicates that 45.9% of the artisans have been practicing their craft for a period ranging from 10 to 25 years. 15.6% of the artisans are relatively new to their craft, with experience ranging from 0 to 5 years. 13.3% of the artisans have extensive experience, with a practice duration spanning from 25 to 50 years. 24.8% of the artisans fall within the 5 to 10 years of practice category.

The data reflects that artisans with 10-25 years and 25-50 years of practice likely possess extensive knowledge and skills in their craft, which can contribute to preserving and advancing traditional techniques.

Figure 5. Size of the family units in the Lahaul-Pangi landscape

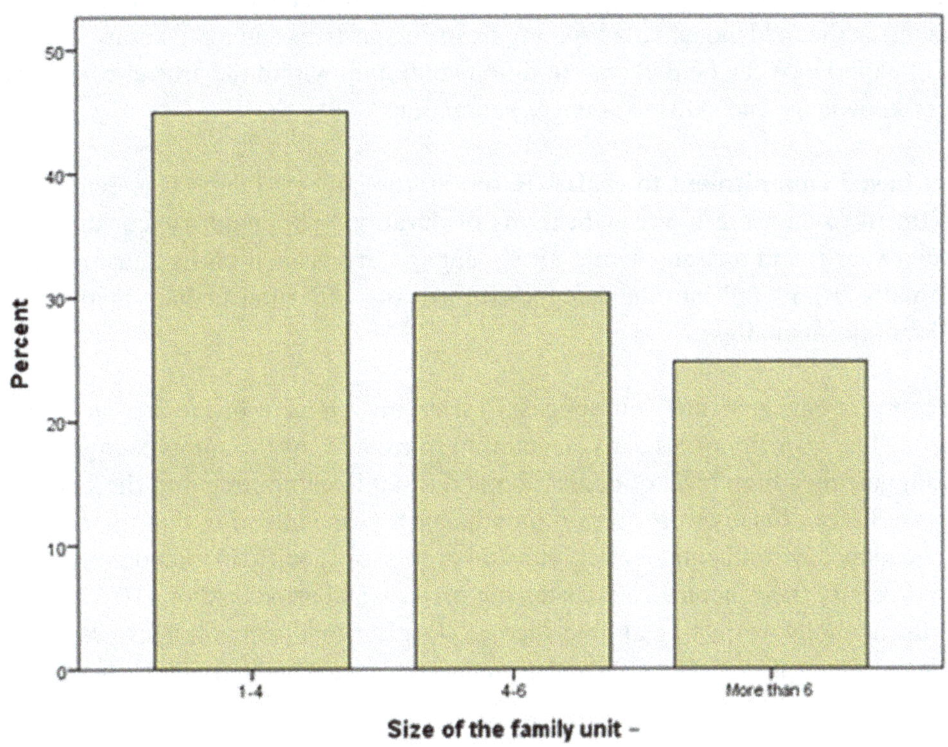

Figure 6. Artisans' experience in practicing their crafts

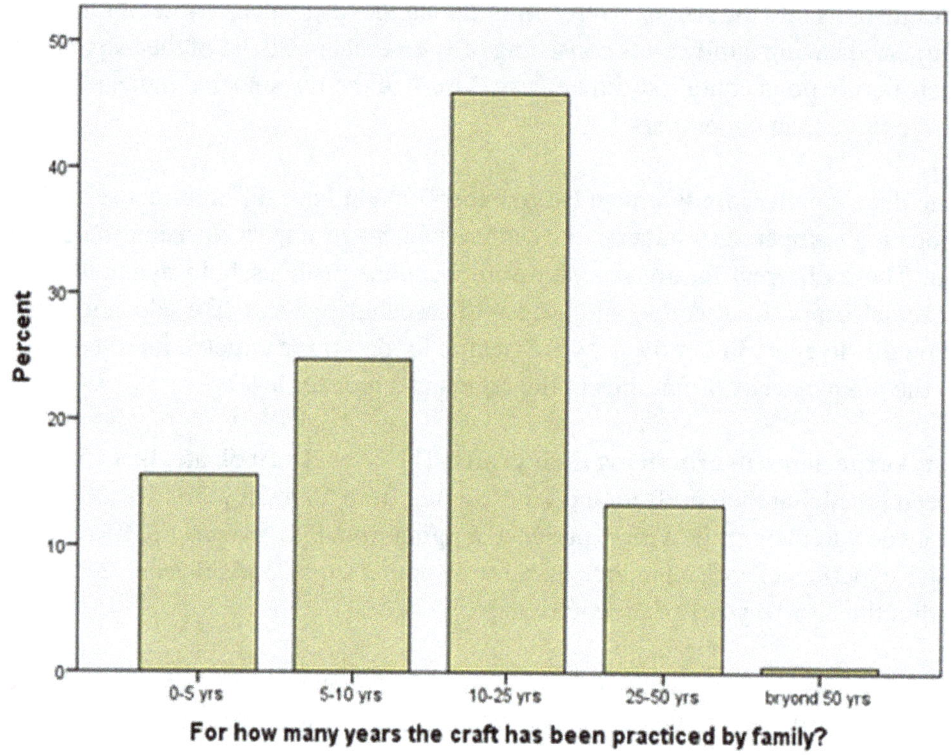

The presence of artisans with 0-5 years and 5-10 years of practice suggests that there is a new generation of artisans entering the field, potentially bringing fresh perspectives and innovations. Understanding the distribution of experience can help in craft tradition continuity within the artisan community, ensuring the transfer of knowledge and skills to younger generations.

7) **Season-based commitment to crafts**: Based on the collected data, it is evident that artisans' commitment to their crafts varies in terms of duration. The majority, constituting 69.3%, are dedicated year-round artisans, while 18.8% engage in seasonal craftsmanship for 3-6 months. Additionally, 10.6% fall into the category of artisans with intermediate involvement, working for 6-9 months annually.

This diversity in engagement levels suggests varied involvement of the artisans concerning time for their crafts. The majority of artisans, accounting for 69.3% of the sample, engage in their crafts year-round suggesting a high level of dedication and a stable commitment to their chosen craft. Approximately 18.8% of artisans work for 3-6 months each year indicating that their craft is seasonal, likely influenced by environmental factors, cultural events, or specific market demand. These artisans may need to diversify their income sources during off-season periods. About 10.6% of artisans work for 6-9 months annually, reflecting an intermediate level of involvement. This group might include individuals who engage in crafts as a part-time pursuit or those who alternate between different crafts or occupations throughout the year

Figure 7. Season-based commitment to crafts

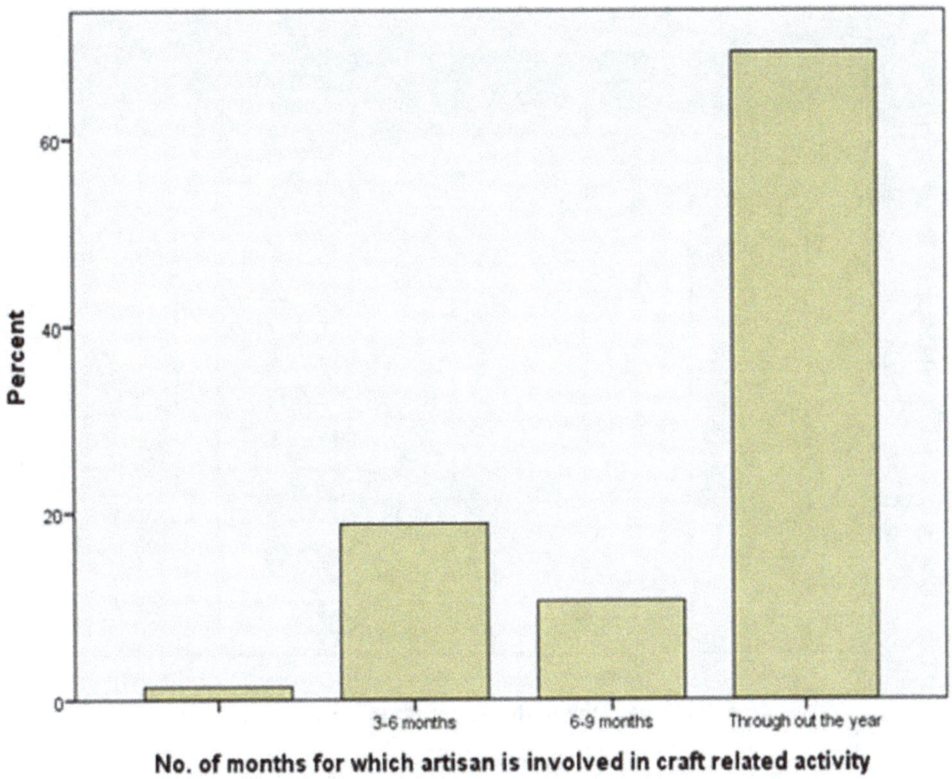

8) **Seasonal patterns of demand**: 33% of the artisans identify the period from October to December as the peak demand season for their products. 27% of the artisans believe that demand for their products remains relatively consistent throughout the year. 29.8% of the artisans highlight the first quarter, from January to March, as a phase with high demand for their products. 7.8% of the artisans indicate that demand is relatively higher from July to September. 2.3% of the artisans note that demand increases during the second quarter, from April to June.

The data indicates that seasonal variations in product demand are significant among artisans. The peak demand in Oct-Dec suggests that festivals and holidays during this period may drive increased sales of artisanal products. High demand in January may be influenced by winter and spring seasons, possibly due to seasonal preferences or special occasions. The substantial percentage (27%) of artisans reporting consistent year-round demand indicates the existence of stable markets for their products. Understanding these seasonal variations is crucial for artisans to fine-tune their craft-making abilities and marketing strategies to meet consumer needs and maximize their income.

9) **New Product development**: The data indicates that a substantial majority, 84%, have not been engaged in the development of new products in the past two years. This suggests a lack of recent innovation or a focus on maintaining existing products or services. With only 15.6% of respondents stating that they have been actively working on new products, it implies that a relatively small portion of the group is prioritizing design innovation.

Figure 8. Seasonal demand for the products

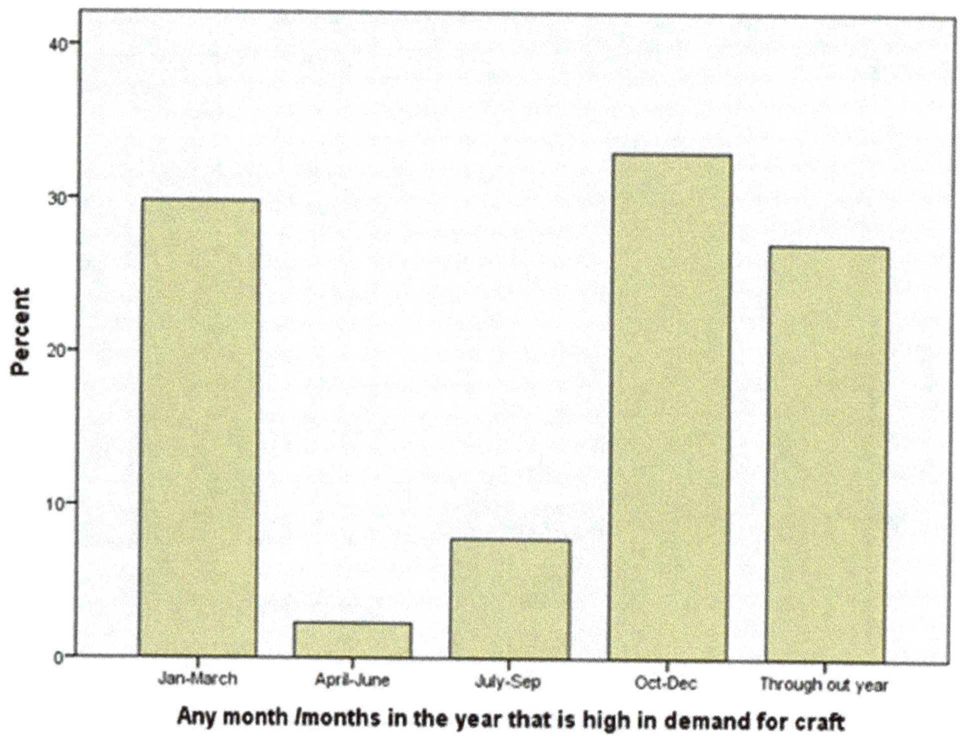

Figure 9. New product development in the artisan community

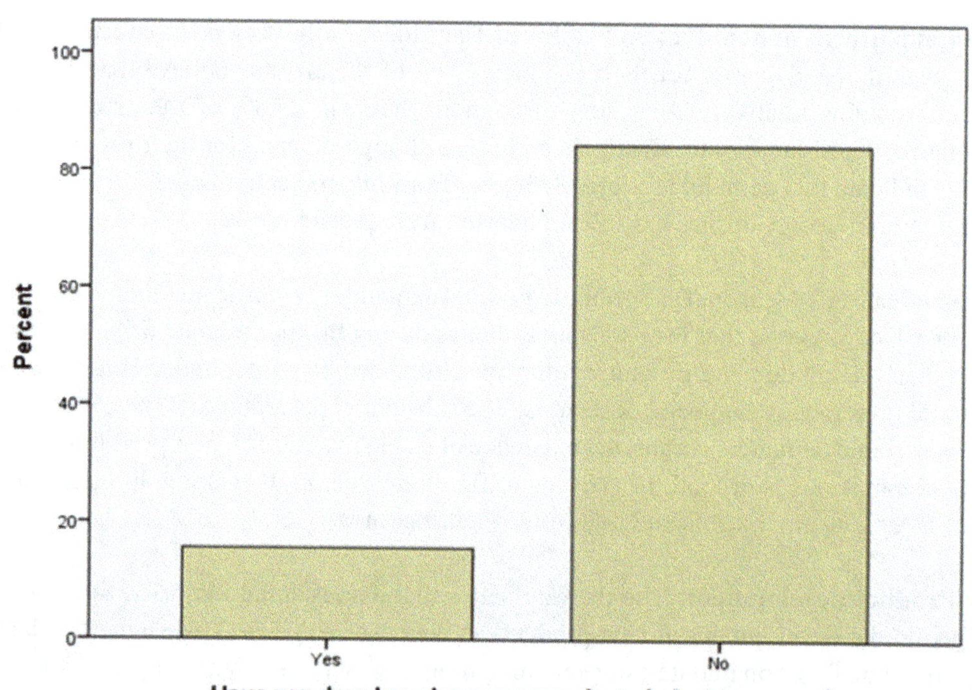

The data indicates that the majority (84%) of respondents or entities in the surveyed group have not developed any new products in the past two years. Only a small minority (15.6%) reported actively working on new products during this time. This suggests a potential gap in new products within the artisan community of the Lahual-Pangi landscape and underscores the importance of assessing fresh design concepts and resource allocation to remain in the market. This may be due to various factors such as listed resources, a cautious approach to risks, or market demand. The low percentage of respondents involved in new product development could imply an untapped income opportunity for their growth and expansion.

10) **Product diversification in Lahual-Pangi landscape**: The data collected from 218 artisans provides information about the types of products they are engaged in making. 76% of the surveyed artisans were engaged in making wool products. 6% of the artisans were involved in making clothing. 10.1% of the artisans were focused on making tailored products. 16% of the total artisans were involved in making products categorized as "other."

The data reveals that a substantial portion of artisans in the Lahaul-Pangi region are engaged in making wool products, highlighting the significance of wool craftsmanship in the local artisan community. Additionally, there is diversity with artisans making clothing, tailored products, and other types of crafts. This information can guide efforts to promote and support these artisans, align production with market demand, and foster income-based opportunities in the region.

11) **Design orientation of women artisans concerning traditional and latest designs**: A substantial majority, approximately 67.4% of artisans, are focused on producing traditional designs. This sug-

Figure 10. Product diversification in the Lahaul-Pangi landscape

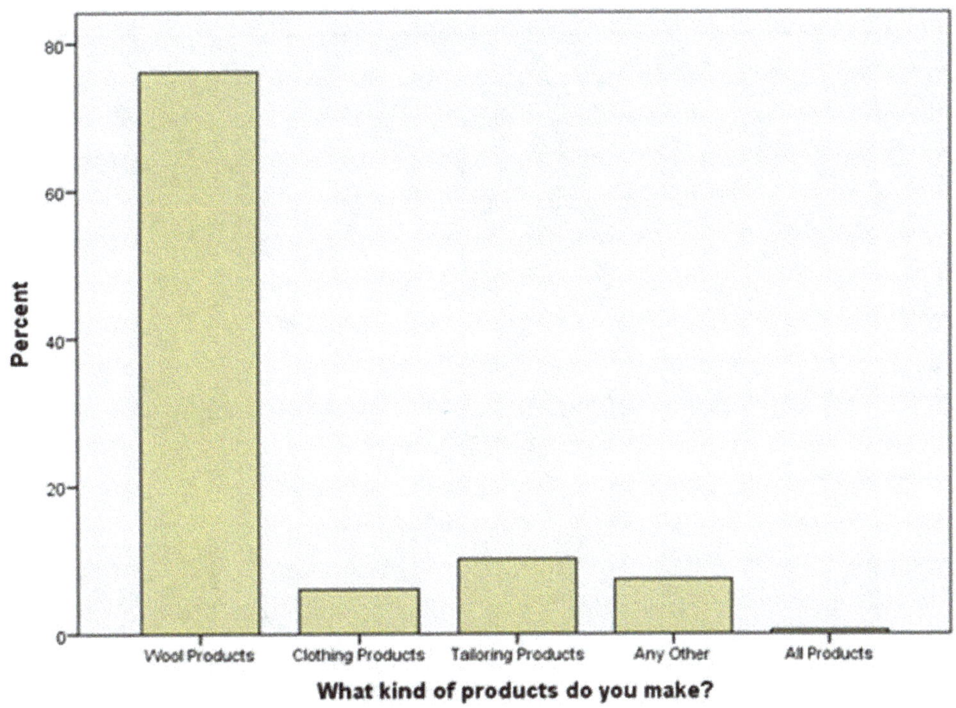

gests a strong commitment to preserving and continuing traditional craftsmanship. About 21.1% of artisans are engaged in creating the latest designs, indicating a segment of artisans who are willing to innovate and adapt to contemporary design trends. Only 11% of artisans are working on designs based on market demands.

This collected data suggests that a relatively small portion of artisans may be actively responding to market trends and consumer preferences. To thrive in the craft sector, a balance between traditional craftsmanship, innovation, and designs that appeal to the market may be necessary. This can help artisans cater to a wide range of customer preferences and remain competitive. In summary, most artisans prioritize traditional designs, while a smaller but significant proportion embrace modern trends or the latest designs but only a few are working as per the market demand. Finding this balance between tradition and innovation is crucial for the sustained presence of artisans and their crafts.

12) **Skill development interest in artisans:** The provided statistics reflect the areas in which artisans wish to improve or strengthen themselves. The majority of artisans (52.3%) express a desire to strengthen their marketing skills. A significant portion (27.1%) wishes to improve their design skills. Artisans may seek to differentiate their products through unique designs. About 14.2% of artisans aim to enhance their production skills. While a smaller percentage (2.3%) expresses interest in strengthening packaging skills, this could still be significant.

The data highlights the aspirations of artisans to improve various aspects of their crafts. The majority want to enhance their marketing skills, indicating a recognition of the importance of effective promotion.

Figure 11. Design orientation of women artisans concerning traditional and latest designs

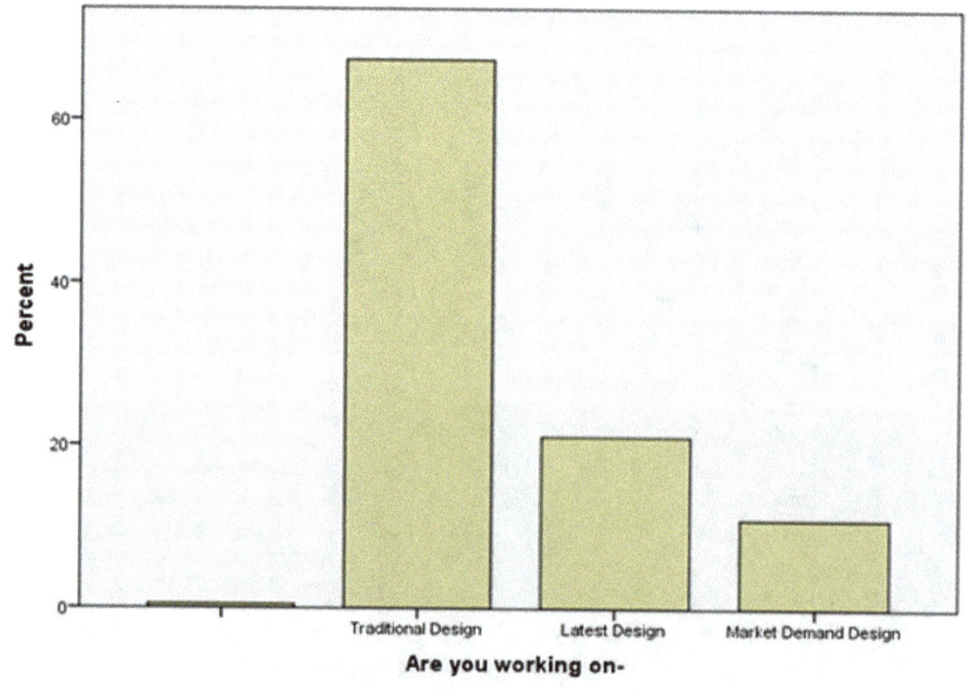

Figure 12. Areas where the artisan would like to strengthen

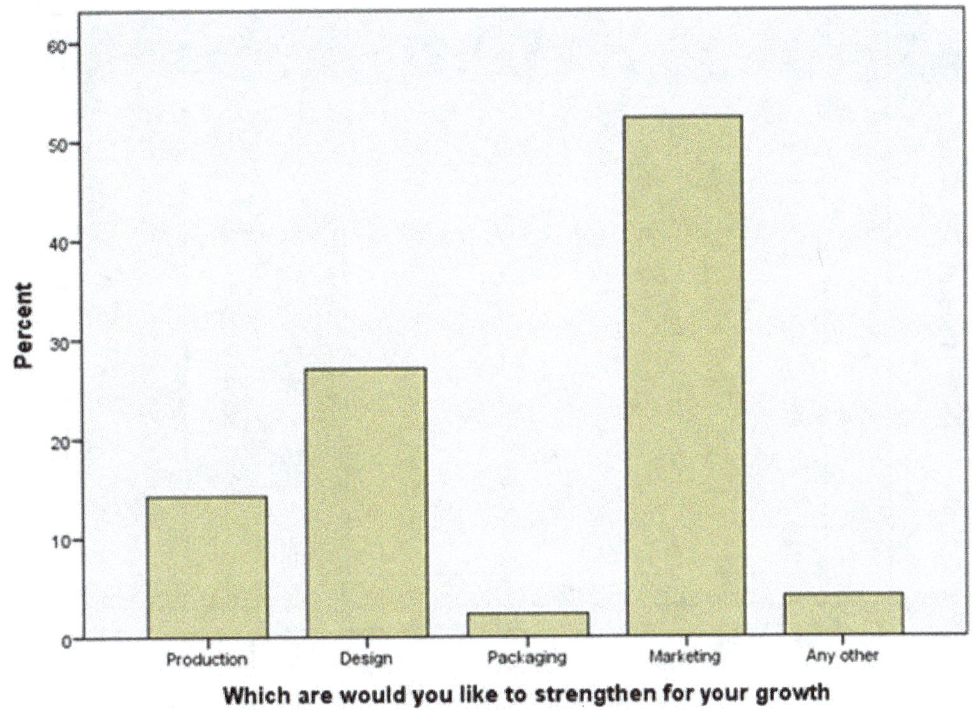

Design improvement is also a significant goal, suggesting a desire to create unique and attractive products. A portion of artisans aims to enhance the efficiency of their crafts, emphasizing the need for streamlined processes. Even a small percentage focuses on packaging skills, acknowledging the impact of presentation on product appeal. The distribution of artisan priorities across marketing, design, production, and packaging implies that artisans recognize the need for a well-rounded approach to succeed in the market.

13) **Challenges faced by artisans to sell their products**: 29.8% of the artisans find the biggest difficulty in selling their product because of distance from their unit to the market, 26.6% find the difficulty because of lack of domestic local markets, 2.3% find it difficult to sell the products because of agents/middlemen.
14) **Monthly income of the women artisans of Lahaul- Pangi landscape**.: A significant portion, approximately 68.3%, of artisans earn a monthly income in the range of Rs 2500-5000 indicating the majority of artisans fall within a relatively low-income bracket. About 26.1% of artisans have a slightly higher income, falling in the range of Rs 5000-7000 per month.

In summary, the majority of artisans surveyed have monthly incomes in the range of Rs 2500-5000, indicating a predominantly low-income group. While a minority earns slightly more, very few artisans enjoy higher incomes. This data underscores the need for economic support, market access, and skill enhancement programs to improve the livelihoods of traditional artisans.

Figure 13. Challenges faced by artisans to sell their products

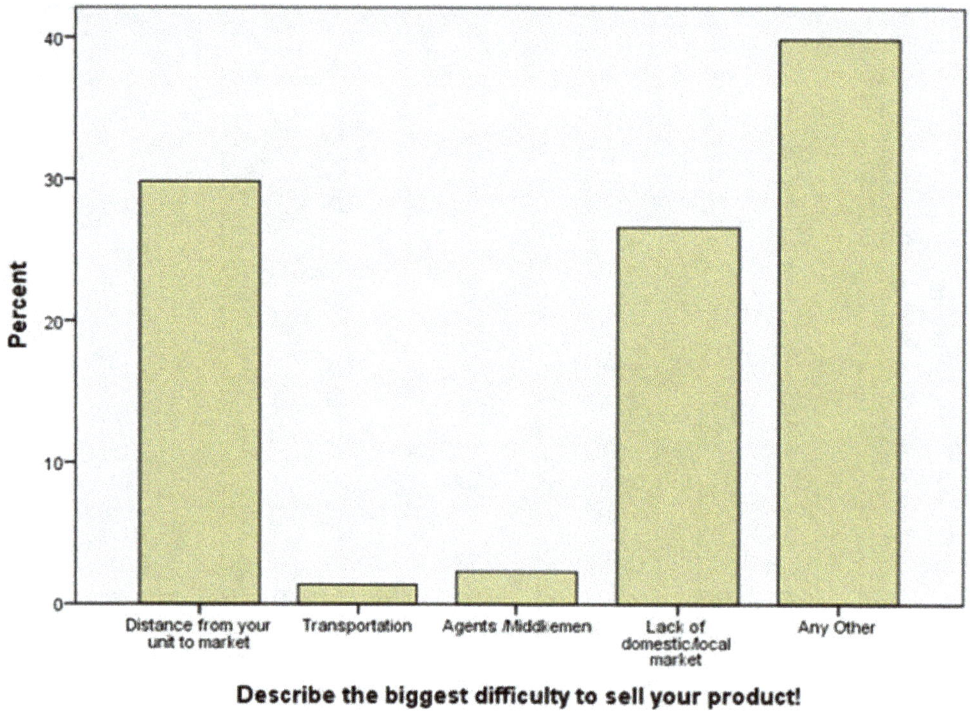

Figure 14. Monthly income of the women arti-sans of Lahaul- Pangi landscape

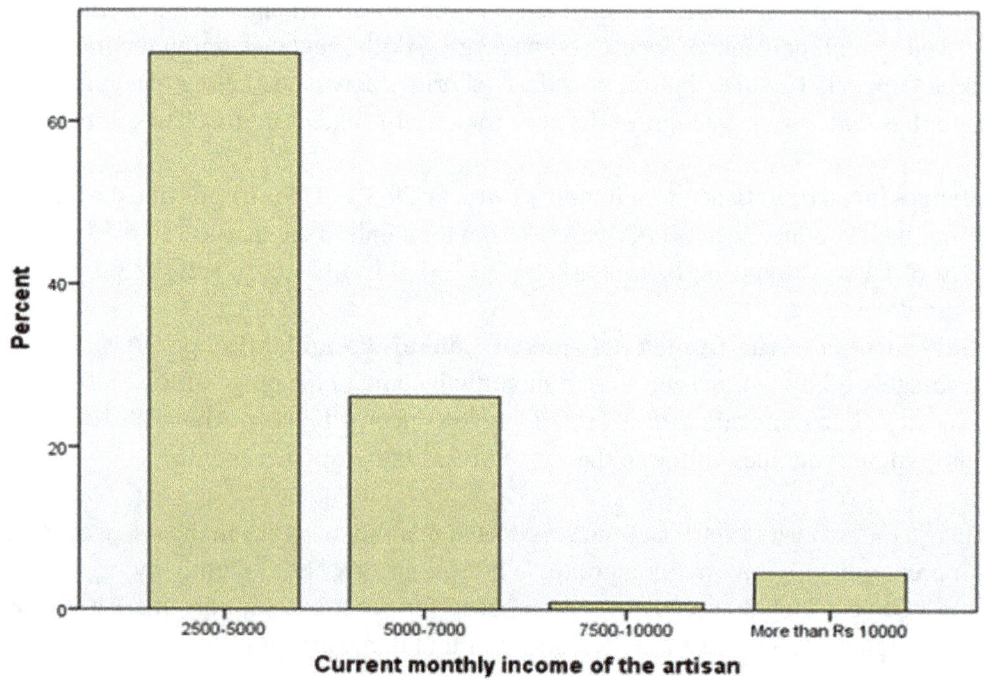

15) **Interest of artisans in passing on the traditional craft-based knowledge:** A majority, constituting 55% of the surveyed individuals, are not interested in training their children in traditional craft practices. However, it's noteworthy that 45% of respondents are still willing to pass down traditional craft practices to their children. The 45% who are ready to train their children in craft practices represent a group dedicated to preserving their cultural traditions.

In summary, the data indicates that a majority of respondents are not interested in training their children in traditional craft practices, highlighting a potential decline in these skills. However, a significant minority is still committed to passing on these traditions, signaling hope for the preservation of cultural heritage. The majority's disinterest in traditional crafts may pose challenges for the continuation of these practices. Efforts may be needed to engage and educate the disinterested majority to ensure the continuity of traditional craft practices.

16) **Training interest of artisans:** A significant majority of artisans, specifically 96.8% of them, have chosen not to undergo any training to enhance or update their skills in the recent past.

This statement summarizes that almost all artisans, at a staggering 96.8%, have declined the opportunity to receive training aimed at improving their skill sets in recent times. This suggests a lack of interest or willingness among artisans to invest in skill development through training programs.

Figure 15. Whether artisans are planning to train their children in traditional craft practices

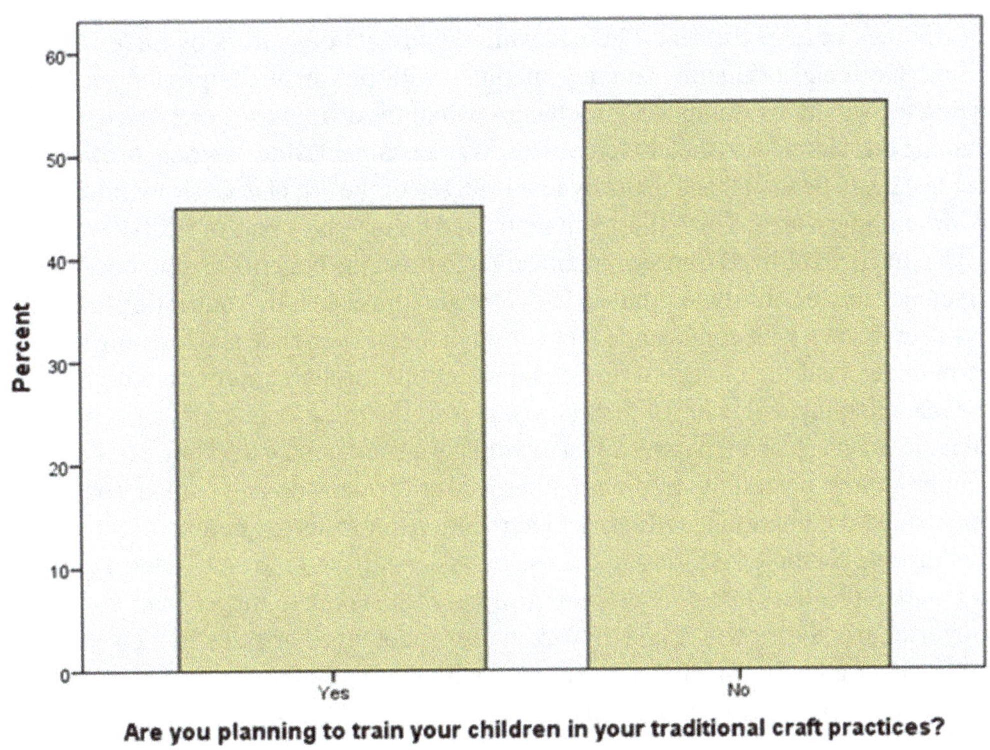

Figure 16. Interest of artisans in training

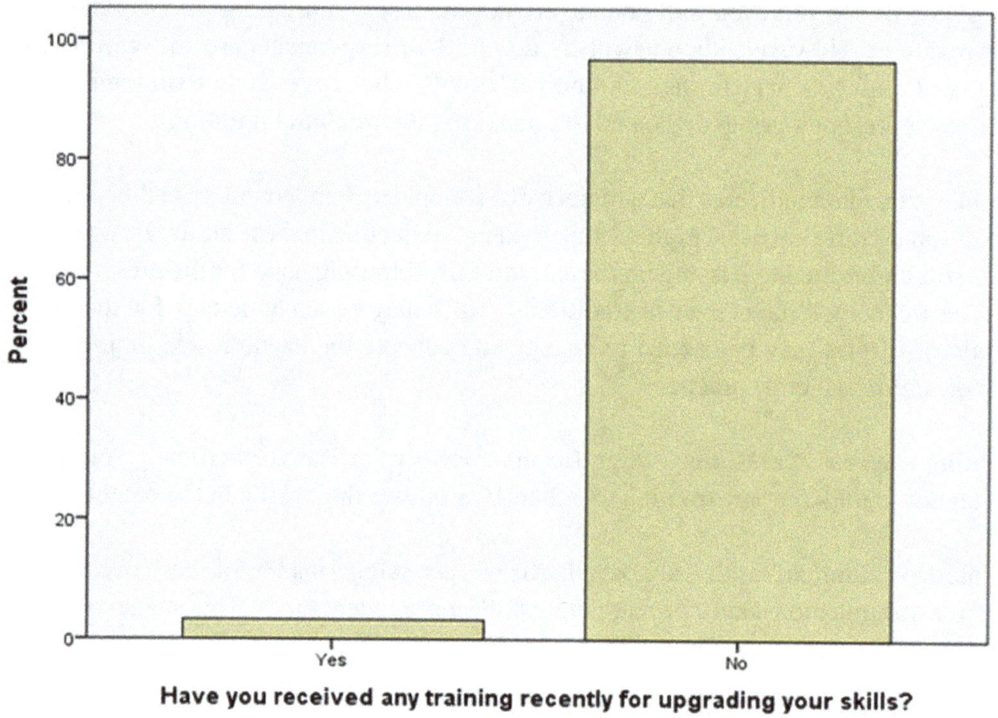

4. DISCUSSION

The survey findings were instrumental in achieving the project objectives by understanding the challenges artisans faced and evaluating training modules with pertinent design and marketing insights. Understanding the variations is important for the growth of the artisans with respect to their availability. Flexible training that can cater to their diverse patterns as sustainable development of craft enhances their income sources. It can be seen that a majority that is 96.8% of the artisans were not interested in training or any skill development area. The NIFT team motivated them to be a part of skill development training programs. The practices utilized were communicating to them the benefits of skill development, such as increased income, and the ability to produce higher quality products; by connecting artisans with experienced master craftsmen who could guide them through their experience and learning process. Artisans were made to understand the impact of design interventions through appropriate training programs to enhance the marketability and value of their crafts, preserving traditional crafts and cultural heritage so that the valuable practice can be passed down to younger generations, etc. They can benefit from these development programs, potentially leading to improved craftsmanship and competitiveness. They were made to understand the importance of design interventions in fostering community capacity building, as they often involve combined efforts and a sense of ownership over the craft's development. Improved designs and market practices can provide rural artisans with access to bigger markets, expanding their base and opportunities for growth. The NIFT team also encouraged artisans to view skill development as sustainable career growth, enabling a community with master craftsmen to inspire and guide others, qualifying for being a part of government subsidies and schemes, etc. By implementing these motivating practices, artisans can be encouraged to invest in their skill development, leading to personal growth, enhanced craft quality, and the continued vitality of traditional crafts.

4.1 Pre-Training Workshops on Design Development of Traditional Designs

Based on the findings of the survey further training programs were designed before field visits. A design workshop was conducted by a team of designers from NIFT and artisans of Lahaul, both online and online mode at NIFT, Kangra. The design workshop held at NIFT, Kangra campus helped in understand the crafts of the Lahaul region in terms of raw materials and designs, before the field visits. The artisans were also invited to the NIFT campus for exposure workshops, demonstration workshops, exhibitions, and craft bazaars where they get an opportunity to interact with the craft experts, upgrade their product and design knowledge, and understand consumers in the urban markets. The workshop included the following points:

i. **Identification of traditional designs:** Interaction with artisans from Lahaul. Understanding of traditional knitted products of the region concerning technique, tools, yarn, color, motif, and design.
ii. **Display and exhibition of product:** Exhibition of traditional products in Weekend Flee marker- "Mind your own business", Dharamshala. Feedback for existing products concerning design & pricing.
iii. **Design isntervention:** Forecast-based design development of various categories of products like apparel, home furnishing, and lifestyle products.

The main objective of the workshop was to identify the traditional design interns of raw materials, motifs, colors, etc. Organizing exhibitions & and sales of products. Further developing contemporary products keeping the essence of the traditional crazy for apparel and lifestyle accessories.

4.2 Field Visits and Training Programs

Field visits and workshops were conducted in all the panchayats in the Pangi region like Sechu, Dharwas, Sach, Killar, Karyas, Kumar, Purthi & Sural. In Lahaul Panchayat field visits and workshops were conducted in Udaipur, Tindi & Tingret. The field workshops were enriched in the content of handling and understanding the traditional crafts, design process, and introduction to digital media for better salability of the products. The sessions conducted in the villages covered the following topics- Introduction to the project, design Intervention workshop & product display and packaging. Considering the skills of the local community, handicrafts are one of the most potential livelihoods. The project team could also work on the various raw materials and tools used for the practicing of these crafts. Preserving traditional elements of existing crafts and at the same time designing contemporary crafts is always a challenge, however, this was dealt meticulously by the NIFT design team by educating artisans on design elements and motifs with contemporary elements to increase their visibility and hence salability. The project team reached out to the local community members in person and imparted awareness about the content of the training modules in 09 panchayats. During the workshops conducted, the artisans were imparted training on awareness of the project and its initiatives, design interventions, product display and packaging, digital marketing, etc. Such kind of knowledge is imperative for making the artisans confident, independent, and well-informed about the existing market trends.

5. STRENGTHENING LIVELIHOODS AND EMPOWERMENT OF WOMEN ARTISANS

Training programs play a pivotal role in strengthening livelihoods and empowering women artisans engaged in wool-based handicrafts in the picturesque Lahaul and Pangi regions of Himachal Pradesh. These remote areas, nestled in the lap of the Himalayas, have long been known for their rich tradition of crafting exquisite wool-based products. However, for the women artisans here, harnessing their skills to improve their livelihoods and socio-economic status was a challenge until the inception of these training initiatives. In these primeval valleys, where the winters are long and harsh, the practice of wool-based handicrafts has been a way of life for generations. Traditionally, women in Lahaul and Pangi have been the custodians of this art, passing down their expertise from mother to daughter. Yet, the absence of formal training often resulted in limited exposure to new techniques and market trends. This made it difficult for these women to compete in the broader marketplace and secure better prices for their products. With the implementation of the training programs designed specifically for these women artisans, confidence building was generated in the community. These programs are a testament to the transformative power of education and skill-building. They provide women with the opportunity to learn new techniques, enhance their existing skills, and gain insights into modern market dynamics. As a result, these artisans become not just producers but also entrepreneurs, capable of making informed decisions about their craft and business. One of the most immediate benefits of these training programs is the improvement in the quality of products that was evident in their products. The rural women artisans could learn about the latest trends in design, color combinations, and product innovation that had helped them to acquire the ability to adapt traditional techniques to contemporary tastes, ensuring that their creations remain relevant and attractive to customers. This has led to increased demand for their wool-based handicrafts in both domestic and international markets. By honing their skills and understanding market dynamics, these artisans are better equipped to negotiate fair prices for their products and had empowered them economically. This financial independence not only improves their standard of living but also gives them a sense of agency and self-worth. In a region where economic opportunities are limited, this newfound empowerment is nothing short of life-changing. The ripple effect of these training programs extends beyond individual artisans and had helped in fostering a sense of community and cooperation. Women who once worked in isolation now come together to share their knowledge and experiences. This collective approach not only enhances their skills further but also opens doors to group enterprises. Women's self-help groups and cooperatives have emerged as powerful entities, pooled resources and marketing their products more effectively. Furthermore, such training programs encourage sustainability and environmental consciousness. As artisans become more aware of the environmental impact of their craft, they adopt eco-friendly practices, such as using natural dyes and sustainable sourcing of raw materials. This not only aligns with global trends but also ensures the preservation of their traditional craft for future generations.

The prime focus of the entire project was to strengthen and empower the women artisans by designing training programs and imparting them further for sustained growth. During the workshops conducted, the artisans were also imparted awareness of the project and its initiatives, design interventions, product display and packaging, digital marketing, etc. Such kind of knowledge is imperative for making the artisans confident, independent, and well-informed about the existing market trends. Equipping the artisans and making them aware of the benefits associated with artisan cards was another exemplary initiative taken for the benefit of the community.

The activities undertaken in each project village include:

i. To identify artisan groups, SHGs, and cooperatives.
ii. To collect a database of artisans & and master craftsmen.
iii. Identify how much they are willing to adapt to new designs and new products.
iv. Identify problems they are facing in developing products (health, material or supply of raw material, production, marketing, etc.).
v. Identify widow artisans for entrepreneur support.
vi. Identify whether the products are detailed and finished for the market or not.
vii. To identify popular hand-knitted products online based on price, quality, and online customer reviews.
viii. Identify brands that are already working on Himachali hand-knitted products.
ix. Identify Master artisans.
x. Gave brief for online marketing and offline marketing and a general overview of marketing channels and what are the possible ways to sell their products according to emerging market trends.
xi. Design intervention workshop.
xii. Demonstration of domestic knitting machines.
xiii. Demonstration of natural dyeing workshop.
xiv. Encouraging artisans to make new designs as per global trends and forecasts.
xv. Discussed the artisan card how to make an artisan card and the benefits of an artisan card.

Such training programs for women artisans practicing wool-based handicrafts in the Lahaul and Pangi regions of Himachal Pradesh are instrumental in strengthening livelihoods and empowering these women. They improve the quality of products, enhance economic independence, foster community cooperation, and promote sustainable practices. These initiatives are not just about skill-building; they are about transforming lives and preserving a cherished cultural heritage. In these remote Himalayan valleys, education and empowerment are indeed the keys to a brighter future for women artisans and their families.

6. CONCLUSION

Considering the skills of the local community, handicrafts are one of the most potential livelihood options in the project landscape that were dealt with by the project team with appropriate interventions. NIFT initiatives included developing the capacities of the communities, utilizing abundant natural resources, and introducing new and innovative designs and interventions at marketing and branding levels. The project has reached out to the local community members in person and imparted awareness about the content of the training modules. Motivating the local community to attend and benefit from these training programs was another colossal task executed by the project team persistently. Training programs lead to women's empowerment by equipping them with appropriate skills for better economic opportunities. Such kind of knowledge is imperative for making the artisans confident, independent, and well-informed about the existing market trends. To impart design interventions effectively by NIFT, it was essential to create comprehensive training modules that cover various aspects of design, including both traditional and contemporary design principles. The broader area of training programs imparted by NIFT included understanding of design; traditional and modern design styles, sourcing of raw materials appropriate for the crafts, techniques for effective marketing, including online and offline strategies participating in craft fairs and exhibitions, etc. The tremendous positive response from these artisans was exemplary and motivational for the project team to carry out further initiatives with more zeal and enthusiasm. The training was imparted to all 218 artisans successfully. In conclusion, NIFT Kangra's efforts in the Lahaul-Pangi region addressed the challenges faced by the local communities by focusing on capacity building, harnessing natural resources, and implementing branding and marketing strategies. By enhancing the skills of artisans and promoting sustainable livelihoods, the study contributes to the conservation of the fragile Himalayan ecosystem and improves the economic prospects of the region's inhabitants. The endeavors undertaken by NIFT Kangra in the enchanting Lahaul-Pangi region stand as an exemplary model of how education, skill-building, and community engagement can breathe new life into remote and marginalized areas. This holistic approach has not only enriched the lives of the artisans but has also acted as a beacon of hope for the entire community. The success of this project is not measured solely by the number of artisans who completed the training but by the profound and enduring impact it has had on the region as a whole. Beyond the acquisition of technical skills, the women artisans have emerged as leaders in their own right. They now have the knowledge and tools to make informed decisions about their craft, their businesses, and their lives. The ripple effect of these initiatives is visible in various aspects of the community. Economic empowerment has led to improved living standards, access to education for their children, and better healthcare. The increased income generated by the sale of their handcrafted products has the potential to break the cycle of poverty that has persisted for generations. Moreover, the preservation of traditional handicrafts is not just an economic endeavor but also a cultural imperative. The Lahaul-Pangi region is home to a rich tapestry of traditions and heritage, and the continuation of these crafts ensures that the legacy is passed on to future generations. It is a source of pride for the artisans, and it connects them to their roots in a rapidly changing world. The sustainable practices introduced by the project team also deserve commendation. As the artisans become more attuned to the environmental impact of their craft, they have adopted eco-friendly methods, including the use of natural dyes and responsible sourcing of materials. This environmental consciousness is not just a trendy concept but a way of life that ensures the delicate Himalayan ecosystem remains unharmed. The success of these training programs has not gone unnoticed. Artisans who once worked in isolation have now formed self-help groups and cooperatives. This collective approach amplifies their voices in the marketplace and provides a platform for shared learning and support. It's not just about competition; it's about collaboration and mutual upliftment. Looking ahead, the Lahaul-Pangi region is poised for continued growth and development. The skills acquired through these training programs are an enduring asset that will continue to benefit the artisans and their families. Moreover, as these empowered women take on leadership roles within their communities, they become agents of change, advocating for the preservation of their cultural heritage and the sustainable development of their region.

In conclusion, NIFT Kangra's efforts in the Lahaul-Pangi region of Himachal Pradesh are a testament to the transformative power of education, skill-building, and community engagement. The impact goes far beyond economic prosperity; it encompasses the preservation of cultural heritage, environmental stewardship, and the empowerment of women who have long been the backbone of their communities. As we celebrate the success of these training programs, we also recognize that they represent a roadmap to a brighter and more sustainable future for the artisans and their breathtaking Himalayan homeland. The Lahaul-Pangi region serves as an inspiring example of how tradition and innovation can coexist harmoniously to uplift communities and preserve the essence of a place and its people.

Acknowledgment: Authors of this work acknowledge the support given by organizations like Global Environment Facility, Himachal Pradesh Forest department, UNDP (under Secure Himalya Project),

NIFT Kangra, and the various rural artisans based in the landscape for their important contribution without which this paper would not have been possible.

REFERENCES

District of Lahaul and Spiti. (n.d.). *About District.* District of Lahaul and Spiti. https://hplahaulspiti.nic.in/geographical-conditions/

Heritage, B. (n.d.). *Traditional Crafts.* Bharat Heritage. https://www.bharatheritage.in/himachal-pradesh/traditional_crafts.html

India Craft. (n.d.). *Crafts of Himachal Pradesh.* India Craft. http://www.india-crafts.com/state_specials/himachal_handicrafts/

Himachal Pradesh Tourism Development Corporation. (n.d.). *Himachal Handicrafts and Handloom.* Himachal Pradesh Tourism Development Coroporation. http://hptdc.in/index.php/himachal-handicrafts-and-handloom/

Himalaya, S. (n.d.). *Himachal Pradesh.* Secure Himalaya. https://securehimalaya.org/himachal/

Shimla. (2021). Tribal crafts fair becomes hit among tourists in Kinnaur. *The Statesman.* https://www.thestatesman.com/cities/shimla/tribal-crafts-fair-becomes-hit-among-tourists-kinnaur-1503026016.html

Shubhyatra. (n.d.). *Arts and Crafts.* Shubhyatra. http://www.shubhyatra.com/himachal-pradesh/arts-crafts.html

Chapter 16
The Sustainable Development of Mountain Tourism at the Hills of Munnar in the State of Kerala:
A Destination Branding Perspective

Prasanth Udayakumar
Indian Institute of Tourism and Travel Management, India

ABSTRACT

India has many mountain destinations to offer to the international mountain enthusiasts. However, these destinations are known to the outside world as holiday, leisure, and recreation destinations. But ample scopes are there to promote them and brand them as "mountain destinations." The Munnar Hills located in the Idukki district of the state Kerala is one such destination which could be promoted as a mountain destination. The goal of this chapter is to examine and map the destination's current resources in order to create a detailed plan for re-branding Munnar as a mountain destination. To obtain information on the expectations of millennial visitors and adventure seekers visiting Munnar Hills, surveys are administered to them regarding their expectations regarding their experiences at mountain destinations. A new product line has been developed based on the data acquired from them, keeping up with global trends in the mountain tourism industry. Input from policy makers, government representatives, and community leaders is being sought for the re-branding of Munnar as a mountain destination.

1. INTRODUCTION

The COVID-19 epidemic in late 2019 had a profound impact on how tourists perceive tourist spots all around the world. Travellers have to reconsider their vacation plans and preferences as a result of social segregation policies, lockdown, and international travel restrictions. Mountain locations became a distinctive niche in the tourism sector as a result of these changes. Mountain regions have grown in appeal

DOI: 10.4018/979-8-3693-0823-3.ch016

as a result of their innate qualities that encourage social isolation. These areas are desirable during a pandemic because they include wide-open spaces, clean air, and the ability to participate in outdoor activities. Mountainous areas have enticed tourists looking to reduce their chance of contracting the illness.

The epidemic has made travel-related health and fitness more important. Mountains provide chances for physical activity, including hiking, skiing, and yoga retreats, which are in line with travellers' preferences for experiences that are wellness-focused. Popular metropolitan areas had trouble controlling crowds and avoiding social awkwardness. Mountain areas were better able to deploy safety measures and provide visitors a sense of security since they had lower population densities. In the post-pandemic era, it is anticipated that demand for mountain destinations would remain high. These places have attracted travellers' attention, and they may do so in the future for the reasons listed above. To keep up with this demand, destinations must adjust to shifting consumer demands and uphold a high level of hygienic practise.

2. LITERATURE REVIEW

Mountain tourism has become increasingly popular in recent years as a result of its distinctive natural sceneries, leisure activities, and cultural attractions. An overview of the most important research findings and developments in the area of mountain tourism is the goal of this survey of the literature. We can learn a lot about the difficulties, effects, and sustainable growth of mountain tourism destinations by studying the existing literature.

"Introduction: Tourism in the Mountain Environments" provide an overview of the tourism industry in mountainous regions (Hall, C. M., and Weiler, B., 1992). The introduction by Hall and Weiler provides a foundational overview of the difficulties associated with mountain tourism while stressing its benefits and drawbacks. It paves the way for further in-depth discussions regarding eco-friendly tourist methods and efficient management plans for mountain destinations later on in the book.

Another work (Price, M. F., and Price, L. W. (2000) present a comprehensive exploration of the geographical aspects of mountain tourism, emphasizing its connection to place and space. They discuss strategies for destination development, marketing, and infrastructure planning of Mountain Destinations, with a focus on responsible and sustainable approaches.

(Mieczkowski, Z., 1995) discusses the concept of tourism climatology, which is the study of how climate and weather conditions influence tourism and recreation activities in Mountain areas. He emphasizes the importance of understanding climatic factors in destination planning and marketing at mountain destinations.

(Gössling, Scott, and Hall, 2013) pinpoints the urgency of addressing water-related issues in mountain destinations and the need for interdisciplinary research and collaboration between the tourism industry, water resource managers, and policymakers to develop sustainable solutions in such destinations. The authors examine how tourism activities in mountains interact with water, the environmental and socio-economic impacts of these interactions, and the challenges posed by managing this critical resource within the context of tourism development.

(Frey, N. M., & George, R., 2019) introduce the significance of mountain destinations in the global tourism industry and highlight the unique environmental and sociocultural attributes of these areas. They emphasize the need for sustainable tourism practices to protect these sensitive ecosystems and communities. They examine the strategies implemented in Jasper National Park, which is a popular mountain destination, to promote sustainable tourism. These strategies include visitor management, environmental stewardship, community engagement, and partnerships with stakeholders.

(The UNWTO 2018) report discusses the role of culture and heritage in sustainable mountain tourism development. It highlighted the need for responsible tourism practices that respect and protect the cultural and natural assets of destinations, especially mountains and other sensitive areas.

Despite the fact that India's numerous and expansive mountain ranges offer enormous potential for mountain tourism, a number of factors indicate that this sector is still in its infancy. There also experience severe dearth in literature confined to Mountain Tourism in India. This study tries to fill the gap by indulging in an exploratory research on Munnar Hill station, which is a globally acclaimed Mountain Tourism Destination

3. OBJECTIVES OF THE STUDY

1. To map the existing tourism products and resources of Munnar
2. To prepare a route map for new product development in the destination
3. To suggest a model for re-branding Munnar as a Mountain Tourism Destination

3.1 Methodology of the Study

The current study is exploratory in character and uses both primary and secondary data because mountain tourism in the nation is still in its infancy. A poll of 200 visitors is conducted at random to gauge the level of interest in the destination's current tourism offerings. To learn more about the values seek out by mountain enthusiasts on their mountain vacations, another 100 millennial travellers were interviewed. Policy makers and community representatives were interviewed about their thoughts on the destination's re-branding through focus interviews and panel discussions.

4. BACKGROUND

4.1 The Scope of Mountain Tourism in India

Mountains are just one example of the many natural treasures that India's varied geography has to offer. Travellers seeking adventure, peace, and spiritual enlightenment have always been drawn to India's mountains because of its diverse landscapes, cultures, and experiences. The mountain tourism business in India is a timeless treasure, spanning the formidable Himalayas in the north to the lush Western Ghats in the south. Mountain destinations in India geographically spread across three regions as follows:

4.1.2 The Himalayan Region

The Himalayan mountain range, sometimes known as the "abode of snow," is the highlight of mountain tourism in India. The Himalayas, which span multiple states and include Sikkim, Jammu & Kashmir, Himachal Pradesh, Uttarakhand, and others, provide an unmatched combination of spectacular scenery, holy places, and adventure options.

Trekking: The Himalayas are home to some of the most well-known trekking routes in the world, including the Roopkund Trek and the Everest Base Camp Trek. Trekkers from all over the world come to participate in these excursions to feel the rush of conquering high-altitude terrain.

Spiritual retreats: Locations like Rishikesh and Dharamshala have developed into spiritual hotspots, drawing yogis and people looking for inner calm. These locations provide yoga and meditation retreats against the lovely Himalayan setting.

Adventure Sports: White-water rafting, paragliding, and skiing are just a few of the adventure sports that the area offers, making it a year-round destination.

4.1.3 The Western Ghats

Along India's western coast, the Western Ghats, a UNESCO World Heritage site, are renowned for their lush vegetation, tumbling waterfalls, and a variety of species. Several states, including Kerala, Karnataka, Tamil Nadu, and Goa, are bordered by this mountain range.

The Western Ghats are an ideal destination for ecotourism. In places like Periyar Tiger Reserve and Nagarhole National Park, tourists can explore national parks and wildlife sanctuaries and get a chance to see rare types of plants and animals. The Western Ghats' hill towns, like Coorg and Wayanad, are well known for their plantations of coffee and spices. Visitors can join guided tours to discover more about the raising of these crops and to take in the tranquil beauty of the area.

4.2.4 The Eastern Himalayas

While receiving less attention than its western counterparts, the Eastern Himalayas are no less magnificent. This area includes Arunachal Pradesh, Assam, and Nagaland in the northeastern Indian states.

Cultural diversity: Numerous indigenous cultures and ethnic groupings can be found in the Eastern Himalayas. While seeing this particular region of India, visitors can immerse themselves in the customs, music, and festivals of the area.

Wildlife and Biodiversity: The area is renowned for its great biodiversity, and national parks like Kaziranga and Manas give wildlife aficionados the chance to see endangered animals like the one-horned rhinoceros and Bengal tiger.

Although mountain tourism is thriving in India, it still confronts a number of difficulties, such as environmental deterioration, overtourism in some regions, and the need for sustainable development. To address these issues, stricter laws, community involvement, and responsible tourism practises are necessary.

Mountain tourism in India has a bright future. The significance of sustainable tourism is being recognised by the government and local communities more and more. The demand on popular locations will be lessened by investments made in the infrastructure, accessibility, and promotion of less well-known locations.

4.2 The State of Kerala and Western Ghats

The Western Ghats, a bio diverse hot spot in India, has attracted considerable attention from researchers interested in understanding its ecological significance, and the role of Kerala, a state in southwestern India, within this system.

Kerala has become a focus of ecological research due to its particular physical location inside the Western Ghats. A microcosm of the entire Western Ghats ecosystem, the state, also known as "God's Own Country," is said to cover a variety of landscapes, from high-altitude forests to coastal regions, according to (Gupta & Nair, 2017). As a result of Kerala's abundant flora and fauna, (Suresh & Menon, 2019) have investigated the Western Ghats' biodiversity in great detail, as well as the issues associated with its conservation.

Kerala's contribution to the Western Ghats' hydrological cycle is a crucial component of the state's function in the ecosystem. The importance of Kerala's rivers in sustaining the Western Ghats' water balance (Krishnan & Rajan, 2018). These rivers, which originate in the Ghats and flow through Kerala, provide water to inhabitants downstream as well as ecosystems. The preservation of these water bodies is essential for the survival of the entire Western Ghats region (Thomas & Joseph, 2020).

In order to preserve biodiversity in the Western Ghats, Kerala's forests are also essential. The state's protected regions, including Silent Valley National Park and Periyar Tiger Reserve, are crucial habitats for a number of endangered species. As a result of Kerala's conservation efforts, the Western Ghats' endemic flora and fauna have been preserved, making Kerala a significant role in the regional conservation of biodiversity.

Furthermore, the ecosystems of the Western Ghats are impacted by Kerala's agricultural practises. The growth of cash crops like rubber and tea in the state has caused habitat fragmentation and a loss of biodiversity (Menon & John, 2017). For Kerala and the entire Western Ghats to plan for sustainable land use, it is crucial to comprehend the ecological effects of various agricultural practises.

In conclusion, Kerala's position in the Western Ghats system is complex and includes its contributions to the hydrological cycle, biodiversity preservation, and land use practises, all of which are related to tourism either directly or indirectly.

4.3 Munnar as a Mountain Tourism Destination

Munnar is a picturesque town located in the Western Ghats of Kerala, a state in southwestern India. Its geographical location within the Western Ghats contributes to its stunning natural beauty and unique climate. This picturesque hill town, positioned at a height of around 5,200 feet above sea level, has long captivated tourists and nature lovers. Munnar is a must-visit location for anyone looking to get away from the rush and bustle of daily life because of its attractiveness, which stems from both its magnificent landscapes and its rich cultural legacy.

5. DISCUSSIONS AND RESULTS

5.1 Mapping of Mountain Tourism Resources at Munnar

Munnar has already established a reputation as a honeymoon and adventure travel destination. Two hundred randomly selected tourists participated in a survey to find out how popular the traditional tourism offerings are at the location. The tourists were asked to rank the tourist activities that are currently offered at the destination based on their personal preferences by the researcher. The gathered data was subjected to factor analysis in order to evaluate the outcome, which is best described as follows:

Table 1. Factor Analysis showing the popularity of current tourism products in the destination among the tourists.

Sl. No.	Factors	Eigen Value	Variance (%)	CV
1.	Adventure Tourism	4.147	31.351	31.351
2.	Rural Tourism	3.872	29.521	60.872
3.	Wellness Tourism	2.327	09.219	70.091
4.	Heritage Tourism	1.282	08.915	79.006
5.	Pilgrim Tourism	1.199	08.361	87.367

According to the data above, adventure tourism is more common among visitors to Munnar. Both rural tourism and wellness tourism draw sizable numbers of visitors. Though in smaller numbers, pilgrims and heritage enthusiasts are also making the trip to the location. A mapping of the destination's current tourism offerings has been completed as below, drawing from both this analysis and the researcher's on-site observations.

5.1.1 Adventure Tourism

From trekking to paragliding, this charming hill station offers an array of thrilling experiences for adventure enthusiasts.

Trekking - A hiker's paradise, Munnar's topography is undulating. The area offers a wide range of trekking options that are suitable for all levels of experience thanks to its lush forests, flowing waterfalls, and expansive tea gardens. The trekking choices in Munnar were clarified through a Delphi analysis conducted with decision-makers and government representatives. There are approximately 10 well-known trekking paths nearby.

Jeep Safari - Munnar provides thrilling vehicle safaris for those seeking excitement without the physical strain of walking. These off-road excursions take tourists deep inside the Western Ghats, where they can explore inaccessible rural locations. The Kolukkumalai Jeep Safari, which takes tourists to the highest tea plantation in the world. Difficult terrain, sheer cliffs, and fantastical panoramic views are typical features of Mountain Destinations which could be explored in Jeep Safaris.

Mountain Biking - Munnar's varied topography has helped make mountain riding more well-liked there. There is a vast network of trails in the area, from simple tracks for new riders to difficult routes for experienced ones. There are countless mountain biking chances in Munnar, whether the tourists choose a leisurely ride among tea gardens or an adrenaline-pumping downhill descent through the hills.

Table 2. Mapping of Trekking paths popular in Munnar

1.	Anamudi Peak	6.	Kolukkumalai Tea Estate Trek
2.	Meesapulimala Trek	7.	Attukal Waterfalls Trek
3.	Chokramudi Trek	8.	Echo Point to Top Station Trek
4.	Lakkam Waterfalls Trek	9.	Kannan Devan Hills Trek
5.	Chinnar Wildlife Sanctuary Trek	10.	Marayoor Sandalwood Forest Trek

Paragliding - The region's favorable wind conditions, varying altitudes, and mesmerizing landscapes make it an ideal destination for paragliding enthusiasts. The service of experienced instructors from the local community is the added advantage of paragliding experience at Munnar.

Rock Climbing - Rock climbers will find an exciting challenge on Munnar's rocky outcrops and cliffs. Munnar has something to offer everyone, whether the tourist is an experienced climber or a novice wishing to try hand at this thrilling sport. Numerous climbing locations provide varying degrees of difficulty, and professional guides are available to assure a secure and pleasurable experience.

White-water Rafting - Munnar is more well-known for its tranquil backwaters, but during the monsoon season, it also provides opportunities for white-water rafting. For daring rafting adventures, adjacent rivers like the Kundala and the Chinnar offer exhilarating rapids and cascades. For fans of water sports, the flowing waters and verdant surrounds make an exhilarating experience.

5.1.2 Rural Tourism

A well-liked rural tourism destination in India is Munnar, which is well-known for its luxuriant tea plantations, wildlife sanctuaries, and cultural diversity. The area offers distinctive activities like tours of tea estates, cultural events, and eco-friendly lodgings as analyzied by Rajasekar & Prasanna (2019). Travellers looking for a connection with environment and the local community have been drawn to Munnar's rural tourism industry. Rural tourists are drawn to Munnar by its traditional social structures, way of life, beautiful surroundings, and patterns of land use. Some of the potentials for rural tourism in the area include learning how to farm, manufacturing handicrafts, tasting local cuisine, participating in pastoral activities, and attending traditional rites and festivals. Despite the benefits, there are still problems including over development, seasonality, and cultural commodification. Munnar can, nevertheless, preserve its allure and originality by diversifying its attractions, planning responsible tourism, and fostering strong community ties as opined by George & Prakash (2021).

5.1.3 Heritage Tourism

Tourists interested in learning about the history, culture, and customs of a place should visit Munnar because of its famed rich cultural and natural legacy. The region has a long history of human settlement and was first inhabited by indigenous people. Over the years, the area was influenced by a number of dynasties and colonial powers, including the British Empire, the Chera dynasty, and the early European planters. The architecture, cuisine, and customs of Munnar are all influenced by the region's rich cultural past. Generation after generation has passed down the traditional agricultural methods of Munnar, particularly the cultivation of tea and spices. The region's folklore, music, dance, and cultural festivals offer a look into its rich history. Many indigenous tribes, such the Muthuvans and Mannans, who each have their own cultural traditions, are also found in the Munnar region.

The development of this industry has been aided by the opening of heritage centres, museums, and cultural events. Tourists frequently visit the Tata Tea Museum, which chronicles the development of Munnar's tea industry, and the Munnar History Museum, which provides insights into the area's cultural legacy. Local communities have actively promoted heritage tourism by educating tourists about their culture and traditions. For tourists, traditional arts like Kathakali and Kalaripayattu are frequently performed as suggested by Anand. A. (2017)

Many Malayalam films have been shot in the scenic mountains and among the rich cultural legacy of the Munnar Hills that explored the potential for film tourism in the mountains. Numerous renowned Malayalam authors have depicted the stunning Munnar Hills in their fictions that explored the potential for literary tourism in the mountains.

5.1.4 Spiritual Tourism

Many tourists travel to Munnar specifically to seek spiritual enrichment, contemplation, or religious experiences. According to Coelho & George (2018) Munnar's spiritual tourism has made eco-spirituality, which emphasises a harmonious link between people and the environment, a central subject. An environment that is favourable for meditation and introspection is created by the verdant greenery, diverse wildlife, and clean air. The mountain town is also home to a number of historic churches, ashrams, and temples, offering a variety of options for spiritual inquiry.

5.1.5 Wellness Tourism

According to (Hall, 2016) the British planted tea and coffee plantations in the area during the colonial era, which is when Munnar first gained notoriety as a wellness resort. It quickly became a popular site for rest and recuperation because to its mild climate, natural surroundings, and tranquil atmosphere. Ayurveda, yoga, and meditation have become increasingly popular in Munnar as it has developed over time.

5.2 New Product Development

A second poll has been done among 100 Millennials who were selected using the snowball sample method, in addition to the one mentioned above. These are youthful travellers with a particular interest in mountainous areas. The purpose of the survey was to gather data regarding respondents' expectations for the value of travelling to mountain destinations. This data is useful for creating new Tourism Products in the area that can support Munnar's re-branding as a mountain tourism destination.

Table 3. Mapping of spiritual spots in Munnar

1.	Attukal Bhagavathy Temple	5.	CSI Christ Church
2.	Aruvikkuzhi Waterfall and Temple	6.	Kolukkumalai Temple
3.	Nallathanni Temple	7.	Pambadum Shola National Park
4.	Mount Carmel Church		

Table 4. Mapping of wellness resources in Munnar

1.	Ayurvedic Retreats	There are many Ayurvedic clinics and spas in the area that provide conventional Ayurvedic medicines and treatments. These treatments are thought to support both physical and emotional wellness.
2.	Yoga and Meditation Retreats	Munnar is an ideal location for yoga and meditation retreats. These antiquated arts can be practised by visitors in the peace and quiet of nature.
3.	Adventure Wellness	Adventure lovers are also catered to in Munnar. Treks, nature hikes, and other outdoor pursuits combine exercise with scenic beauty to promote wellness holistically.

Table 5. Factor analysis of value perception of millennial tourists

Sl. No.	Factors	Eigen Value	Variance (%)	CV
1.	Educational Value	3.036	20.240	20.240
2.	Experiential Value	2.761	18.410	38.650
3.	Unique Value	1.216	08.108	46.758
4.	Socialization Value	1.171	07.804	54.561

The aforementioned component analysis demonstrates that travellers to mountain areas are mostly interested in experiential and educational values.. Additionally, people are searching for unique values in the destination that they won't discover anywhere else. Many tourists also want to socialize with the unique mountain community, which has many unique characteristics.

The researcher had two panel discussions, one with local leaders and the other with policy makers in the destination. These discussions shed insight on creating a new product portfolio that would satisfy the values that mountain enthusiasts have in mind. To learn more about the tourism offerings in other mountain destinations across the nation, the researcher also looked at secondary data. Accordingly a new product portfolio has been suggested as follows:

5.2.1 Star Gazing Observatories

The Top Station in Munnar is an ideal location for Star Gazing and Astro-photograhy. The atmosphere of this place is cold and dry, lending the perfect setting for unobstructed stargazing. The local villagers who run Homestays near by Top Station could be trained in star gazing with telescopes, there by offering astro-tourists a stunning nocturnal experience at their homestays. The Starscapes, the first chain of Astronomy Observatories in India, has set up stargazing observatories in locations such as Kausani and Bhimtal in Uttarakhand and Ooty in Tamilnadu. The scope of setting up similar observatories in Munnar could not be ruled out. Special astronomy events and festivals could also be organized during International Dark Sky Week which falls between 15 - 22 April every year, like the kind of one which is celebrated at Mukteshwar in Uttarakhand.

5.2.2 Promotion of Pastoral Tourism

The Pastoral Communities across the globe follows unique lifestyles and habits which could be sources of tourist attractions. In Munnar, where rural and agricultural activities are prominent, pastoral tourism opportunities are abundant, allowing visitors to immerse themselves in pastoral activities such as cattle rearing, veterinary care, breeding and reproduction of cattle, feeding cattle, shelter making etc.

Around 11 varieties of indigenous pastoral communities are found in Munnar whose way of living and pastoral activities are sources of tourist attractions. These pastoral communities have diverse customs, languages, and practices, contributing to the cultural richness of the destination.

Table 6. Mapping of pastoral communities found in Munnar

Pastoral Communities found in Munnar			
1.	Kurichiyas	7.	Vettuvan
2.	Mannan	8.	Boyas
3.	Kurumbas	9.	Dheevaras
4.	Paniyas	10.	Narikuravas
5.	Cholanaikans	11.	Uralikuravas
6.	Kadar		

Table 7. Mapping of pastoral activities in Munnar

Pastoral Tourism Activities in Munnar		
1.	Grazing	Cattle are led by shepherds to graze in open fields and hunt for food. They make sure there is sufficient grass and plants to meet the nutritional requirements of the herd.
2.	Watering	For the cattle's health and wellbeing, it is essential to provide access to sources of fresh, clean water. Shepherds frequently direct their flocks to drinking holes or rivers in the area. This catches the attention of tourists who spend their time near such water bodies for boating or any other purpose.
3.	Herding	Cattle are kept in groups by shepherds, who also restrain them from wandering too far. To help with herding, they use additional livestock guardian animals in addition to trained dogs. In pastoral lands, the sight of shepherds moving away with a neatly arranged herd of cattle is a touristic attraction.
4.	Fencing and Enclosure Maintenance	To keep the cattle safe from predators or to manage grazing areas, shepherds may need to erect temporary enclosures or maintain existing fences. Shepherds use their creativity while fencing which often makes use of locally available materials. Tourists viewing this and helping shepherds could be possible and can provide opportunities for experiential tourism activities.
5.	Veterinary Care	Shepherds monitoring the health of the cattle and provide basic veterinary care and treating injuries or illnesses could be interesting activities for the tourists.
6.	Breeding and Reproduction	In pastoral areas tourists could witness communities indulging in cattle management activities such as controlling breeding seasons, ensuring that cows and bulls successfully mate, and supervising the calving or birthing procedure.
7.	Hay making	Hay may occasionally be given to cattle as supplemental feed, particularly during droughts or when grasses are scarce. Making hay could be an attraction in Pastoral Tourism.
8.	Dairy Farming	In several pastoral settlements in Munnar, dairy cattle are raised. In such communities, milking is a regular activity, and they also operate small cottage industries that produce cheese and other dairy products.
9.	Marketing and Sales	Shepherds frequently have to sell their cattle or the goods they produce, like meat, milk, or wool. This entails tasks like driving cattle to marketplaces and haggling over sales. Tourists can also visit local cattle markets in the destination where pastors indulge in trading of cattle and cattle products.
10.	Tradition and Culture	The pastoral communities of Munnar have a wealth of cultural practises and traditions related to raising cattle. These could consist of ceremonies, music, and storytelling.

5.2.3 Trekking Through Flower Valleys

In Sikkim's Barsey Rhododendron Sanctuary, Maharashtra's Kaas Plateau, Nagaland's Dzüko Valley and Uttarakhand's Valley of Flowers, trekking through flower valleys is already very popular. The Valley of Flowers in Uttarakhand is well known for its alpine flora, which includes calendulas, primroses, daisies, and poppy flowers. Nagaland's Dzüko Valley is unique for lily, which grows only in this pocket behind the Japfu mountain range. Similar to this, Munnar is well known for its Neelakurunji flowers which bloom once in every twelve years. To commemorate the bloom of this exceptional flora, the

Department of Tourism, Government of Kerala, could hold the "Neelakurunji Fest" and could organize unique Trekking Trails once every twelve years.

5.2.4 Tea Tourism Experience

In India's mountains, tea plantations provide tourists a chance to explore local culture and learn about rare tea varietals. Since more people are interested in learning about the history and culture of tea as well as savouring the distinctive flavours and smells of many types of tea, tea tourism has grown in popularity in recent years. Hill towns all around the country, from Kangra to Munnar, are characterised by undulating hills covered in vivid green tea leaves. Additionally, a few tea estates and historic bungalows in these locations allow visitors to experience life as a tea planter.Tea aficionados from all over the world frequent the Makaibari Estate, Glenburn Estate, and Teanest Bungalow in Darjeeling and Coonoor, respectively, to get a firsthand look at the process of getting tea from the field to the cup.

Similar to the ones in Darjeeling and Coonoor, the Tata Tea Museum in Munnar, which is housed in a colonial home, may also be publicised. This museum showcases the development of Munnar's tea industry and features a collection of equipment and artefacts used in the industry. Additionally, visitors can learn about the finer points of tea brewing and taste.

5.3 Re-branding of Munnar as a Mountain Destination

A strong brand enables destinations to withstand the test of time and remain competitive while remaining innovative, dynamic, and agile in their approach as the travel industry changes at an unprecedented rate along with constantly changing traveller behaviour, the impact of sustainability, and living in a hyper connected world. Munnar is a well popular destination which has been branded by highlighting its climate, topography and other factors. The youthful generation now has more spare cash and vacation time because to the growing IT sector and the availability of remote employment options. The younger generation can learn more about new places to visit thanks to technological advancements. Due to all these benefits, travellers are drawn to less-traveled but no less intriguing locations rather than overly famous ones. Munnar is not an exception either. The allure of this well-liked destination is eroding as younger travellers, with greater disposable incomes and access to modern amenities, can now afford other, comparable locations. In light of this, re-branding Munnar as a mountain destination would be appropriate.

5.4 Attributes of Branding

Locational Advantage - The Himalayan Mountains are too far away for those who want to climb mountains from India's southern and western coasts. They have easier access to Munnar, which is located in the Western Ghats. The destination's re branding may benefit from this advantageous location. Additionally, Munnar might provide all the entertainment and attractions found in other Himalayan mountain destinations.

Proximity to Beaches and Backwaters - Kerala's unique landscape, which includes mountainous terrain, backwaters, and coastal plains, attracts a wide variety of tourists. The benefit of visiting Munnar, a mountain destination, is that visitors may also explore the 150 kilometres nearby beaches and backwaters. Travellers would have the chance to combine mountains, beaches, and backwaters in one package—an opportunity they might not have while visiting Himalayan mountain destinations.

Accesibility - Munnar is reachable by every kind of transportation. The state boasts good rail and road networks, as well as four airports. The Munnar district of Idukki lacks a railway network. Nonetheless, travellers may disembark at the 150-kilometer Kottayam or Ernakulam railway stations, from where they could continue their journey to Munnar by road. Munnar is hence more accessible than other mountainous areas in the nation.

5.5 A Branding Model for the Destination

It might be argued that Kerala Tourism was the driving force behind the nation's destination branding movement. The state's numerous advertising campaigns, with their breathtaking imagery, made other state tourism boards envious and helped to create the history of destination marketing in India.

Ten policy makers from the state tourism department participated in a focus group interview to provide their input on the creation of a new branding strategy for Munnar. Based on the knowledge gained from them, the researcher created the model that is shown below.

5.5.1 Co-Branding With Responsible Tourism Mission

The Kerala government established the Responsible Tourism Mission as the central organisation to promote and carry out the concepts and programmes of responsible tourism throughout the state. It is impossible to completely rule out the idea of co-branding Munnar's mountain tourism with the Responsible Tourism Mission. The Responsible Tourism Mission, which operates under the slogan "Making better places for people to live in and better places for people to visit," has already made a name for itself because it is a special programme that is exclusive to Kerala. Given that mountains are ecologically sensitive regions where responsible tourism practices are suitable, promoting mountain tourism in Munnar under the Responsible Tourism Mission brand identity would make sense. Eight destinations are currently included in the RT Mission: Kumarakom, Thekkady, Kovalam, Wayanad, Vagamon, Bekal, Kumily, and Muzuris. The mission's goal was to promote a variety of destinations within the state that would emphasise responsible tourism. Rebranding Munnar as a "Responsible Mountain Destination" may be aided by its inclusion on the list.

5.5.2 Co-creation Experiences Through Website

A brand is characterized as a social process wherein customers, stakeholders, and service providers collaborate to create value that is added. Companies and customers co-create value through their brands in communication venues. This includes what are referred to as "co-creation experiences," or points of interaction, which can be found both offline (stores, products, and visual identities) and online (websites and social networks). At these points of contact, the brand creates and shares information and opinions on items that make up the identity and perceptions that the brand will acquire on the market with staff, stakeholders, brand communities, and other customers.

Participation in online co-creation activities is positively impacted by the perceived quality of a destination's website. With its appealing design and audio-visual components, Kerala Tourism's website won numerous international awards for best tourism website. This could be used as an advantage for re-branding the destination under study. Redesigned Kerala Tourism website featuring photos and material showcasing Munnar's mountainous beauty might maximize co-creation experiences, which in turn could aid in the destination's re-branding.

5.5.3 Visual Branding Through Short Video Clips

Kerala Tourism has a long history of using video clips to promote its travel destinations. The state's tourism attractions are portrayed in theme-based commercials such as "Your Moment is Waiting" and "The Great Backwaters," which are masterful combinations of words and eye-catching imagery. Stark Communications is the company behind these advertisements.

The short film "Your Moment is Waiting" masterfully depicted a woman's voyage through a variety of events in the state, including sailing in a house boat, experiencing shirodhara (a type of Ayurvedic therapy), watching a Kathakali dancer practise without his usual makeup, and more.

The short film "Your Moment is Waiting" masterfully depicted a woman's voyage through a variety of events in the state, including sailing in a house boat, experiencing shirodhara (a type of Ayurvedic therapy), watching a Kathakali dancer practise without his usual makeup, and more. The "The Great Backwaters" campaign, which was equally remarkable, was captured utilising cutting-edge aerial photography techniques. This single short film elevated the state's backwaters to the same level as the Grand Canyon and the Great Wall of China.

A short film of a similar nature, titled "Mountains calling you," might be produced to present the new product line for Munnar's mountain tourism.

5.5.4 Bus and Taxi Branding

Kerala Tourism was the pioneer in investigating the potential of branding their travel products through the use of public transport in foreign nations. Double-decker buses with 'bus branding' featuring the famous houseboats and scenic Kerala backwaters on one side and Kathakali on the other, are operated out of five depots in Central London. A 'Taxi Branding' campaign was launched in Dubai a few years ago by Kerala Tourism, aimed for Arab tourists, and featured the state's backwaters and waterfalls on taxis operating in the busy city.

It might be feasible to do bus and taxi branding of a similar nature for Munnar. The buses and taxis of selected overseas countries might carry the picturesque tea plantations, neelakurunji flowers, and cattle-rearing shepherds.

6. CONCLUSION

It is well deserved that Munnar has been upgraded to a Mountain Tourism Destination. The vast array of tourism products that are already available may be marketed under the heading of Mountain Tourism. The study's recommended new product development approach also fits the goal. To this extent, the government's, the community's, and the tourists' combined efforts would be advantageous. The study's suggested re-branding tactics make sense in light of the evolving tastes of younger travellers and the exigencies of the moment. Support from the local community will inevitably be needed during the implementation phase. When the new products recommended by the study are developed, new opportunities will present themselves to the youth of the area, which could be tapped only through the support of the Government in the form of skill development and capacity building.

REFERENCES

Anand, A. (2017). Tourism Development in Munnar: A Socio-Cultural Perspective. In S. Das & R. N. Sahoo (Eds.), *Tourism Marketing* (pp. 301–311). Springer.

Coelho, R. A., & George, B. P. (2018). Spiritual tourism in Kerala, India: An analysis of foreign tourist perspectives. *Journal of Heritage Tourism*, *13*(2), 107–121.

George, M., & Prakash, G. (2021). Rural Tourism in Munnar: Challenges and Opportunities. *Journal of Sustainable Tourism*, *29*(8), 1090–1106.

Gössling, S., Scott, D., & Hall, C. M. (2013). Tourism and water: Interactions, impacts, and challenges. *Annual Review of Environment and Resources*, *38*, 1–28.

Gupta, A., & Nair, S. (2017). Kerala's Western Ghats: A Biodiversity Hotspot. *Journal of Ecological Studies*, *10*(2), 78–91.

Hall, C. M. (2016). Wellness tourism: A new paradigm for growth in the tourism and hospitality industry. *Journal of Hospitality and Tourism Management*, *27*, 67–73.

Hall, C. M., & Weiler, B. (1992). Introduction: Tourism in the mountain environments. In Tourism in the Mountain South: A Double-Edged Sword (pp. 1-15). University of Tennessee Press.

Krishnan, R., & Rajan, S. (2018). Hydrological Significance of Kerala's Western Ghats Rivers. *Water Resources Research*, *20*(3), 134–147.

Menon, P., & John, K. (2017). Impact of Cash Crop Cultivation on Biodiversity in Kerala. *Environmental Science & Policy*, *12*(5), 260–273.

Mieczkowski, Z. (1995). The tourism climatology: The state of the art. *Tourism Recreation Research*, *20*(2), 71–75.

Price, M. F., & Price, L. W. (2000). *The geography of tourism and recreation: Environment, place, and space*. Routledge.

Rajasekar, R., & Prasanna, A. (2019). Munnar: An Emerging Rural Tourism Destination. *Tourism Planning & Development*, *16*(1), 53–68.

Suresh, M., & Menon, A. (2019). Protected Areas in Kerala: Biodiversity Conservation and Challenges. *The Journal of Wildlife Management*, *45*(4), 321–335.

Thomas, A., & Joseph, B. (2020). Kerala's Involvement in Western Ghats Conservation Initiatives. *Environmental Management*, *25*(1), 87–98.

UNWTO. (2018). *UNWTO/UNESCO World Conference on Tourism and Culture: Building a New Partnership*. UNWTO & UNESCO.

Chapter 17
Tourism Development and Environmental Conservation in the Himalayas:
Challenges and Opportunities

Lalat Indu Misra
Faculty of Hospitality and Tourism Management, Siksha 'O' Anusandhan, India

Ansuman Samal
Faculty of Hospitality and Tourism Management, Siksha 'O' Anusandhan, India

Susanta Ranjan Chaini
Faculty of Hospitality and Tourism Management, Siksha 'O' Anusandhan, India

Santosh Bisoi
Faculty of Hospitality and Tourism Management, Siksha 'O' Anusandhan, India

Dankan Gowda V.
https://orcid.org/0000-0003-0724-0333
BMS Institute of Technology and Management, India

ABSTRACT

In recent years, the Himalayas have seen an increase in tourism, which has benefited both the local economy and the region's indigenous population. Rapid tourist expansion has prompted calls for appropriate conservation measures to mitigate its effects. This research examines the pros and cons of protecting the Himalayan ecology while simultaneously growing tourism there. The course starts out with a quick survey of the Himalayas' many historical and cultural treasures. Concerns attributed to tourism's fast rise include increased resource consumption, environmental damage, waste management challenges, and the loss of indigenous knowledge and practices. The research highlights a number of socioeconomic issues encountered by local communities, including changes in individual lifestyle and the loss of a shared cultural identity. Opportunities for the preservation of the Himalayan ecology brought about by increased tourism are also highlighted in the chapter. Opportunities include the promotion of sustainable tourism practices and community-based initiatives, as well as nature-based tourist activities that raise awareness and generate support for conservation efforts. The research highlights the significance of collaborative problem-solving, community involvement, and successful policy frameworks.

DOI: 10.4018/979-8-3693-0823-3.ch017

1. INTRODUCTION

The Himalayas, which are frequently referred to as the "abode of snow," span about 1,500 miles from west to east through the countries of India, Nepal, Bhutan, China, and Pakistan. This mountain chain is well-known not just for its stunning scenery, but also for the rich variety of habitats it contains, from tropical rain forests to high alpine meadows. Tourism in the Himalayas has a long history, dating back to ancient pilgrimage routes and more recently evolving to include adventure tourism, such as trekking, mountaineering, and river rafting, among others. With more than 40,000 tourists visiting regions like Ladakh in India and an increasing number of trekkers heading to Nepal each year, the tourism industry has become a significant economic force in the region.

Tourism stands as a cornerstone in the global economy, exhibiting robust growth and significant contributions across various sectors. According to a 2016 report by the World Travel & Tourism Council (WTTC), the industry injected an impressive $7.2 trillion into the world's GDP in 2015, accounting for nearly 10% of global economic output. The employment landscape also reaped substantial benefits, with the sector responsible for creating 284 million jobs worldwide, 108 million of which were direct employment opportunities.

The United Nations World Tourism Organization's 2015 data underlines the industry's progressive expansion, highlighting a 4.4% growth rate in international tourist arrivals, which totalled 1.184 billion. Geographically, Asia and North America led the growth charts, with Europe, Africa, and the Middle East trailing behind. More granular data reveals that in terms of sub-regions, Southeast Asia and South Asia were at the forefront, posting tourism GDP growth rates of 7.9% and 7.4% respectively in 2015. Other notable contributors included the Caribbean at 5.1%, Sub-Saharan Africa at 3.3%, and North America at 3.1%, while South America and North Africa lagged behind with growth rates of 1.5% and 1.4% respectively.

Furthermore, the tourism industry plays an indispensable role in international trade, constituting 5.4% of global exports and generating a revenue of $1.3 trillion. The sector is among the top five export earners in over 150 countries and stands as the primary export for 60 nations. For developing countries and Least Developed Countries (LDCs), tourism is particularly vital. It serves as the principal source of foreign exchange for one-third of developing countries and for half of the LDCs, where it can contribute up to 40% of their GDP, as cited by the United Nations Conference on Trade and Development (UNCTAD).

When discussing tourism, the phrases "sustainable," "eco," and "natural" are frequently used to characterize policies or activities that try to lessen the impact of tourism on natural resources, local

Figure 1. Tourism intersections and synergies

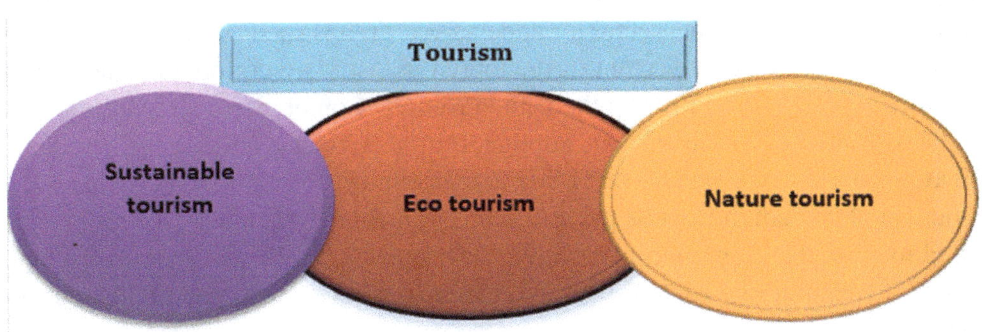

economies, and cultural traditions. The meanings of these terms vary, yet they all have some things in common. Figure.1. presents the tourism Intersections and Synergies. Ecological, social, and economic considerations are all integral to the notion of sustainable tourism. It aims to provide economic advantages to local populations without negatively impacting the culture or nature of the area. UNWTO defines sustainable tourism as tourism that "takes full account of its current and future economic, social, and environmental impacts, addressing the needs of visitors, the industry, the environment, and host communities."

Key Objectives: Sustainable tourism is about the long haul, using resources judiciously, distributing wealth fairly, and treating visitors with respect.

Examples: Hotels that use sustainable energy, hire locals, handle cultural artifacts with respect, and provide educational events for visitors all get high marks.

Eco-tourism is a kind of sustainable travel that aims to protect natural environments. Its goals include enlightening tourists, bolstering local economies, preserving biodiversity, and promoting mutual respect among the world's peoples and their natural surroundings.

Key Objectives: Benefits to local communities, conservation of natural areas, and the dissemination of information about ecosystems in the area.

Examples: Activities such as wildlife viewing, guided treks, bird watching, and conservation missions.

The primary focus of natural tourism is on recreation and aesthetic appreciation of natural environments, rather than on conservation or environmental education. Even if it is done in a manner that tries to lessen its effects, it doesn't always mean it contributes to larger sustainability objectives or serves as a teaching moment.

Key Objectives: Activities like hiking, snorkeling, and camping are examples of ways in which natural environments may be discovered and appreciated.

Examples: Outdoor pursuits such as kayaking, scuba diving, and exploring national parks.

Intersections and Synergies: Note that the categories are not exclusive of one another. If eco-tourism helps local economies and safeguards historic sites, it might be considered a type of sustainable travel. Sustainable practices and an awareness of the environment are also compatible with natural tourism. By segmenting the tourist industry in this way, stakeholders would be better able to tailor laws and procedures to the specific demands and objectives of their respective fields. Community-based strategies in sustainable tourism, for instance, may assist in more fairly distributing economic rewards, while a certification procedure for eco-tourist providers can assure adherence to environmental standards. Finally, dividing tourism into "sustainable," "eco," and "natural" categories is crucial for gaining a deeper knowledge of, and better control over, the business as a whole, and better preparing it to respond to changing social and environmental conditions.

As a highly labor-intensive industry, tourism provides a wealth of opportunities for people of many skill sets. As a result, it is an essential sector for underserved groups such as women, young people, and those with lower incomes. Seventy percent of the tourist industry's workforce is made up of women, according to a study published in 2010 by the United Nations Conference on Trade and Development (UNCTAD). On the foreign exchange front, Foreign Tourist Arrivals (FTAs) reached 8.02 million in 2015, contributing an estimated US$ 21.06 billion to the country's foreign exchange reserves.

Notably, outbound tourism from India has also been burgeoning. It stood at 18.3 million in 2014 and was anticipated to cross the 20 million mark in the subsequent year. Domestic tourism, too, is on the upswing, with 1,281.95 million visits to states and Union Territories recorded in 2014, as reported by the Ministry of Tourism.

Such growth positions tourism as a cornerstone of India's tertiary sector and a significant contributor to its robust economy. WTTC projections even predict that by 2026, India will ascend from its 2015 rank of 12th to a noteworthy 7th position globally in terms of contribution to GDP. Various factors fuel this exponential growth, including the rising middle class, increasing foreign visits, and a surge in domestic tourism. This has likely been catalyzed by greater travel awareness, the desire for respite from routine, personalized travel experiences, and advancements in transportation and communication infrastructure.

The World Tourism Organization further corroborates the sector's optimistic outlook, forecasting that by 2030, India will emerge as the tourism leader in South Asia, with approximately 8.9 million arrivals.

Tourism is not merely an industry focused on leisure and recreation; it also serves as a multifaceted catalyst for economic, socio-cultural, and ecological development. Recognizing these comprehensive benefits, planners, developers, and policymakers are increasingly integrating tourism into strategies for rural upliftment. Given that approximately 75% of the world's impoverished population resides in rural areas, the transformative potential of tourism as a means of poverty alleviation is substantial.

India serves as a case in point. While agriculture is often considered the backbone of rural India, tourism presents an equally compelling opportunity for economic diversification. In terms of employment generation, tourism outperforms traditional sectors. For every one million invested in the tourism industry in India, 47.5 direct jobs and an additional 85-90 indirect jobs are created. This stands in contrast to agriculture, which generates 44.6 direct jobs, and manufacturing, which accounts for 12.6 direct jobs for the same investment.

To create a tourist industry that is environmentally responsible and socially responsible is a challenging enterprise that calls for multi-pronged strategies to assure its viability over the long run. This requires striking a balance between the competing needs of the economy, society, and the environment, as well as being flexible in the face of shifting conditions and advances in technology (Figure.2). The development of sustainable tourism is the shared duty of a diverse group of stakeholders, including local communities, governments, non-governmental organizations, and the commercial sector. This responsibility must be shared in order to be successful. It is essential that they meaningfully participate in the decision-making processes in order to guarantee that the interests of all parties are taken into consideration. It is impera-

Figure 2. Principles of development of sustainable tourism

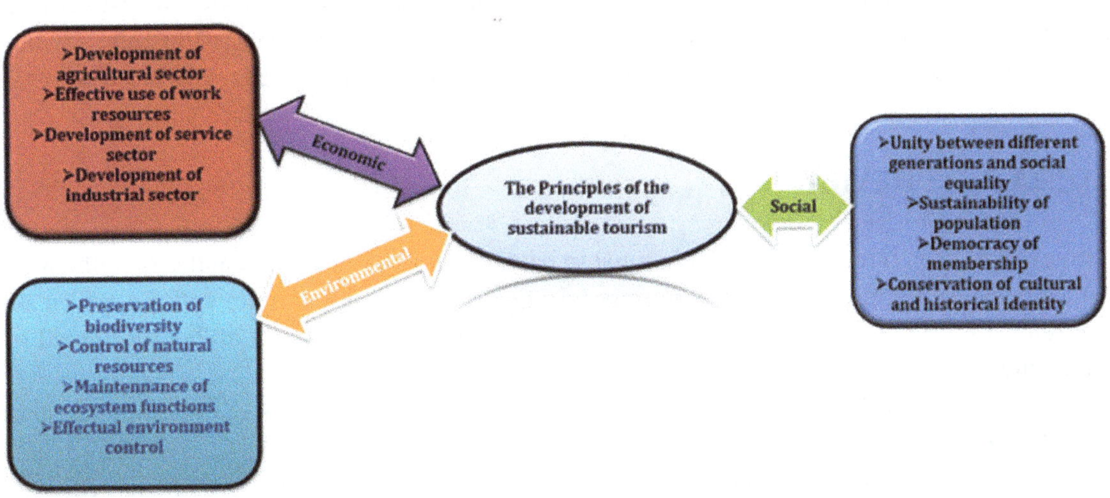

tive that the communities in the area benefit economically from tourist activities in the form of job development, possibilities for business, and revenue-sharing arrangements. For the sake of alleviating poverty and strengthening social cohesion, the benefits of economic growth should be allocated fairly. Cultures and traditions must be respected in order for tourist growth to be environmentally sustainable. This involves educating tourists about local customs and heritage and ensuring that tourism activities do not dilute or commoditize unique cultural attributes. Ecological integrity is crucial for the long-term viability of tourism destinations. The tourist development plan should include conservation efforts, resource efficiency, waste management, and the appropriate use of natural assets. A long-term perspective is necessary for sustainable tourism, one that accounts for the carrying capacity of the ecosystem and the need of constant evaluation and adaptation. Transparency in planning, decision-making, and monitoring is essential, thus strong governance structures should be in place. Achieving accountability may be facilitated by using standards, certification programs, and performance indicators. The tourism business and its stakeholders need to be taught the value of eco-friendly methods. Responsible tourism may be promoted via advertising campaigns, training courses, and educational campaigns. The goal of sustainable tourism is to lessen the industry's ecological, financial, and societal footprint. This includes preventing the exploitation of local people and reducing the amount of visitors in vulnerable places, as well as reducing the environmental impact of hotels. Instead than trying to increase the total number of visitors, the emphasis should be on luring those who are serious about ecotourism and willing to respect local traditions and customs.

The specific aims of this chapter are as follows: To provide an in-depth analysis of the historical and current state of tourism in the Himalayas. To evaluate the economic, social, and cultural impacts of tourism on the region. To explore the challenges posed by tourism to environmental conservation efforts. To identify opportunities for sustainable tourism practices that can mutually benefit both the tourism industry and conservation efforts. To provide examples of both effective and ineffective strategies for reconciling tourism with environmental protection.

In order to provide readers a complete picture of the topic at hand, this chapter is broken down into many major sections: An overview of the context, justification, goals, and organizational framework. A review of the relevant literature provides a synopsis of previous studies. The Economic, Social, and Historical Consequences of Himalayan Tourism Development. The importance of environmental conservation in the Himalayas, ongoing initiatives, and obstacles are discussed. Conflicting policies, limited resources, and pressing socioeconomic problems all provide formidable obstacles. Possibilities: Investigating paths for responsible travel and civic engagement. Case studies are actual events used to illustrate a point. The last section of a research paper should summarize the results and provide suggestions for further study.

2. LITERATURE REVIEW

Academics have been interested in the relationship between mountain tourism and environmental protection for decades. Following is a brief summary of the literature on the topic, with an eye on studies that are relevant to the Himalayas. The economic, social, and ecological will all be considered in this literature study.

Economic Impacts of Tourism: The importance of tourism to the economies of mountainous areas has been emphasized by a number of studies. (Abdollahzadeh, G., & Sharifzadeh, A.2014) highlighted the importance of tourism as a driver of economic growth and employment creation in underserved mountain areas. More recent research by (Bagri, S. C., & Kala, D. 2016) quantified the financial benefits accruing to local communities in the form of revenue from lodging, food services, and local guides. However, these studies also note that excessive focus on economic gains can overshadow environmental considerations.

Social and Cultural Impacts: Research focusing on the social and cultural impacts of tourism is more heterogeneous. Some scholars argue that tourism can help preserve cultural heritage by adding economic value to traditional practices (Bahrami, R., & Noori, K. 2013). In contrast, (Blesic, I., Pivac, T., S. 2014) warned that the influx of external influences could disrupt traditional ways of life and exacerbate social inequalities. When planning a trip to the culturally varied Himalayas, these factors become extremely important.

Environmental Concerns: Several studies have examined how mountain tourism affects the local ecosystem. The ecological footprint of tourism in China's mountainous areas was studied by (Brida, J. G., Disegna, M. 2014), and they found that it may cause environmental deterioration if it were left unchecked. Meanwhile, (Chand, R., & Leimgruber, W 2016).

investigated the potential of local conservation efforts to provide a long-term framework for mountain tourism. Because of the Himalayas' ecological vulnerability, these issues are especially pertinent.

Sustainable Tourism Models: There has been a lot of study done on the topic of sustainable tourism in the setting of hilly areas. Ecotourism, cultural tourism, and adventure tourism are only a few examples of sustainable travel that were discussed in detail by (Das, M. B., Kappor-Mehta, S, 2015). Tourism in fragile ecosystems like the Himalayas, they concluded, can only be sustained in the long run via an integrated strategy that prioritizes both economic growth and environmental protection.

Comparative Studies: The Himalayas have been compared to the Alps, the Rockies, and other mountain ranges in a small number of studies. Waste management, resource depletion, and community relocation are just a few of the issues that have been brought into the spotlight by comparative studies, which have also highlighted the specific geographical and cultural variables that need regional solutions.

Policy and Governance: The role of government policies and governance in balancing tourism and environmental conservation has been the focus of research by scholars like (Golzardi, F., Sarvaramini, S. 2012). They stressed that effective governance models should include multi-stakeholder involvement, from local communities to international organizations, to ensure that tourism development is both economically beneficial and ecologically sustainable.

Tourism is undoubtedly one of the most rapidly expanding industries globally, serving as a veritable boon not only for tourists but also for local communities. It brings a plethora of advantages including, but not limited to, employment generation, infrastructure enhancement, an increase in per capita income, foreign exchange earnings, a rise in the standard of living, international camaraderie, heritage preservation, and nature conservation. These diverse benefits have led multiple governments—both in developing and developed nations to embrace tourism as a tool for comprehensive economic, socio-cultural, and ecological development.

However, it is imperative to acknowledge that tourism is not an unmitigated good. Alongside its manifold benefits, the industry also has the potential to produce a variety of negative consequences, such as seasonal employment, subpar wages, environmental degradation, social discord, and moral quandaries like prostitution and gambling. Moreover, it can lead to the erosion of cultural values, vandalism, xenophobia, loss of biodiversity, and land expropriation.

In view of these challenges, there is an increasing focus on principles like sustainable development and community-based approaches to tourism. Alternative tourism models are being explored to ensure that tourism evolves in a manner that aligns with the long-term well-being of all stakeholders involved tourists, local communities, and the environment. The Brundtland Commission's 1987 definition of Sustainable Tourism Development serves as a useful guidepost in this context: "Development that meets the needs of the present without compromising the ability of future generations to meet their own needs."

Therefore, tourism should be seen not merely as an industry but as an instrument of development and change, warranting thoughtful planning and ethical considerations. If tourism initiatives are developed within the carrying capacity of the environment, the resource base, and the local community, the industry can continue to flourish as a renewable resource, offering sustained benefits across various dimensions of human well-being and ecological balance.

Figure.3. elucidates the complex interplay of factors involved in the development of tourism, reinforcing the need for a holistic approach to its management and growth.

Alternative tourism operates under the overarching paradigm of sustainable development, emphasizing the protection and conservation of both the environment and local resources of tourist destinations. According to Moscardo, as cited in Christou's work, alternative tourism aims for a multifaceted approach that not only safeguards the long-term attractiveness of a tourist locale but also mitigates adverse impacts

Figure 3. Scheme of tourism development

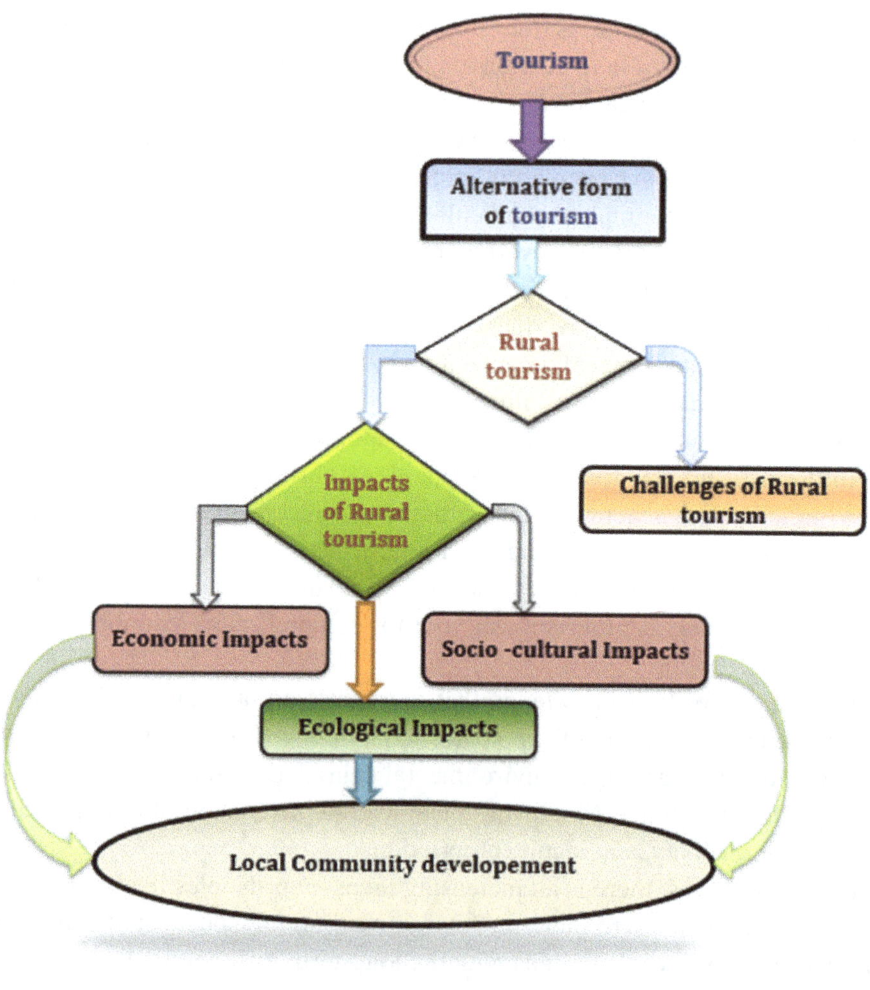

on the destination, its environment, and its population, all while providing socio-economic benefits to the community.

Butler further delineates specific attributes that characterize alternative tourism. First and foremost, it is small-scale in nature, ensuring that the environmental footprint is minimized. Additionally, it is predominantly owned and operated by local communities, thereby ensuring that the financial benefits are retained within the locality. Alternative tourism often involves travel to remote natural areas, places where the emphasis is on preserving the natural environment rather than exploiting it. Another key tenet is its focus on the sustenance and well-being of local communities, affirming that tourism should not be a zero-sum game where gains in one area come at the cost of losses in another. Above all, sustainable development serves as the guiding principle, ensuring that tourism practices are both ecologically viable and socially equitable.

This elucidation serves to comprehensively outline the concept of alternative tourism, situating it within a framework that prioritizes long-term sustainability, local ownership, and equitable socio-economic benefits. By adhering to these principles, alternative tourism offers a viable path toward realizing tourism's full potential as a catalyst for positive change, without compromising the needs of future generations.

Rural tourism, a subset of alternative tourism, adheres to the principle of sustainable development and has evolved as an initiative to bolster rural economies that are often remote, inaccessible, non-industrialized, and primarily agrarian. The motivations for fostering rural tourism vary between developed and developing countries, reflecting their disparate socio-economic contexts. In developed nations, the emphasis often lies in offering travelers an authentic "country experience" and directing tourist flow towards rural settings, thereby diversifying the types of experiences available to tourists and alleviating the pressure on more traditional, urban tourist destinations.

Conversely, in developing and least developed countries (LDCs), rural tourism is seen as a mechanism to uplift communities plagued by poor socio-economic conditions. The focus here is on leveraging the innate but underutilized potential of these regions to function as viable tourism destinations. The ultimate objective is to stimulate rural economic growth, create employment opportunities, and enhance income levels through the vehicle of tourism.

Agriculture, traditionally the mainstay of rural livelihoods, has increasingly shown signs of stagnation and decline worldwide. Given this context, the urgency to identify alternative or supplementary income-generating activities has escalated. Tourism, in this regard, serves as a promising avenue for diversification. Unlike more traditional sectors like agriculture, tourism is a multi-faceted industry that can provide a broad range of vocational opportunities. These can include everything from hospitality and tour guiding to handicrafts and local food production, each contributing to both the social and economic upliftment of rural communities.

Thus, rural tourism is not merely a novel concept but a strategic approach aligned with broader development goals. It encompasses a myriad of initiatives that are both ecologically sustainable and socio-economically beneficial. It offers a lifeline to rural areas that are struggling with the challenges posed by a globalizing world, including the deterioration of traditional forms of livelihood. By placing the well-being of local communities and environmental conservation at its core, rural tourism embodies a sustainable and ethical model for development, offering a mutually beneficial relationship between tourists and host communities.

The conceptualization of rural tourism as a significant area of academic and practical interest traces its roots back to the 1970s and 1980s. This period marked the beginnings of substantial growth in tourism activity in the rural sectors of Western Europe, a trend that subsequently expanded to include the

Americas, Australia, Japan, and eventually Eastern Europe. However, it would be remiss to categorize rural tourism as a strictly modern phenomenon; the aesthetic and intrinsic appeal of rural landscapes has long been celebrated in various cultural mediums including painting and literature. Such artistic reflections contributed to a romanticized perception of the countryside, imbuing it with a sense of allure that predates contemporary tourism initiatives.

The attractions of rural areas, such as picturesque landscapes, diverse flora and fauna, water bodies like rivers and lakes, mountainous terrains, and breathtaking natural phenomena like sunsets, are timeless features that have captivated human imagination for centuries. These features form the foundational elements around which rural tourism has been developed.

The motivations and perspectives surrounding rural tourism are influenced by a host of variables and can significantly differ from one country to another. In developed nations, rural tourism often serves as an alternative to traditional tourism destinations, offering the promise of tranquillity, a return to simplicity, and a respite from urbanized lifestyles. In developing nations and LDCs, the motivations might be more economically driven, leveraging tourism as a strategic tool for rural development, economic diversification, and poverty alleviation.

This multiplicity of perspectives enriches the concept of rural tourism, allowing it to be tailored to meet the specific needs and potentials of different geographic and cultural contexts. Consequently, rural tourism can be viewed not only as a mere industry but also as an adaptable framework for sustainable development. Whether serving as an instrument for economic revitalization or as a conduit for cultural preservation and ecological stewardship, rural tourism stands as a versatile and important phenomenon with manifold implications for societies globally.

Tourism in rural areas presents a multi-faceted opportunity for sustainable development, offering a spectrum of benefits that can profoundly impact local communities. Entrepreneurial prospects emerge as local artisans, farmers, and service providers find a ready market for their goods and services. Infrastructure development often follows, catering to the needs of tourists but also significantly improving local amenities. A surge in employment opportunities becomes likely, often in roles requiring a diverse range of skill sets—from hospitality management to tour guiding to craft making.

Particularly noteworthy is the opportunity for social empowerment. Women, who constitute a substantial part of the workforce in many rural settings, find avenues for economic independence, thereby contributing to broader gender equality objectives. Youth employment also sees an upswing, reducing the allure of urban migration for better prospects. Environmental preservation and heritage conservation become central tenets of rural tourism, as these are often the very attributes that attract tourists in the first place. The integrated approach often allows for the inclusion of marginalized communities, thereby contributing to a more equitable socio-economic landscape.

From the perspective of the tourist, rural tourism offers an array of unique experiences. Whether it is engaging in agritourism activities at working farms, enjoying thematic rural experiences in the form of "Touri-agriculture," or simply soaking in the tranquillity of the countryside, the options are diverse and cater to a range of preferences. Activities can vary from adventure sports to immersive cultural experiences, each fulfilling the growing demand for more experiential forms of tourism.

It is important to recognize, however, that many rural areas are culturally and environmentally fragile. Therefore, careful tourist planning and administration is essential. The help of the host country in this case is crucial. The local customs, dynamics, and ecological constraints are all well understood by the populace. Involving locals in tourism planning not only improves the quality of the visitor's experience but also makes for more sustainable and culturally appropriate development.

Thus, rural tourism should be seen as more than just a money-making opportunity; it's also a multifaceted kind of environmental, social, and economic responsibility. It calls for a holistic and integrated strategy that takes into account both the financial needs of the tourist industry and the moral commitments to environmental protection and social cohesion. Achieving this equilibrium is crucial for rural tourism's potential to spur sustainable economic growth.

In essence, the current body of literature shows a rich tapestry of economic potential and problems connected to the growth of mountain tourism. It also underscores the necessity for environmentally responsible tourism practices, especially in fragile ecosystems like the Himalayas. By exploring these various facets in depth, this chapter aims to build upon existing knowledge and offer new insights into the sustainable development of tourism in the Himalayas. The concept of urgency (Figure.4.) in sustainable tourism development refers to the critical need to implement sustainable practices as rapidly as possible, given the accelerating impacts of tourism on environments, cultures, and local economies. As tourism continues to grow globally, its adverse effects—such as carbon emissions, resource depletion, cultural erosion, and social inequalities—become more pronounced, necessitating immediate action. A model of urgency in this context would incorporate rapid assessment, strategic planning, and immediate action on multiple fronts. The urgency model underscores the need for swift, collaborative, and adaptive action to mitigate the adverse effects of tourism. It is an iterative model, continuously adapting to new data and outcomes, as the situation demands quick yet informed decisions for effective sustainable tourism development.

The scholarly examination of rural tourism reveals its growing importance across nations at various stages of development—from advanced economies to emerging and underdeveloped countries. However, the characteristics, forms, and driving factors behind rural tourism differ significantly depending on the geographical context. The very concepts of what constitutes 'rural tourism' and 'rurality' can vary from one location to another. In the specific case of Himachal Pradesh, there exists a substantial body of research on tourism in general, but studies specifically focusing on the specialized area of rural tourism are notably scarce.

Figure 4. Model urgency sustainable tourism development

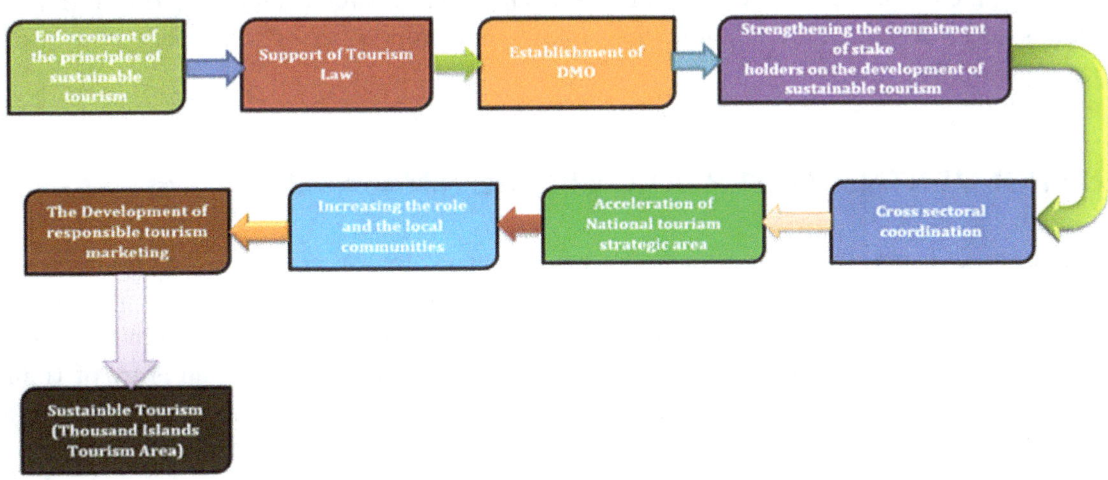

The prevailing research landscape on rural tourism is markedly skewed toward the experiences of tourists, often overlooking the perspectives of local communities. While some studies have delved into individual cases within rural tourism, a holistic exploration of the state of Himachal Pradesh remains conspicuously absent. Instead of probing the consequences of tourism on local communities, many inquiries primarily aim to enhance tourist attraction and diversify tourism offerings. Furthermore, neither state nor federal tourism departments have maintained comprehensive statistics on rural tourism or officially sanctioned rural sites, hindering empirical research. Even though economic advantages are a frequent topic in the discourse on rural tourism, there's an apparent void when it comes to examining the quality of these economic benefits. Equally concerning is the minimal attention given to the socio-cultural and environmental implications of rural tourism in existing studies. Moreover, the dearth of research focusing on local perceptions across economic, socio-cultural, and ecological dimensions in Himachal Pradesh impedes a well-rounded understanding of rural residents' views, thereby affecting the overarching objectives of rural tourism initiatives.

There exists an imperative need for additional scholarly work that focuses on the viewpoints and experiences of local residents to comprehensively evaluate how integrated rural tourism can be administered within a sustainable paradigm. The impact of tourism at the community level must be scrutinized not solely through the lens of visitor numbers or tourism-related businesses, often referred to as demand and supply variables. Rather, it should also take into account the assessments of the local rural populace concerning the advantages and disadvantages that tourism brings to their lives. This balanced approach will enable a more nuanced understanding that encompasses economic, social, and environmental dimensions, thereby facilitating the attainment of long-term sustainability goals in rural tourism.

The majority of scholarly investigations into rural tourism predominantly employ economic metrics as the principal yardstick for evaluating success, thereby sidelining the equally significant criteria of social equity and the growth of social, cultural, and ecological capital. Moreover, the regular oversight of tracking the influence of tourism activities on local livelihoods and the sustainable utilization of ecological resources constitutes a notable lacuna in existing research. Investigations that capture community perceptions can offer invaluable insights to planners, policy makers, and developers for crafting more sustainable strategies that address the interests of tourists without overlooking community-specific concerns. The complex nature of rural tourism presents unique challenges that necessitate nuanced approaches. The local community plays an indispensable role in this context, and research aimed at understanding their attitudes towards tourism is crucial. Such research should aim to instil a positive outlook among locals regarding rural tourism, rather than merely adhering to conventional, business-centric or imperialistic models.

3. ENVIRONMENTAL CONSERVATION IN THE HIMALAYAS

Importance of Conservation: The Himalayan region is an ecological powerhouse of immense global significance. This mountainous area hosts multiple biodiversity hotspots and serves as a source of numerous major river systems that provide water to nearly a quarter of the world's population. The region's diverse ecosystems range from subtropical jungles to alpine meadows, hosting an array of flora and fauna, some of which are critically endangered. Moreover, the Himalayas act as a 'climate regulator,' influencing monsoon patterns and serving as a massive carbon sink. Given this ecological richness and its global environmental implications, conservation in the Himalayas is not merely a regional concern but a matter of international priority.

Current Efforts: Several conservation initiatives have been instituted in the Himalayas, both at the governmental and non-governmental levels. Protected areas, such as national parks and wildlife reserves, have been established in countries like India, Nepal, and Bhutan. For instance, the Great Himalayan National Park in India and Sagarmatha National Park in Nepal are UNESCO World Heritage Sites committed to biodiversity conservation. Local non-profits and international organizations like the World Wildlife Fund (WWF) also contribute through community outreach, wildlife conservation projects, and environmental education.

Policy initiatives aimed at environmental conservation have seen varying degrees of success. Legislation regarding waste management, restrictions on trekking routes, and quotas on tourist numbers have been implemented to minimize environmental impact (Iorio, M., & Corsale, A. 2010). However, the effectiveness of these policies is often compromised by lax enforcement, insufficient funding, and sometimes, local resistance due to economic dependencies on tourism.

Impacts of Tourism on Environment: Tourism development has led to both direct and indirect environmental degradation in the Himalayas. Direct impacts include deforestation for infrastructure development, waste accumulation in trekking and camping sites, and pollution of local water sources. In areas like Everest Base Camp and popular trails in Himachal Pradesh, waste mismanagement has become an urgent issue, with litter and non-biodegradable materials posing long-term environmental risks.

Indirect impacts are equally troubling and include increased carbon emissions from transport, the depletion of natural resources to cater to tourists, and disturbances to local wildlife. Additionally, the influx of tourists puts pressure on local resources like water and firewood, which exacerbates existing challenges related to resource scarcity.

Himachal Pradesh is globally celebrated for its breathtaking landscapes and serves as a year-round haven for tourists. Beyond its natural allure, the state is equally revered for its rich tapestry of culture, encompassing customs, festivals, and traditional crafts. This cultural richness is most vividly manifested in its rural locales, be it Naggar with its arts and crafts, Garli-Paragpur with its heritage, or Baroh exemplifying Gurukul culture (Jalilian, M. A., Danehkar, A., 2012). Despite this, numerous rural areas in the state remain untapped in terms of their tourism potential. To address this, the Himachal Pradesh government introduced the Home Stay Scheme in 2008, aiming to offer tourists an authentic rural experience. However, the more holistic approach to rural tourism was actualized with the "HarGaon Ki Kahani" initiative in 2010. This ambitious project identified one village in each district to promote based on its unique stories, folk traditions, and historical significance, providing essential amenities to improve tourist experience. A considerable budget of Rs 20,243.92 crore has been allocated for the initiative, focusing on specific villages like Shamsher in Kullu, Salasi in Bilaspur, and Bela in Hamirpur, among others, to develop and promote them as tourist destinations.

Approximately 89.97% of Himachal Pradesh's population resides in rural areas, a statistic that clearly outlines the state's predominantly agrarian character. The economic backbone of the state is largely supported by sectors like agriculture, horticulture, forestry, and hydropower generation. Furthermore, Himachal Pradesh is a significant market for arts and handicrafts, an industry that could play a role in the diversification of the rural economy. Despite these assets, a multitude of rural locations remain largely unexplored in terms of their tourism potential (James, G. 2015). This indicates a vast scope for rural tourism to be a catalyst for sustainable development, as well as a mechanism to augment economic benefits for rural communities. Currently ranked 13th in the nation for tourist visits, the state has an excellent opportunity to improve this standing by effectively leveraging its resources in rural tourism.

Tourism in Himachal Pradesh accounts for an estimable 7.2% of the state's Gross Domestic Product, showcasing the sector's considerable economic influence. Given the state's abundant natural and cultural assets, there exists immense potential for further expansion in the tourism industry. For the fiscal year 2016-17, the state budget allocated a substantial INR 5,273.01 lakh specifically for tourism development. This reflects the government's prioritization of the sector, with an emphasis on bolstering infrastructural assets to facilitate tourism. Such enhancements include upgrades in public utilities, transportation networks, communication systems, and aviation facilities (Katoch, A. 2014). The Economic Survey of Himachal Pradesh for 2015-16 reports the presence of 2,416 registered hotels with a total bed capacity of 67,097, along with 662 home stays providing 1,838 rooms. Additionally, the Asian Development Bank has sanctioned funding for 20 projects amounting to US$ 95.16 million. Among these are projects aimed at fostering tourism skills in rural clusters like Dhameta, Kangra-Pragpur, Naina-Devi, and Shimla. As a result of these concerted efforts, the state has witnessed a notable uptick in tourist inflow, as depicted in Figure.5.

Rural tourism in Himachal Pradesh is exhibiting promising growth and holds significant untapped potential, particularly given the state's rich cultural fabric, which encompasses traditions, culinary arts, festivals, folklore, handicrafts, and heritage sites. Additionally, the state's natural beauty offers yet another layer of allure. In an era marked by globalization, urbanization, and economic restructuring, countries and regions are grappling with the challenge of reinventing their economies (Katoch, A., & Gangotia, A. 2014). Within this context, rural tourism emerges as a viable option to address multiple dimensions of rural development—economic, sociocultural, and ecological. However, it is critical to be mindful of the potential adverse effects of tourism, necessitating comprehensive research to facilitate an informed and balanced understanding of the sector.

Sustainable rural tourism aims to achieve a synergistic relationship with local communities, focusing on cultural preservation while minimizing negative impacts across economic, sociocultural, and ecological dimensions. The framework for such tourism encourages local participation in planning and decision-

Figure 5. Tourist arrival (in Lakhs) in Himachal Pradesh

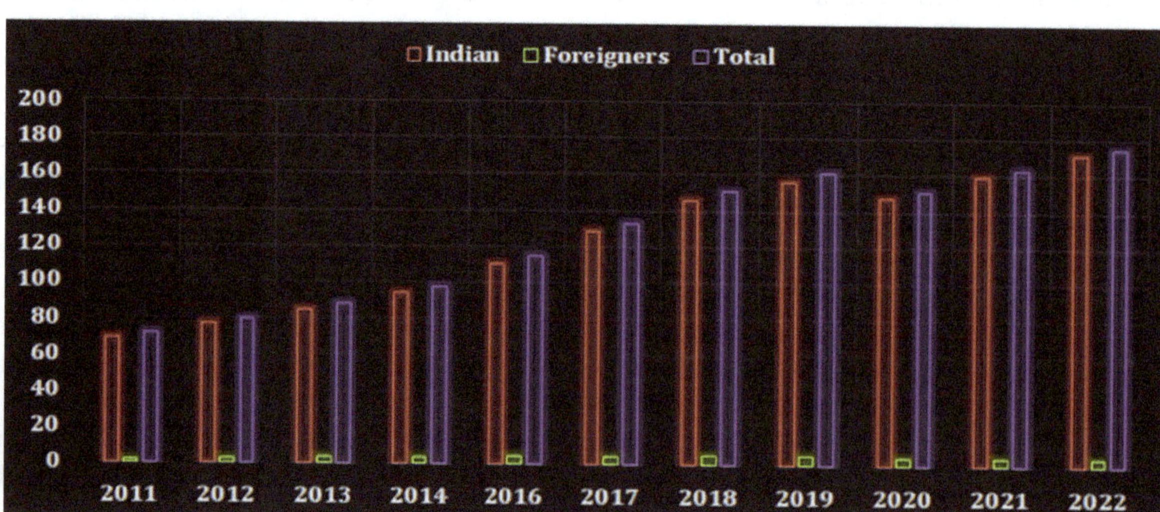

making processes; adopts a discerning marketing strategy; fosters local entrepreneurship; maintains high-quality tourist experiences; supports rural economies and agricultural sectors; and lays emphasis on achieving long-term developmental goals. When orchestrated with care and foresight, tourism has the capability to revitalize declining rural regions by creating job opportunities, stimulating income, and facilitating capital inflows, thereby contributing materially to rural development.

Rural tourism, despite its potential benefits, faces a complex array of challenges that cannot be overlooked. Among these are issues related to the quality of tourism offerings, the adequacy and robustness of local infrastructure, and ease of access to rural destinations. Furthermore, the presence of skilled labor capable of maintaining and elevating the tourism experience is a significant concern (Khakpour, A. 2013). Equally important is the engagement and support of local communities, as well as the interest of investors who are essential for catalysing development. The challenges are exacerbated in developing and underdeveloped nations, where political and institutional barriers add another layer of complexity to the already multifaceted problem set facing rural tourism.

The genesis of tourism in Himachal Pradesh can be traced back to 1955, when a tourism information center was formally relocated to the state by the Government of India. Subsequent to the territorial reorganization of Punjab in 1966, additional areas were annexed to Himachal Pradesh, prompting the separation of the Department of Tourism from Public Relations that same year (Hernandez, J. M., & Casimiro, L. A. 2011). This organizational shift marked the beginning of a remarkable upswing in tourism activity, as evidenced by the subsequent rise in tourist inflows. To institutionalize and accelerate this progress, the Himachal Pradesh Tourism Development Corporation was inaugurated in 1972. Cementing tourism's stature within the state's economic portfolio, it was officially recognized as an industry on August 1, 1993. Over the years, targeted tourism policies have been enacted to drive sectoral growth, focusing on key industry aspects and providing fiscal allocations for diverse tourism-related activities. In a further move to stimulate investment and entrepreneurship in the sector, tourism offices have been strategically dispersed across the state.

The strategic roadmap for tourism in Himachal Pradesh was initially laid out in the 2000 tourism policy. Aimed at stimulating employment, the policy set a target of creating 47 jobs for every investment of 10 lakh INR. The policy's mission statement was multi-faceted, advocating for sustainable tourism development, promoting responsible tourism practices, and prioritizing job opportunities in rural and tribal areas (Li, P., Ryan, C., & Cave, J. 2016). It also aimed to extend the average duration of tourist stays through activity-based tourism and to elevate adventure tourism to global standards. Among its other focal points were the mitigation of seasonal variations in tourism traffic, the development of tourism in rural and tribal regions, and the enhancement of the five 'A's (Accessibility, Accommodation, Attractions, Amenities, and Activities) for pilgrimage sites. In pursuit of these objectives, the policy outlined the need for robust infrastructure—ranging from basic to specialized tourism infrastructure, entertainment facilities, and lodging options. To operationalize this vision, tourism clusters were identified, and a tourism development board, chaired by the Chief Minister, was constituted to formulate policy guidelines (Himachal Pradesh Development Report, 2002).

Subsequent policies, like the 2005 Tourism Policy, expanded the scope by incorporating more stakeholder participation, including NGOs and private entities, and identifying thrust areas such as rural tourism, ecotourism, and pilgrimage tourism, among others. It also conducted a SWOT (Strengths, Weaknesses, Opportunities, and Threats) analysis of the state's tourism landscape. The most recent iteration in 2013 builds upon these foundations but places a pronounced emphasis on sustainable development (Mukherjee, D. 2012). Its vision is to position Himachal Pradesh as a global sustainable tourism destination by the year 2029. Furthermore, it seeks to leverage sustainable tourism as an economic engine to provide ample employment and entrepreneurial opportunities to residents while preserving the state's natural and cultural heritage, thus ensuring long-term prosperity and an elevated quality of life for local communities.

The consequences of this environmental degradation are multi-faceted. First, it undermines the very ecological richness that attracts tourists, thereby threatening the long-term sustainability of the tourism industry itself. Second, it disrupts the delicate balance of local ecosystems, which can result in a loss of biodiversity, increased vulnerability to natural disasters, and compromised water quality. Finally, these environmental issues risk marginalizing local communities, who rely on the natural environment for their livelihoods, but have limited capacities to adapt to rapid ecological changes.

In conclusion, while the Himalayan region holds significant ecological importance, the current efforts at environmental conservation are faced with challenges stemming from rapidly growing tourism. The environmental impacts of tourism, both direct and indirect, have begun to take a toll on the fragile ecosystems, indicating an urgent need for more effective and sustainable management strategies.

4. CHALLENGES

The burgeoning interest in rural tourism presents a double-edged sword for communities in peripheral areas. On one hand, this form of tourism represents a boon for local economies, offering an alternative revenue stream to traditional agricultural endeavours. On the other, it threatens the ecological and social fabric of these often fragile environments. To dissect these challenges and potentials in a more granular fashion, one can look at various dimensions.

Economically, rural tourism often enhances the financial resilience of peripheral communities. It acts as a counterweight to the income volatility typically associated with agrarian economies. This economic windfall can be especially meaningful for rural youth and women, offering opportunities for employment closer to home. Local artisans and craftsmen benefit as well, finding new markets for their goods among tourists eager for authentic experiences.

However, the allure of quick economic gains must not overshadow the need for responsible tourism development. Infrastructure development, often accelerated to meet the demands of increased tourism, can exert undue pressure on the local environment. Improperly managed construction activities might cause irreparable damage to geological and soil profiles. Such environmental degradation is not just ecologically irresponsible but also undermines the very features that make rural areas attractive to tourists in the first place.

The concept of "carrying capacity" becomes highly relevant here. Carrying capacity refers to the maximum number of individuals that a specific environment can sustain without suffering degradation. If the number of tourists exceeds this capacity, the area is prone to a host of negative externalities ranging from littering and noise pollution to more severe ecological impacts such as the destruction of local flora and fauna. Thus, the management of tourist flows becomes a critical concern. Possible strategies could include the implementation of visitor quotas during peak seasons, or the promotion of off-season travel to balance the yearly load.

Given these complexities, a multi-stakeholder approach becomes essential for sustainable rural tourism development. Local governments, community leaders, and environmental organizations must collaborate to ensure that tourism initiatives align with long-term sustainability goals. Community-based participatory planning can offer valuable insights into the unique needs and limitations of each rural area, thereby ensuring that the benefits of tourism are reaped while minimizing the downsides.

Therefore, the task ahead is not just to promote rural tourism, but to do so in a manner that respects the delicate ecological balance and cultural heritage of rural communities. An approach rooted in sustainability not only serves to preserve the environment but also ensures that the tourism sector remains a viable economic driver for these areas over the long term.

The socio-cultural ramifications of tourism in rural areas indeed represent another facet of the complex interplay between local communities and the tourism industry. When tourists flood rural areas in search of authentic experiences, they often unintentionally catalyze alterations in the local cultural landscape. While this interaction may bring some level of novelty and economic benefit to the community, it frequently leads to adverse effects that deserve nuanced analysis.

One of the key concepts at play here is “demonstration effects,” where the values, lifestyles, and material conditions of tourists become aspirational for local communities. This can result in a myriad of consequences, including but not limited to, the transformation of traditional crafts into commoditized products tailored for tourist consumption. Such commoditization can dilute the rich cultural heritage of these areas, turning age-old traditions into mere commercial transactions.

Similarly, the phenomenon of “staged authenticity” presents another ethical quagmire. Here, cultural practices and traditions are intentionally modified to meet the expectations or preconceptions of tourists. Such alterations may provide short-term economic gains but can lead to long-term erosion of the cultural fabric of the community. The cultural legacy becomes more of a choreographed performance rather than a lived experience, essentially altering its very essence.

Moreover, the relationship between the host community and tourists can go through several phases, ranging from initial euphoria to eventual antagonism or even apathy. This dynamic is a consequence of how tourism is managed and can manifest in different forms including social conflict, and the alteration of social norms and values. These phases are indicative of how locals’ views of tourism change over time, from initial enthusiasm to growing disillusionment with the industry as its cultural and environmental consequences become more obvious.

An method based on shared decision-making and moral accountability is necessary to tackle these societal and cultural difficulties. Communities at the grassroots level should be given agency over the commercialization of their heritage. Local governments, community groups, and individual individuals should all have a voice in the decision-making process. More sustainable and considerate conduct from both visitors and locals might result from educational campaigns that raise awareness of the possible drawbacks of cultural exchanges.

Policy Conflicts: The Himalayas face a delicate balancing act between tourist expansion and environmental protection, with the two often at odds with one another due to competing regulations. On the one hand, tourism is seen as a boon by national and regional governments in terms of tax income and employment opportunities. Infrastructure development, such as the building of hotels, roads, and other tourist services, is therefore often prioritized in public policy. Conservation policies, on the other hand, push for curbs on human activity in environmentally delicate areas. This includes, but is not limited to, tourist limits, waste management regulations, and building moratoriums.

The contradictory scenario caused by these competing policy objectives is that the short-term economic benefits from tourism impair the long-term goals of environmental protection. Furthermore, inadequate coordination among different stakeholders—local communities, government agencies, NGOs, and tourism operators—complicates the effective implementation of balanced policies. Such conflicts raise questions about governance models and the need for integrated, multi-stakeholder approaches to reconcile the two objectives effectively.

Resource Scarcity: The rise in tourism puts a considerable strain on already scarce natural resources in the Himalayan region. Water is perhaps the most pressing concern. Increased tourist activities lead to higher water consumption for a variety of uses, from drinking to sanitation. This increased demand exacerbates existing water scarcity issues, affecting both humans and local ecosystems. Forest resources also experience pressure due to tourism-related construction and the consumption of firewood for heating and cooking in hotels and lodges. Moreover, land use changes driven by tourism development often result in soil erosion, habitat fragmentation, and an increased risk of landslides, further escalating the natural resource depletion crisis. This scarcity not only challenges the sustainability of the tourism industry but also has long-term implications for the ecological balance of the region.

Social and Cultural Erosion: Tourism, particularly when it becomes a dominant economic activity, has the potential to impact the social fabric and cultural traditions of local communities. In the Himalayas, many communities have lived in harmony with their environment for generations, practicing sustainable agriculture and resource management as a part of their cultural identity. The influx of tourists and the commodification of local culture can result in the erosion of these traditional practices.

In conclusion, while the potential economic benefits of rural tourism are considerable, they come with associated risks to both the environmental and socio-cultural integrity of local communities. Balancing these competing demands necessitates a sophisticated, multi-stakeholder approach that is committed to long-term sustainability and ethical responsibility. Only then can rural tourism fulfill its promise as a force for positive economic and social change without sacrificing the unique cultural and environmental assets that make these communities worth visiting in the first place.

Increasing hotel and motel building in rural locations is a common response to the influx of visitors to these places. However, this is made more difficult by the fact that tourism is seasonal. Peak seasons see a large influx of tourists to rural tourist hotspots, while off-seasons see a significant drop in activity. Because of seasonality, hotels may have empty rooms for long stretches of the year, generating little economic benefits.

In addition, many rural inhabitants who profit from tourism are unqualified to successfully manage and maintain their operations due to a lack of professional training and expertise. Due in part to a lack of knowledge and experience, many businesses in rural regions fail. Capacity development and skill training programs are essential components of an integrated strategy for overcoming these obstacles. Such efforts need to be modified to suit the specific requirements of rural areas.

For example, traditional crafts and art forms might be modified to cater to tourist tastes, losing their original significance and complexity. Festivals and religious practices may also become commercialized, turned into spectacles for tourist consumption rather than important cultural and religious events. Additionally, social structures can be altered as younger generations are drawn away from traditional occupations towards service jobs in the tourism sector, sometimes leading to generational tensions.

In summary, the challenges of reconciling tourism development with environmental conservation in the Himalayas are manifold. Conflicting policies, resource scarcity, and the erosion of social and cultural fabrics pose significant obstacles to achieving a harmonious balance. Addressing these challenges requires a multi-faceted approach that incorporates economic, ecological, and social dimensions, recognizing that each element is deeply interconnected with the others.

5. OPPORTUNITIES

Sustainable Tourism: Amidst the challenges of tourism development and environmental conservation in the Himalayas, several models of sustainable tourism present opportunities for harmonious coexistence. One such model is eco-tourism, which promotes responsible travel to natural areas that conserves the environment and sustains the well-being of local people. Another is cultural tourism, aimed at protecting and preserving local traditions while providing an authentic experience to visitors. Adventure tourism, with its focus on outdoor activities, can also be organized in a manner that minimizes environmental impact. These models prioritize local sustainability and low impact on natural resources. For example, the principles of 'Leave No Trace' can be integrated into trekking and mountaineering activities. By adapting these existing models to the unique Himalayan context, stakeholders can create a tourism industry that is economically viable, environmentally responsible, and socially equitable.

Community Involvement: Community-based tourism has shown promise in various global contexts and could be particularly effective in the Himalayan region. In this model, local communities are involved at every stage of tourism development, from planning to implementation. By giving locals a stake in tourism operations—be it through ownership, employment, or decision-making—this approach can mitigate some of the social and cultural erosions discussed earlier.

Moreover, when communities are actively engaged in tourism, they are more likely to implement and adhere to sustainable practices. For instance, locals can serve as guides who educate tourists about responsible behavior in sensitive ecosystems. They can also benefit from supplemental income through the sale of local crafts, food, or traditional medicines, thereby adding economic value to cultural and natural conservation efforts. In essence, community-based tourism can create a symbiotic relationship between preserving the environment and promoting economic growth.

Government and NGO Role: Governmental and non-governmental organizations play a critical role in achieving a balanced approach to tourism and conservation. First and foremost, they can act as regulators, setting guidelines for sustainable practices and ensuring compliance through regular monitoring and penalties for violations. For example, permits for trekking in certain regions could be issued based on environmental impact assessments. Furthermore, government bodies can incentivize sustainable practices through grants, tax breaks, and other financial incentives for tourism operators who adhere to eco-friendly operations. At the same time, they should invest in creating and maintaining necessary infrastructure like waste management systems that are environmentally compliant.

NGOs, on the other hand, can fill gaps in community education, research, and advocacy. They can collaborate with local communities to implement conservation projects and promote sustainable tourism. By acting as mediators between different stakeholders—government agencies, tourism operators, and local communities—NGOs can facilitate dialogues to develop holistic solutions to the complex challenges the Himalayan region faces.

In conclusion, the Himalayas, though fraught with challenges, offer ample opportunities for sustainable tourism development that benefits both the environment and local communities. By adopting models of sustainable tourism, involving communities in tourism operations, and leveraging the regulatory and facilitative roles of governmental and non-governmental organizations, there exists a viable pathway to a harmonious and sustainable future for this majestic region.

6. CASE STUDIES

6.1 Successful Attempts

Bhutan's High-Value, Low-Impact Tourism Model: Bhutan offers an intriguing case study in balancing tourism and environmental conservation successfully. The kingdom's "High-Value, Low-Impact" model prioritizes quality over quantity, drawing a limited number of tourists willing to pay a premium for a sustainable and culturally enriching experience. This approach has several benefits, including lower environmental degradation and a higher level of engagement with local communities. The revenues generated are channeled back into conservation projects, making it a self-sustaining model.

Annapurna Conservation Area, Nepal: The Annapurna Conservation Area in Nepal has effectively utilized community-based tourism. Local people are involved in managing trails, educating tourists, and maintaining the natural environment. The project has not only resulted in better conservation practices but also improved livelihoods for the local communities. The Annapurna model shows how community involvement can be successfully integrated into conservation efforts while promoting tourism.

6.2 Unsuccessful Attempts

Solid Waste Management in Everest Base Camp: Despite the high volume of visitors that make the climb to Everest Base Camp every year, there has been no improvement in the handling of trash left behind. Plastics, cans, and even oxygen cylinders often litter the trails and campgrounds, inflicting long-term ecological harm despite several regulations and attempts to clean up the region. This case study shows how difficult it may be to monitor and enforce environmental rules in less accessible regions.

Unregulated Tourism in Manali-Leh Highway, India: Due to better road conditions and more reasonably priced travel packages, tourism along India's Manali-Leh Highway has increased dramatically in recent years. Waste management problems, soil erosion, and water contamination are only some of the environmental difficulties that have arisen due to a lack of regulation. In this scenario, the ecosystem is degrading because of an overemphasis on tourism without sufficient consideration of its environmental effect.

These case studies demonstrate that while some initiatives have been successful in balancing tourism and conservation, others have faltered, often due to poor regulation, lack of community involvement, or a singular focus on economic gains at the expense of environmental sustainability. The success stories offer valuable lessons in the importance of integrated planning, stakeholder involvement, and regulatory oversight, while the failures underline the consequences of negligence and the complexities inherent in managing both tourism and environmental conservation in the fragile ecosystems of the Himalayas.

Beyond the well-trodden tourist paths lie villages and landscapes where one can truly engage with the essence of Himachal Pradesh. Here, the quintessential features of the state, ranging from its scenic vistas to its unique customs and traditions, are preserved and vividly displayed, far from the hustle and bustle of urbanized tourist hubs. study's observation regarding the nationality distribution among tourists visiting these rural locations is notable. A significant majority, 68.7%, are domestic tourists, while a substantial 31.3% are international visitors. This data implies several crucial aspects: The local Indian populace has a considerable interest in experiencing the deeper, more authentic facets of their own country, reflecting a domestic appetite for 'roots tourism.' The high percentage of foreign tourists, although less than the domestic figures, is nevertheless significant. It points towards an international interest in the 'unfiltered' experiences that Himachal's rural areas can offer. The balance between domestic and international visitors also suggests that the tourism product of rural Himachal Pradesh has a broad-based appeal, indicating that it meets various cultural expectations and interest levels. Given these findings, one could argue that the management of rural tourism in Himachal Pradesh faces the critical challenge of catering to diverse demographic groups while maintaining the authenticity and sustainability of the rural experience. Decisions about infrastructure, community involvement, marketing, and sustainability would need to be finely calibrated to serve both domestic and international tourist populations without diluting the unique attributes that make rural Himachal Pradesh so appealing in the first place.

The challenge lies in striking the right balance between modernity and tradition, accessibility and preservation, and economic benefit versus social and environmental costs. study offers a valuable empirical foundation for understanding these dynamics and could serve as a critical resource for policymakers, planners, and local communities to collaboratively develop a sustainable model for rural tourism in Himachal Pradesh

The breakdown of the domestic tourist profile, as delineated by Figure 6, presents a compelling narrative about the regional appeal of rural tourism in Himachal Pradesh. A detailed analysis of the percentage contributions from various states reveals insights into cultural, geographical, and possibly economic factors shaping tourism dynamics. **Punjab (20.39%):** Given Punjab's geographical proximity to Himachal Pradesh and the common cultural threads shared between the two states, it is not surprising to find a significant number of tourists hailing from Punjab. **Gujarat (18.44%):** The sizable representation from Gujarat, a state that is not geographically proximate, might indicate a particular tourism market that values the kind of experiences rural Himachal Pradesh offers. This could relate to the natural beauty, the quieter atmosphere, or perhaps even specific tourism packages that resonate with Gujaratis. **Uttar Pradesh (13.59%):** Similar to Punjab, the relative geographical proximity and shared cultural factors can make Himachal Pradesh a natural getaway choice for residents of Uttar Pradesh. **Haryana (8.73%),**

Figure 6. The percentage of domestic tourists visiting rural areas

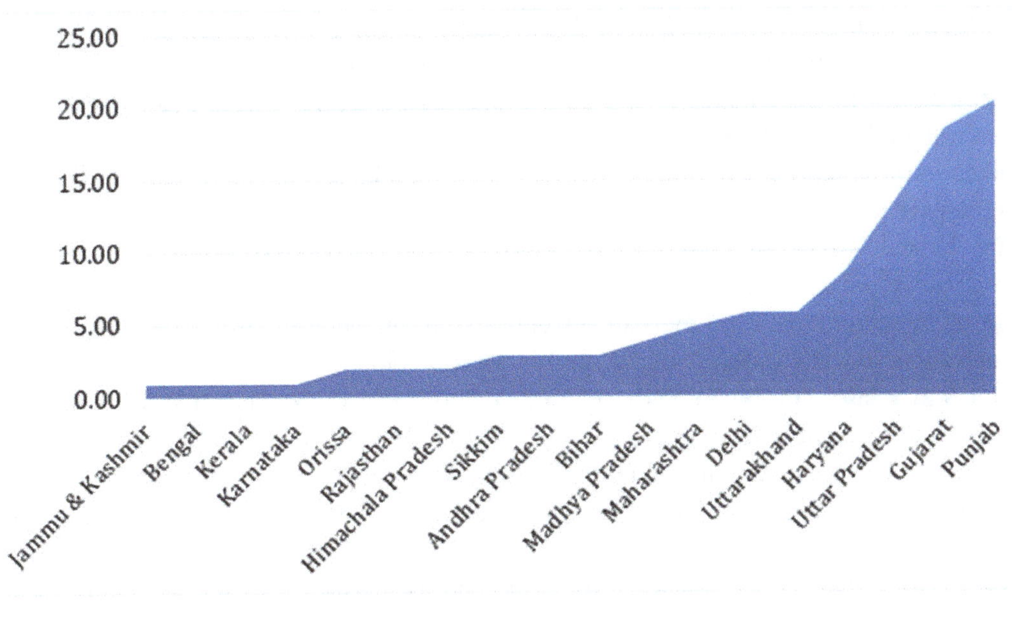

Uttarakhand (5.82%), and Delhi (5.82%): These states and the national capital also show substantial numbers, likely due to their geographical closeness and the ease of travel.

The data suggest several implications for the management and marketing of rural tourism in Himachal Pradesh:

Regional Targeting: Since the states from which most tourists come are identified, marketing campaigns can be regionally targeted to maintain or even increase visitor numbers from these places.

Cultural Sensitivity: Understanding the cultural background of the majority tourist populations could help in tailoring experiences that are more aligned with their expectations, thereby enhancing tourist satisfaction.

Resource Allocation: Knowing where the majority of your tourists come from can help in the logistical planning, from determining where to set up information kiosks to figuring out which regional cuisines might be more in demand.

Expansion Opportunities: The data can also identify states that are underrepresented, which could be potential markets for expansion.

Sustainability: Last but not least, these insights could be instrumental in creating sustainable tourism models that cater to the largest demographics without compromising the local ethos and environment.

Understanding the profile of domestic tourists is thus not just an academic exercise but can have far-reaching implications in the planning, management, and sustainable development of rural tourism in Himachal Pradesh.

The profile of foreign tourists visiting rural areas in Himachal Pradesh, as illustrated by Figure.7, provides a nuanced perspective on the international appeal of the region's rural tourism offerings. This profile also possesses significant implications for the tourism development strategy for the area. **United Kingdom (22.53%) and Russia (22.28%):** These two nations dominate the foreign tourist inflow. The cultural and possibly historical affinities between these nations and India could be factors that make rural Himachal Pradesh an attractive destination. **New Zealand (19.15%):** The sizable representation from New Zealand suggests an opportunity to tap into markets that value rural and nature-oriented tourism experiences. **Australia, Norway, France, South Korea, Italy, and Canada (each 4.26%):** The diversity in this segment could point to the universal appeal of the natural beauty and cultural richness offered by rural Himachal Pradesh. **Targeted Marketing:** Understanding the countries of origin can help tailor promotional campaigns, possibly even partnering with travel agencies in these specific countries. **Customization:** The diverse nationality of the tourists implies a need for varied cultural experiences and possibly multilingual guides. **Policy Tailoring:** Given the variety of countries represented, visa facilitation and other entry requirements could be streamlined for residents of these countries to boost tourism.

Comprehensive analysis underscores the multifaceted nature of tourism in Himachal Pradesh, spotlighting the state's intricate blend of natural allure, cultural richness, and historical significance. Importantly, you've pointed out that this isn't merely a haven for traditional tourism; it's also emerging as a hub for niche sectors like rural tourism, adventure tourism, ecotourism, and geotourism. The state government's proactive response, as manifested in various policies and schemes, suggests a positive trajectory. However, as you astutely observe, the inclusion and well-being of the local community are instrumental for a sustainable tourism model.

Job Creation: The influx of tourists, particularly in the rural sectors, can lead to increased employment opportunities for local communities, ranging from hospitality to tour guiding and artisanal crafts.

Local Entrepreneurship: The popularity of unique, localized experiences like home-stays can empower residents to become entrepreneurs, leveraging their local knowledge and resources.

Figure 7. The percentage of foreign tourists visiting rural areas

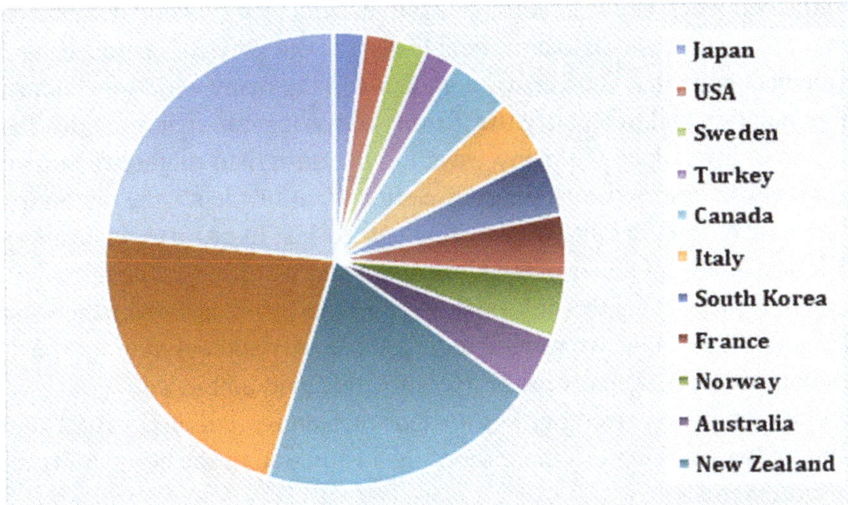

Cultural Exchange: The presence of diverse tourists can foster a mutual exchange of traditions, arts, and practices, enriching the local culture while adding layers of depth to the tourism experience.

Preservation of Traditions: The appeal of authentic experiences can incentivize the preservation of traditional practices, crafts, and art forms that might otherwise be endangered.

Tourism in Himachal Pradesh is undeniably a sector with prodigious potential, and the state has made significant strides in recognizing and harnessing this potential. However, for tourism to be a true catalyst for comprehensive development, an integrative approach is necessary—one that takes into account economic gains, cultural preservation, ecological balance, and above all, the welfare and active participation of the local community. This holistic view can ensure that tourism serves not just as a vehicle for economic growth but also as a tool for sustainable and inclusive development.

The purpose of this study was to examine how rural tourism has affected the local population in this beautiful part of Himachal Pradesh and how it has impacted their progress toward economic and social progress. The effects on the economy, society, and the environment have all been included into the analysis. In addition, our research delved deep into the complex profiles of rural visitors, investigating their choices in vacation spots and measuring their levels of contentment with those spots' features. This academic research also evaluated barriers to rural tourism with the goal of developing realistic approaches to fostering this industry in the study region for the long term. Six different tourist villages in the beautiful Himachal Pradesh landscape (Pragpur, Naggar, Udaipur, Brua, Hewan, and Baggi) were surveyed extensively for this research. The study goals were achieved by the careful use of two separate survey tools for data gathering (Figure.8). The first survey was made for locals to fill out, while the second was made for visitors to fill out, both domestic and international. Through a multi-stage sampling process, we were able to recruit 450 participants for this investigation. Of these respondents, 300 were from the neighborhood and 150 were from out of town. There were a total of 75 respondents questioned throughout all three rural areas; 50 were from the local communities being studied, while the remaining 25 were tourists. The local community survey was intelligently broken down into three components covering demographic information, the effects of rural tourism, and obstacles to its growth. Similarly, the survey instrument for visitors was divided into three clear sections covering respondents' demographic information, the specifics of their trips to rural regions, and their general impressions of such trips. After that, we subjected the collected data to a thorough examination, using several mathematical and statistical methods (such the Chi-square test), to draw conclusions about the data's significance. These analytic procedures were the backbone that allowed us to understand the many ways in which rural tourism has affected the local population and to identify the distinct traits of rural visitors in this fascinating area.

The study sets out to investigate the many ways in which tourism might affect a community, both positively and negatively. In order to investigate these effects in more depth, the research smartly grouped the variables linked with each effect into overarching categories. In order to determine the relationship between these variables and rural tourism, the Chi-square test was used to the data.

The research not only aimed to gather information from locals, but also from the visitors themselves. As a result of taking a holistic view, we were able to get a deeper appreciation for the "5 A's" of a successful tourist destination: attraction, accessibility, ambience, and ambience.

In addition, a comprehensive inventory assessment of features unique to rural tourist settlements was done. This finding helped provide a more thorough evaluation of the community assets and tourist potential of these settlements.

Challenges to rural tourism were examined as a key part of the research. By identifying these challenges, the research was able to proffer pragmatic recommendations aimed at fostering sustainable development in the context of rural tourism in Himachal Pradesh. This insight has proved invaluable in outlining the key dimensions of rural tourism in the region and highlighting the pivotal role played by the local community in its inception, growth, and future trajectory.

Moreover, the research study has been instrumental in pinpointing the bottlenecks within the rural tourism policy framework in Himachal Pradesh and, by extension, in India. It underscores the collective responsibility of all stakeholders involved in the tourism sector. However, it particularly underscores the paramount importance of the local community. The success of rural tourism hinges on the effective transfer of socio-economic benefits to the local populace and their active engagement in the management and organization of rural tourism initiatives. Ultimately, the study underscores the interconnectedness

Figure 8. Impact of tourism development

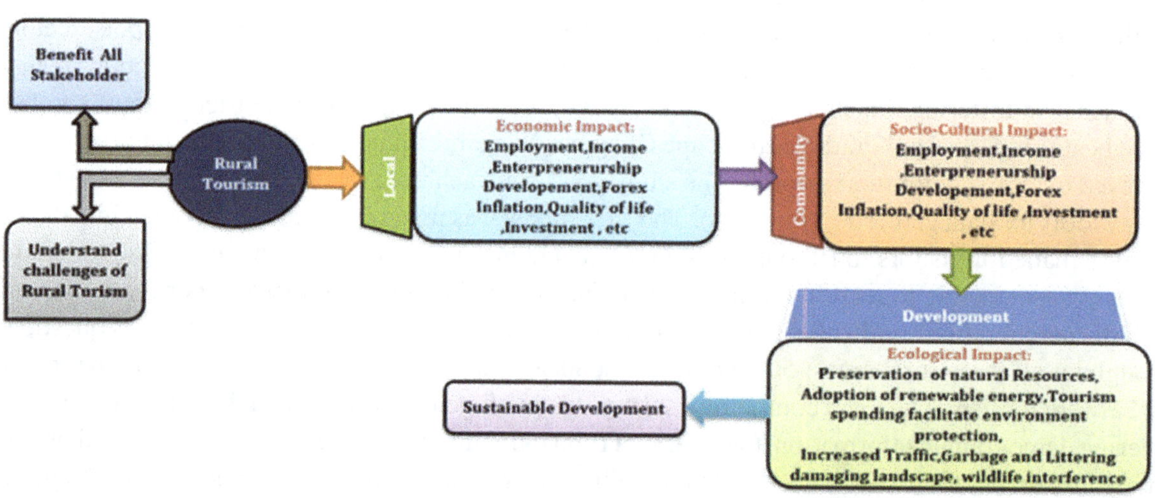

of tourism, local communities, and sustainable development, offering a roadmap for the harmonious coexistence of these elements in the context of Himachal Pradesh's rural tourism landscape.

7. CONCLUSION

The central tension elucidated in this chapter resides at the intersection of tourism development and environmental conservation in the Himalayas—a region of unparalleled ecological and cultural significance. The challenges are manifold, including policy conflicts, resource scarcity, and social and cultural erosion. These challenges are often exacerbated by inadequate governance, insufficient community involvement, and the complexities of managing a fragile and diverse ecosystem.

However, challenges also beget opportunities. Sustainable tourism models such as eco-tourism, cultural tourism, and adventure tourism offer avenues for a more balanced approach. Community-based tourism, where local communities are intimately involved in the planning and management of tourist activities, provides a promising paradigm for economic empowerment without compromising environmental integrity. Furthermore, governmental and non-governmental organizations possess the authority and expertise to implement and regulate policies that can navigate the precarious balance between development and conservation.

7.1 Recommendations for Future Research

Integrated Policy Analysis: Future research could delve into the development of integrated policy frameworks that reconcile the conflicting goals of tourism promotion and environmental conservation.

Socio-economic Impact Studies: Comprehensive studies are needed to understand the long-term socio-economic impacts of different models of tourism on local communities in the Himalayas.

Environmental Monitoring: Research should focus on establishing robust environmental monitoring systems to gauge the real-time impacts of tourism activities on local ecosystems.

Cultural Preservation: Further studies could explore the dynamics between tourism and cultural erosion, potentially leading to the formulation of policies that protect indigenous cultures while allowing for their responsible sharing.

7.2 Policy Formulation

Stakeholder Involvement: Policies should be formulated with active involvement from all stakeholders, including local communities, tourism operators, and conservation organizations.

Strict Regulation and Monitoring: Strict criteria for tourist activities, supported by a strong monitoring and punishment system, must be enforced by regulatory organizations.

Sustainable Practices: Waste management systems, renewable energy utilization, and ethical trekking rules are all examples of sustainable practices that the tourist sector should be incentivized to adopt.

Education and Awareness: Perhaps as part of the application process for hiking permits or guided tours, both visitors and locals might be given information on the environmental value of the place and the effect of irresponsible tourism.

Ultimately, the difficulties of striking a fair balance between tourism and environmental protection in the Himalayas are not insurmountable, despite their complexity and variety. A future where human and non-human societies may both thrive can be achieved via the use of an interdisciplinary strategy that incorporates sustainable tourist models, community engagement, and efficient government.

REFERENCES

Abdollahzadeh, G., & Sharifzadeh, A. (2014). Rural Residents' Perceptions Toward Tourism Development: A Study from Iran. *International Journal of Tourism Research, 16*(2), 126–136. doi:10.1002/jtr.1906

Bagri, S. C., & Kala, D. (2016). Residents' Attitudes toward Tourism Development and Impacts in Koti -Kanasar, Indroli, Pattyur Tourism Circuit of Uttarakhand State, India. *Pasos (El Sauzal), 14*(1), 23–29. doi:10.25145/j.pasos.2016.14.002

Bahrami, R., & Noori, K. (2013). Analysis Of The Role Of Tourism And Its Impact On Rural Development (Case Study Of The Central Part Of Marivan). *Technical Journal of Engineering and Applied Sciences, 3*(12), 1074–1080.

Blesic, I., Pivac, T., Dordevic, J., Stamenkovic, I., & Janicevic, S. (2014). Cultural Events As Part Of Cultural Tourism Development. Case Study: Sombor And Apatin (Serbia). Geografski Zbornik. *Acta Geographica Slovenica, 54*(2), 381–390. doi:10.3986/AGS54406

Brida, J. G., Disegna, M., & Osti, L. (2014). Residents' perceptions of tourism impacts and attitudes towards tourism policies. *Tourismos:An International MultiDisplinary Journal of Tourism, 9*(1), 37–71.

Chand, R., & Leimgruber, W. (2016). *Globalization and marginalization in Mountain Regions: Assets and Challenges in Marginal Regions*. Springer. doi:10.1007/978-3-319-32649-8

Das, M. B., Kappor-Mehta, S., Tas, E. O., & Zumbyte, L. (2015). *Scaling the Heights: Social Inclusion and Sustainable Development in Himachal Pradesh. The World Bank*. International Bank for Reconstruction and Development / The World Bank.

Golzardi, F., & Sarvaramini, S., sadatasilan, K., & Sarvaramini, M. (2012). Residents Attitudes towards Tourism Development: A Case Study of Niasar, Iran. *Research Journal of Applied Sciences, Engineering and Technology*, •••, 863–868.

Hernandez, J. M., & Casimiro, L. A. (2011). Simulation model for a joint mass / rural tourism system. *Tourism and Hospitality Research, 12*(1), 5–14. doi:10.1177/1467358411429525

Iorio, M., & Corsale, A. (2010). Rural Tourism and livelihood strategies in Romania. *Journal of Rural Studies, 26*(2), 152–162. doi:10.1016/j.jrurstud.2009.10.006

Jalilian, M. A., Danehkar, A., & Ali Fami, H. S. (2012). Determination of Indicators and Standards for Tourism Impacts in Protected Karaj River, Iran. *Tourism Management, 33*(1), 61–63. doi:10.1016/j.tourman.2011.01.024

James, G. (2015). Rural Tourism and Biodiversity:A win-win Relationship. *Kurukshetra-A Journal on Rural Development*, 17-21.

Katoch, A. (2014). *The impact of rural tourism on local community development:A study of Himachal Pradesh, India.* [PhD Synopsis, Central university of Himachal Pradesh, Dharamshala].

Katoch, A., & Gangotia, A. (2014). Rural Tourism: An instrument for local development in Hamlets around Dharamshala, Himachal Pradesh, India. AVAHAN-. *Journal of Hospitality and Tourism*, *2*(1), 12–24.

Khakpour, A. (2013). Methodology of comparative studies in education. *Contemporary Educational Researches Journal*, *2*(1), 20–26.

Li, P., Ryan, C., & Cave, J. (2016). Chinese rural tourism development: Transition in the case of Qi-yunshan, Anhui.-2008-2015. *Tourism Management*, *55*, 240–260. doi:10.1016/j.tourman.2016.02.007

Mukherjee, D. (2012). Rural Tourism Path to Economic & Regional Development in India. Kurukshetra, A Journal on Rural Development, 62 (10), 3-6.

Sharpley, R. (2014). Host perceptions of tourism: A review of the research. *Tourism Management*, *42*, 37–49. doi:10.1016/j.tourman.2013.10.007

Chapter 18
Tourism Development in the West Jaintia Hills District, Meghalaya, in the Foothills of the Eastern Himalayas:
Prospects and Challenges

Manroilang Dkhar
Assam University, Silchar, India

ABSTRACT

The Himalayas' most distinctive characteristics are the snow-covered mountains, the diverse flora, and wildlife and the breath-taking scenery. Shillong, the capital of Meghalaya, is a hill station. It is a part of the Eastern Himalayas foothills. From Shillong, Meghalaya, on bright autumn and winter mornings, one may view the Himalayan Range in the direction of north. The easternmost district in the State of Meghalaya, Jaintia Hills, is one of the States 12 Districts. East and West Jaintia are the two districts that make up Jaintia Hills. Jaintia hills is a popular tourist destination with a lot of natural, peaceful, and picturesque views. It is situated in a mountainous area. Compared to East Jaintia Hills, West Jaintia Hills District has a higher level of tourism activity. This chapter is designed based on secondary data and observation method which attempts to investigate the tourism development in West Jaintia Hills District, Meghalaya, in the foothills of Eastern Himalayas.

1. INTRODUCTION

1.1 An Overview of the Global Tourism

Today, tourism is regarded as one of the largest developing sectors and also acts as a major source of income for many countries. As we all know, that tourism involves travelling from one place to another

DOI: 10.4018/979-8-3693-0823-3.ch018

either nationally or internationally for various reasons such as recreation, rest, relaxation, leisure, adventure, religious, family or business purposes for a limited duration.

The United Nations World Tourism Organization (UNWTO) defines tourists as "people travelling to and staying in place outside their usual environment for not more than one consecutive year for leisure business and other purposes." Tourism growth and development has led to an upward trend because it allows tourist to see and visit the world, engage in different activities, experience, and share knowledge about cultures and meet new people. The tastes and preferences of the tourists are likely to grow in the coming years, as they shift their interest from the earlier concept of mass tourism to an innovative alternative tourism which will directly and indirectly contribute to the growth and development of the travel and tourism industry worldwide. Therefore, customer satisfaction, safety and enjoyment are the focus of any tourism business.

1.2 India Tourism

India is surrounded by water on three sides i.e., the Arabian Sea on the West, Indian Ocean on the South and Bay of Bengal in the East which provides various tourists attraction and services to tourists. The Himalayas in the North acts as a backdrop or a hideaway, which offers tourists to choose the destination of their choice for rest, relaxation or to simply rejuvenate their body and soul. India is also a land of various spiritual and cultural history, there are many temples, towns of various architectural styles and world-famous monuments.

Himalaya is a Sanskrit word which implies "Abode of Snow". HIMA which means Snow, and ALAYA which means Abode, are supposed to have inspired the word by early Indian pilgrims who trekked through these mountains. The world's highest mountain chain runs across five countries: India, Pakistan, China, Nepal and Bhutan. It is the principal source of the worlds river system, including the Indus and Ganga-Brahmaputra basins. The Himalayas are home to iconic animals such as the Snow Leopard, Bengal Tiger, and one-horned rhinoceros, along with millions of people. Tourists have been visiting the region since ancient times due to its mesmerizing perspective and appeal to many climbers and hikers from all over the world. The Himalayas reflect nature's power, beauty and grandeur. The Himalayas are a mountain range in Asia that not only serves as a physical barrier dividing the Indian Subcontinent from the Tibetan plateau, but also as a climate-specific waterway and a cultural division. Although the term plural Himalayas is commonly used, its official name is Himalaya. Many people have crossed and left traces of their cultures on the Tibetan Plateau, which was home to people who established their own distinct culture influenced by Buddhism. Tibet received the religion in two waves: in the seventh century from India, Nepal and China, and again in the eleventh century from India.

1.3 The Himalayan Tourism

The Himalayas, a mountain range in Asia where Earth meets Sky, are more than just a physical topographical landmark, they represent the identity of millions of people's civilizations dating back thousands of years. Without these massive mountain ranges, rainfall rushing up from the Indian Ocean would pass across the Indian Subcontinent and into Central Asia, leaving it a desolate wasteland. The Himalayas are an amazing natural wonder that has drawn tourists since time immemorial. Its enormous snow-clad mountains, breathtaking beauty, and plentiful flora and fauna have captivated people all over the world. But, among all of these attractions, its people are a standout. The people, that live in the Himalayan regions are communities that rely on nature and have strong religious and ethnic beliefs in their daily lives. They are friendly and welcoming communities of people who are extremely appreciative of and concerned about the environment in which they live. Even though they live in tough conditions, this does not deter them from continuing and appreciating life. The numerous Himalayan communities are noted for their skilled craftmanship and traditional expertise for producing medicines, crop cultivation, and masonry. They are classified as nomadic pastoralists or subsistence farmers based on their location and climatic conditions. Agriculture and animal husbandry provide a living for the Himalayan people.

The Purvanchal are low hills in the Eastern Himalayas that share borders with China, Bhutan, Myanmar, and Bangladesh. The region spans several geographical States and is well recognized for its deep gorges, terrain that is rocky and untamed beauty, making it a calm and secluded haven for nature enthusiasts and adventure travellers. Along with this, distinctive and different lively cultural practices can be found in each States in the Eastern Himalayas region, where the people of the communities practice their unique beliefs and customs during any festivals. The people who live here come from several ethnic communities and have a reputation to be courteous and warm, sharing an insight of their rich heritage through music, dance, art and handicrafts, traditional costumes and cuisine. Purvanchal Hills offers a variety of natural tourism attractions that appeal to a wide range of interests. From tranquil and enchanting landscapes for nature lovers to exhilarating adventures to immersing in historical aficionados, there is something for everyone. Purvancahl has numerous attractions for all types of visitors.

Some of the important Hills that make up the Purvanchal are:

1. The Patkai Bum and Naga Hills
2. The Mizo Hills and Manipur Hills
3. The Khasi, Jaintia and Garo Hills along Meghalaya,
4. The Dafla Hills in the North.

1.4 North East Tourism, in the Eastern Himalaya Region

North-Eastern region part of India lies on the Eastern Himalayas popularly known as Seven Sisters which comprises the States of Assam, Arunachal Pradesh, Manipur, Meghalaya, Mizoram, Nagaland, Tripura and One Brother which include the State of Sikkim can best describe as pristine, picturesque, wild and untouched natural scenic beauty that each States possesses. With all these the North-Eastern States offers the potential areas of the country in respect of tourism because of its diversity in landscapes and traditions. Considering the development like roads, cultures, forest, employment and so on, it can play a huge important role for the success of the total tourism product of the States. The tribes of India are spread across the country and have yet to break free from their traditional lifestyle, where little is known about them, and link with the social conventional set up. These tribes in India have separated cultures and traditions that have yet to be identified. Most tribals in India are concerned about the natural environment, and they are hostile to protect the rivers, lakes, farmland, and forests that are their primary source of income. Because of this bond between tribal people and nature, the tribal has an attractive, interesting diversity. Gonds, one of the world's largest tribal communities, are primarily found in Andhra Pradesh, Gujarat, Uttar Pradesh, West Bengal, Odisha, Madhya Pradesh, Chhattisgarh, and parts of Maharashtra. Santhals are India's oldest tribal people, predominately found in Jharkhand, Odisha, and West Bengal. Similarly, the North-Eastern tribes of India are quite innovative and creative in constructing objects, accessories, head gears, and other items out of bamboo, stone and clay. Their way of life is mirrored in their everyday representation of traditional games, agricultural practices, hunting and fishing traditions. They are particularly skilled in crafts like as pottery, woodwork carving, textile weaving, as well as bamboo and cane handicrafts. The tribes have distinct dance and songs with intricate rituals. The tribal tourism in the Nort East of India, if well managed and adequate attractions plan to aware the tourists through comprehensive measures, can enhance tourism for an inspirational, reviving, and enlightening experience of the roots of tribal life, art, culture and legacy. The different States on the Eastern Himalayas region feature a rugged and tough environment with high altitude slope and hills, wild flora and fauna, and variety of agriculture crops and methods for cultivation. The diverse of each States location on the Eastern Himalayas have resulted in varying climate conditions ranging from per humid to humid or receiving an average rainfall of roughly 2450mm each year. Mawsynram, lies on the Eastern Himalayan State of Meghalaya, India, receives the most rainfall, averaging 11,873mm per year.

1.5 Tourism in Meghalaya, in the Foothills of the Eastern Himalaya Region

Shillong the capital city of Meghalaya situated on the Eastern Himalayas, is blessed and covered with the gift of nature. Shillong Peak, the tallest peak in Meghalaya at 6449 feet and around 1450 feet above the capital city for about 6m, spreads on a higher tract where the snow-clad Eastern Himalayas Mountain ranges can be seen located in Bhutan and west of Arunachal Pradesh. It is a rare view to have a glimpse of the snow-clad Himalayan mountains in Meghalaya. However, the view is possible just after the monsoon, when the concentration of the dust particles in the atmosphere is low. Another factor for the visibility of the Eastern Himalayas from Meghalaya is the low altitude of the entire Brahmaputra Valley's geographical location. The State of Meghalaya on the foothills of Eastern Himalayas has 12 Districts namely East Garo Hills, South Garo Hills, West Garo Hills, North Garo Hills, South West Garo Hills, East Khasi Hills, West Jaintia Hills, East Jaintia Hills, West Khasi Hills, South West Khasi Hills, Eastern West Khasi Hills and Ri-Bhoi District. Jaintia Hills is divided into two Districts- East Jaintia Hills and West Jaintia Hills District. Also known as the "Abode of Clouds" and receives the highest rainfall during monsoon season, Meghalaya is bestowed with diverse flora and fauna, deep gorges and canyons, magnificent waterfalls, caves, sacred groves and crystal-clear water bodies. Meghalaya geographic location on a hilly area and its ideal moderate climate makes it the perfect destination for tourist attraction. Apart from this, the beauty of Meghalaya lies on the rich cultural heritage and traditions, beliefs and customs, languages, and numerous dialects. The State of Meghalaya comprises a population made up of three primary Tribes: Khasi, Jaintia, and Garo, each with its own distinct culture and traditions. While the Khasi and Jaintias are thought to have migrated from South East Asia, they are both socially and lingually connected to the larger Mon Khmers group. The Garos, on the other hand, are considered to have originated in the Tibeto- Burma region, while the Bodos are thought to have originated in Tibet. The matrilineal system is a tradition that unites all three tribes, in which property is inherited through female lineage (maternal side), with the youngest daughter inheriting the holdings. The Khasi make up about half of Meghalaya's population, and their towns are spread throughout the State. The Jaintias live in the Jaintia Hills, while the Garo live in the Garo Hills in Meghalaya. Aside from that, the Jaintias are known by several names "Syntengs, Jayantias and Pnars. For a long time, the British controlled the Jaintia Hills until 1972, when Meghalaya declared an autonomous council under all three tribal councils.

In accordance with experts in history, the Jaintias are descendants of the Indo-Chinese Mon Khmer people. However, it is stated that that the Jaintias initially came to Assam from South East Asia and later travelled to their current settlements in the beautiful foothill Eastern Himalaya State of Meghalaya. Jaintia

Hills is the easternmost part of the State of Meghalaya. The mountainous region derives its name from the Jaintia Kingdom and the local Pnar people who are also known as the Jaintias. Bordered with Assam, Jaintia Hill is truly a wonderful tourist destination to explore. The majesty of these hills is something everyone must experience in their lives. In order to have an authentic experience while visiting Jaintia Hills, tourists will have to travel to the rural heartlands and wild spaces because the destinations are spread across the different districts of the State. One of them is West Jaintia Hills District in Meghalaya which offers several tourist attractions of its natural landscapes such as various waterfalls over weathered rocks in a secret forest, caves that lies deep inside the heart of the dense forest and the famous boats floating on crystal-clear water. Beside this, the tribe in West Jaintia Hills District are well-known for their rich cultural legacy, which is evident in their music, dance, rituals, and crafts. The tribes are well-known for their beautiful weaving of bamboo carpets, mats, and cloth, particularly traditional garments used during festivals and worn by both men and women of the district. The tribal people of the district are incredibly creative, in order to pass on talents passed to them by their four fathers, such as crafting various musical instruments, jewellery, pottery and wood carving. Their religion is influenced by Hinduism, where the three predominant principles promulgated by God are "kamai yei hok" (to strive for righteousness and justice), "tipbru tipblai" (to know man, to know God) and "tipkur tip kha" (to know one's clansmen, to know one's kinsmen)". Rice is the tribe's essential meal, and different rice snacks are consumed with tea. Rice (red rice) is served with curries made from various indigenous vegetables, as well as chutney made from fermented dry fish with native fruits and vegetables. "Jaktung" (Ja-rice, Ktung-dry fish) is a traditional rice dish offered at child naming ceremony, while "Jatoh" (Ja-rice, toh-beans) is a dish in which several types of beans are cooked along with rice. Each of their dance and festivals is conducted with great zeal and intricate customs.

2. LITERATURE REVIEW

It is very important to understand the tourism industry of a particular country and State, where several research articles have been studied. The opinions of various authors have been pointed out here in review of literature.

Yogesh,E B Khedkar, Snehal Pawar (2019) stated that the tourism industry in India has seen a tremendous growth and development during the recent years, it includes a diverse range of establishment and services such as railways, accommodation, airlines, cruise, hotels, resorts and restaurants. On the other hand, the challenges faced by the tourism industry in India are safety, security, human resources, marketing and promoting issues, environmental, financial and customer related issues. The solution to this is requiring immediate action which include development of effective policies and guidelines and effective management which will help to mitigate the challenges.

Nimit Chowdhary, Suman Billa and Pinaz Tiwari (2022) indicates that the government initiative in the marketing and promoting the country under the slogan "Incredible India" by the Ministry of Tourism, Government of India as a tourist destination has emerged India as one of the fastest tourist destinations in the world. India as a country is known for its diversity in culture, language, temples, forts, natural landscapes and wonderful cuisines. Moreover, the growth of tourism has led to various initiatives like cleanliness drive, e-visa facility, adoption of heritage sites to stakeholders' participation for better management and creating infrastructure along various tourist destination. Indeed, it is noted that the scope and challenges faced by any tourism destination of all countries are different. Therefore, not only the marketing or promotion is required in tourism but also the rejuvenation strategies need solution. **Sidharth Shankar Raju (2019)** the author talks about the scope of tourism courses in India. To develop tourism in any destination of a country we need to understand what the destination attracts or possess such as the socio-cultural, political background, environment, and demographic. By keeping the tourists needs and desires and the target market in mind this should match with the potential of the destination of the society and eventually help to develop a strategic marketing plan. If the destination can attract or the inflow of tourists is increasing, then this will lead to employability and creation of job to many labours and skilled workforce. With the gap in demand and supply between labours in the tourism sector and supply of skilled labours in the tourism sector, there is a need for training centres of that skill. Therefore, the government should encourage various tourism related course in order to promote the scope of tourism of the country and create skilled employees in the workforce according to the tourism sector needs.

Honoria Samson (2017) investigated that social media has been playing a crucial role in everyone's life. It helps us to make travel decision based on our preferences and choices. People share their thoughts, experiences and reviews on social media and this effect will act as a source of information made available for people to make their travel choices. Many tourisms industry is using social media, as this helps in advertising and marketing which have increased the total revenue of the firm as well as provide information which is more real and relevant. Therefore, social media has helped in influencing the buying behaviour of the customers which will hugely impact the tourism industry.

Purusottam Nayak, Sudhanshu K. Mishra (2013) indicates that the State of Meghalaya is endowed with the gift of nature with an ideal pleasant climate. Because of its natural scenic beauty and diversity in culture the State have the potential of tourism. Air and road connectivity, infrastructure facilities at tourist sites in terms of accommodation, transport, banking use, drinking water, sanitation, health care is either lacking or not developed. There is need to train and recruit skilled staffs who are professional and welcoming in the workforce of the tourism industry of the State.

Gourab Bera (2015) investigated that due to the lack of proper infrastructure in the State of Meghalaya in most of the tourist destination there is no organized tour facility conducted to visit or know the real beauty of the tourist spot. Since many conflicts arises between the local people and the promotion of tourism in the State neither the government nor the municipality has taken any steps or measures to promote the tourism of the State and remained invincible and hidden.

Benjamin Lyngdoh (2022) the author explains that tourism development in Meghalaya can be viewed from the last 45 years with nature-based beauty being the unique selling product. The article demonstrates the three stages of tourism development in Meghalaya classified as Inception Stage (1980s-1990s) tourism can be traced from Sohra, Setting-up Stage (2000-2014) tourism extended to other parts of Meghalaya with a sense of professionalism, Take-off Stage (2015-2019) tourism took off to other nature-based attractions which became very popular and resulted to the growth of tourism entrepreneurship in Meghalaya. The way forward laid focus on sustainable tourism, development of community-based tourism, protection of natural resources, infrastructure development and a stress on service quality. The rules and regulations are very important as well as require the active participations of various stakeholders.

Saibal Kumar Saha, Sawarisa Suiam, Ankita Sarangi (2022) this article demonstrates that the people of Ialong Village, West Jaintia Hills District, Meghalaya believed that with the impact of rural tourism in the village it has provided and improved facilities like electricity, hygiene, toilets, accommodation and have attracted more tourist to the region. However due to lack of advertisement and promotion of the different villages about the attractions or offerings, it has limited the growth of potential in tourism of these villages.

East-Himalaya.com (2002) indicates there are many tourism tourist attractions in Jaintia Hills, where tourism is still at its emerging stage.

Saibal Kumar Saha, Tushar Chandra Babai, Sangita Saha (2018) argues that the reasons behind the tourist not visiting some unknown attractive places in Meghalaya are due to transportation, lack of time, lack of information and age problems. Therefore, all these problems can be mitigated through promotion of those unknown tourist destinations and support the local bodies of different villages, NGOs to take up necessary steps to get that unconstructed place constructed for better transport and communication.

Arup Barman, Ranjit Singh, Venkata Rao Yedia (2010) stated that in order to be sustainable, cultural tourism should incorporate local people of tribal backgrounds in the development process and community-inspired ideas. If this is achieved, the tribal culture will be conserved and preserved, as will economically progress.

Padampat Singhania (2022) "This Study has been conducted to acquaint the growth prospects in the form of social, economic, environmental and facility aspects and challenges of sustainable tribal tourism in India".

Tamish Agarwal, Rshita Jain (2023) stated that exploration of tribal places will aid in the economic development and expansion of local tribes while also assisting them in gaining an awareness of their culture. Along with this information, tourists that visit these indigenous cultural communities will learn about the local people's lifestyle and traditions. As a result, in addition to tourist visits, they would spend their money on momentous and local goods from communities nearby. As a result, Indian culture would be incomplete without the distinct lifestyles of the adverse tribes. Tribal tourism in Rajasthan will help to boost the regions tourism economy while also maintain the regions original cultural heritage.

Hsiao-Ming Chang, Ching- Hui Lin, Ying- Wan Chuang (2021) mentioned that people of Taiwan value and appreciate the connection between sustainable development and tribal tourist operations with the preservation of aboriginal culture. The development of the tribal tourist industry's indigenous peoples has relied heavily on industrial development. Government support and cooperation, as well as investment from diverse stakeholders, must be a crucial motivator for the long-term development of tribal tourism. Thus, Economic, Society and Environmentally Sustainable Development are the three factors which will contribute to the tribes' long-term development.

Suneet Verma, Madhu Murdia (2017) examines that tourists are constantly on the lookout for new adventurous escapes and experiences. It is common for tourists to believe that they do not wish to experience new cultures or face new problems from tribes. Furthermore, it is the potential of tribal tourism (history, culture, costumes, cuisines etc) that is accountable for the effective development and investigation of any lovely unexplored sites of the vastly diverse State of Rajasthan. As a result, a plan of action for a new tourism product to draw and meet visitor expectations would be devised in Sothern Rajasthan.

Divesh Kumar Paraste (2021) has focused the recognition of various indigenous resources seeks management and promotion of them on the prospects of the indigenous community. Management of indigenous resources is a method of making indigenous tourism available and providing service to tourists so that indigenous tourism is known, experienced and enjoyed. Promotion of indigenous tourism is the process of advancing of indigenous tourist product information through multiple forms of publicity. The impact of indigenous tourist promotion on the community is favourable since the people care about their culture.

Gour Krishna Saha (2015) the author discusses how, in the case pf the North Eastern region, no effective actions have been taken to boost tourism despite the region's immense prospects for natural scenic splendour, rich bio diversity, and a house of extensive and diverse cultural ethnic history. All if this creates a potential tourist's destination with fully interrelated tourism promotion methos in the North Eastern Region.

John Mang Muan Lian Zou, G.P Prasain (2017) this paper stresses the various tourist spot in India's Northeastern Region, which have stayed unexplored for numerous years. To bring these tourist spots to the forefront using multiple channels and approaches and to entice tourists to visit the region, the brand identity "Paradise Unexplored" has been sold in the global marketplace tailored specifically for North East India.

3. THE SCENERIO OF TOURISM DEVELOPMENT IN INDIA

Tourism has recently emerged as the world's most rapid and most significant service industry. It contributes to a country's numerous development processes by creating revenue and improving the local communities. When it comes to the Indian tourist setting, tourism is rapidly developing and fiscally significant, as it has aided in the country's numerous growths. Because of India's richness in culture, traditions and customs, as well as its geographic location in the tropical zone, more tourists are visiting the country to enjoy the variety of amenities and services provided by the tourism and hospitality business. Cultural tourism, tribal tourism, beach tourism, adventure tourism and many more forms of attractions entice tourists to visit India. Foreign tourists visit India for a variety of reasons be it for leisure, medical, business, education or jobs on foreign deputation etc.

"According to the estimates from the United Nations World Tourism Organisation (UNWTO) for 2023, international tourist arrivals could reach 80 to 95 percent of pre pandemic levels depending on the extent of the economic slowdown, travel recovery in Asia-Pacific and geographical tensions, among other factors".

3.1 Tourism Sector Highlights

"According to the Ministry of Development of North Eastern Region the following highlights of the Indian tourism scenario are projected as follows:

1. By 2030, India is projected to generate $250 billion in GDP from tourism, creates 137 million jobs in the sector, earn $56 billion in foreign exchange, and attract 25 million foreign arrivals.
2. Indian travel market is expected to grow from an estimated $75 billion in Financial Year (FY)20 to $125 billion by FY27
3. It is projected that the Indian tourism and hospitality industry will earn $50.9 billion from visitor exports by 2028, a substantial increase from the $28.9 billion earned in 2018. Additionally, the number of international tourist arrivals is expected to reach 30.5 million by 2028.
4. It is estimated that outbound trips from India will touch 29 million by 2025 and cross the $24 billion mark by 2024.

5. The tourism sector in India provided employment to 39 million people, which accounted for 80% of the country's total employment in FY20. It is estimated that this sector will create around 53 million jobs by 2029".

(Source: Ministry of Development of North Eastern Region)

3.2 Tourism Development In Northeast India

The Northeast States are located on the easternmost Himalayan Range and are made up of seven Sisters States that is Assam, Arunachal Pradesh, Manipur, Meghalaya, Mizoram, Nagaland, Tripura and one Brother State Sikkim that share boundaries with one or five nations to the east of the country. The region is known for its steep slopes and green valleys, rolling hills, Himalayan Mountain range, rich cultural legacy and various tea gardens that spread over different States that provide a breath-taking and mesmerizing view of the region. Each States tourist attractions make it a suitable hideaway for nature tourism having distinct identity expressed in their culture and traditions which are portrayed in festivals, music, dance and cuisines. Because of its unrivalled attractions, the natural scenic splendour of Northeast India is a combination of adventure and eco-tourism. The region has various national parks and wildlife sanctuaries, including the well-known UNESCO World Heritage Site Kaziranga National Park in Assam, which is famous for its one-horned rhinoceros. The fourth highest waterfall in India Nohkalikai Falls, is situated in Meghalaya, while the Rainbow Falls in Arunachal Pradesh is a famous picnic destination with picturesque landscapes. The oldest and the most venerated shrines in India situated in Assam the Kamkhya Temple and the famed Living roots bridges of Meghalaya, which serves as unique attractions due to the local people's inventiveness.

Assam is carrying out well in the development of tourism in Northeast India, tourism was recognized here in 1958. During the British period, Shillong gained an attractiveness for outdoor recreation, which led tourists to visit and explore Kaziranga, Kamakhya Temple, and Shillong in the 1930. "According to the provisional data by the Ministry of Tourism, Government of India, 118.45 lakh domestic and 1.04 lakh foreign tourists visited the Northeastern States during the year 2022."

The number of tourists visiting the Northeast States has steadily increased during the last decade. Domestic tourist arrivals increased from 3.13 million in 2003 to 95 million in 2019, representing an 8% Compound Annual Growth Rate (CAGR) throughout the time. As a result, tourism provides a unique opportunity for operators, hotels, resorts and infrastructure developers to capitalize on Northeast India's potential that remains untapped.

3.3 Tourism Development In Meghalaya

Tourism in Meghalaya has evolved since the 1970s. it has evolved as a potential State destination for both domestic and international tourists. Meghalaya is located in the Eastern part of the Himalayan region, its physical natural attractions include abounding and rich flora and fauna, canyons, deep gorges, lush green hills, caves, waterfalls, and crystal-clear water. Apart from festivals, cultural heritage and cuisines, these nature-based features comprise the States distinctive selling point by contributing to the region's economic and social growth. Khongtong Village, also known as Whistling Village, is a valley region in East Khasi Hills that was selected by the UNWTO as the "Best Tourism Village" in 2021. The village inhabitants assign a unique tune to call each other that acts as a distinctive identification. Another site that has just been added to the tentative UNESCO list is the Living Root Bridges, which are made up of interconnected roots and have been created for ages by the native inhabitants of the region (Khasi and Jaintia). The Meghalaya Tourism App, that launched in 2020, allows users to effortlessly arrange their travels to the State by reserving lodging, taxis and guides. Within the course of next five years, the government intends to double the total number of lodging units in the State, with 50% of those being of excellent quality (tariff greater than Rs 5000). To improve and advance the States cultural history, the government will collaborate with local communities to set up destinations in which tourists are able to reside, acquire information, and experience the traditional way of life. In this regard, pottery making in Larnai village (West Jaintia Hills District), silk making in Umden(Ri Bhoi District), beekeeping and brewing of bitchi (local rice beer) in the Garo Hills.

The Meghalaya Tourism Department is aiming to develop conditions that would draw people from all over the world to experience the regions distinctive culture. According to a tourist industry study, one of the key reasons for the gradual development of the tourism business is a lack of a place to live. The government is also attempting to develop infrastructure fir the tourism by fostering and promoting locally produced products, steps such as ecotourism, and the sustainable operation of tourism and hospitality facilities. The administration is organizing several events to draw tourists to the State, including Megha Kayak Festival which feature kayakers from all around the world, the well-known Cherry Blossom Festival, and the Shillong Literary Festival. In addition, it is expected that the tourism sector provides approximately 4.1% of the State's GSDP (Gross State Domestic Product)

"In 2019, the tourist footfall in Meghalaya stood at about 12.7 lakhs (including 25,000 foreigners). Among the North-eastern States, this is highest after Assam and Sikkim. The footfalls have grown steadily over the last decade and are again witnessing a recovery post the COVID pandemic. It is expected that the annual tourist footfall will cross15lakhs by 2024".

4. OBJECTIVE OF THE STUDY

- To study the tourism potential of West Jaintia Hills District in the State of Meghalaya.
- To find the prospects of tourism in West Jaintia Hills District.
- To analyse the challenges faced by the tourism in West Jaintia Hills District.
- To discover solutions and make suitable suggestions for difficulties encountered by the tourism industry in West Jaintia Hills District.

5. RESEARCH METHODOLOGY

The study is descriptive in nature and is primarily based on secondary data available and method of observation have been conducted.

5.1 Research Location

Located on the mountainous region Jaintia Hills tourism attraction is full of breath-taking serene natural spots, picturesque views, and numerous other tourist spots one who seek peace and tranquillity. There is more tourist activity in West Jaintia Hills District compared to East Jaintia Hills District. The reason being, West Jaintia Hills District is blessed with mesmerising waterfalls, crystal clear river, living root bridge, enormous caves, natural sceneries it possesses and a diverse cultural heritage of the tribes living here. Even though tourism is still at its emerging stage, West Jaintia Hills District offers many natural tourist attractions to name a few the Krangsuri Waterfall, Nartiang Village the summer palace of the Jaintia Kings, Nartiang Monoliths a collection of mighty stones, Umngot crystal clear river, homestay, resorts, camping, adventure activities, Ialong Park etc. Its cultural heritage, festivals, inheritance and the innovative and skillful craftmanship by the tribal communities of the district in cane and bamboo tasks include the construction of bamboo vessels for storing water, long and large baskets (khoh-tkot) or panniers (polo) trays, mats and basket of various sizes and shapes. Jowai is the headquarter of West Jaintia Hills District.

5.2 Methods of Data Collection

1. **PRIMARY DATA COLLECTION:** For the purpose of the study primary data is collected through observation.
2. **SECONDARY DATA COLLECTION**: For the purpose of the study secondary data is gathered from several research articles, journals and other electronic websites.

6. PROSPECTS OF TOURISM IN WEST JAINTIA HILLS DISTRICT

CULTURAL HERITAGE: West Jaintia Hills District is the sub tribal home of one of Meghalaya's major tribes, the Jaintias or Pnars. They are thought to be the descendants of the first Mongolian influx into India, and they speak the Austro-Asiatic language family's "Pnar". The tribe's cultural legacy is expressed in their traditions, dances, art and craft. They are skilled in weaving, wood carving, cane and bamboo, pottery, making and designing traditional musical instrument and jewellery for various festivals and ceremonies.

While the majority of tribal people in this district professed Christian faith, the rest practiced the old tribal religion known as "Ka Niamtre". They considered that their religion was given by God rather being formed by man and emanated from God's orders. As a result, Niamtre emphasizes the importance of giving equal weight with other humans so as to achieve God's fulfilment.

INHERITANCE: The tribes here practice a rare and unique matrilineal society in which children derive their identity and surname only from their mother. While this is fortunate for women because they are treated fairly, males also have rights because the father is always the head of the family. Immediately following the mother is the maternal uncle, who has complete control over his sister's family. Among this tribe, the youngest daughter, who will inherit and have custody of all the possessions and property, is responsible for caring and protecting the family. Rites and rituals are demanded of her. Marriage between members of the same clan is absolutely forbidden. After marriage, the groom lives with his wife.

TRADITIONAL DRESS: The traditional garment worn by the tribes of this District is created by talented local tribal weavers. During festivals, both men and women dressed up from head to toe in elaborate attire. Most tribal women traditionally wear gold and silver jewellery such as Khaila (earrings), kpein ksiar (gold necklace), sahkti (rings), khadu (bangles), paila (string of threaded beads) and pangsngiat (headgear). U Kynji (chain) is the only ornament men wear across the shoulder down to the waist.

FOR MEN: The tribal men's traditional attire comprises yuslein(dhoti), jymphong (sleeveless coat), yoospong (turban) and paila (string of threaded beads). The Doloi (Chief) are required to wear the jain-spong (turban), which represents their social position.

FOR WOMEN: This District tribal women own Khyrwang and Thohsaru made of eri silk found in the area. They hide their bodies with a velvet blouse and the lower part Thoh Khyrwang wrap around skirts up to the ankle. Then a cape-like muga silk material is worn across one shoulder and tucked with a safety pin. Kyrchah (cloth having checks pattern) is worn everyday which acts like an apron or worn over to cover the head post-harvest seasons.

6.1 Dance and Music

LAHO DANCE: It is a popular dance done undertaken by both young male and female dancers. While male wear limited jewellery but the female dressed in their best traditional attire with gold and silver ornaments. The traditional attire endures to be elegant and colourful and the attire continues to enchant the audience, particularly tourists. The dance begins with young men on either side of a women dancing in steps while holding arms together in coordination and harmony. No instruments are employed here, instead, a man with a wonderful recitation voice recites hilarious coupling rhythms to the delight of the audience, who bursts into laughter.

CHAD SUKRA: It is a joyful heart dance that is performed before sowing seed, i.e., the beginning of the spring season. The Tribal people who practiced the Ka Niamtre religion believed that before planting crops, one should pray to God to protect them from natural disasters. This practiced has been followed since time immemorial to pay homage to Ka Bei Rymaw (Mother Earth), who protects and feeds everyone on this planet in order for them to survive. Farming and agriculture are the main occupation of the tribe, but those who succeed in other sectors of employment gather together to celebrate life with dance and joy. This dance also indicated by sending signals and emotions to other villages, towns and cities that seeding activities will begin in their respective fields.

Chad Sukra provides an opportunity for both young and old people to showcase their talents in restoring cultural legacy such as dance, singings and drama. All of these abilities from the tribal people of this district will be showcased in keeping with the legacy of "U KHON YNNIAW WASA YNNIAW WASUNG YNNIAW TREP U YNNIAW SKUM" translates as "Sons and Daughters of the Seven Huts". This dance is an occasion with morally sound lessons and a message to everyone to start shaping their lives, to start fresh, a new beginning and a more successful happy year ahead by expelling negative energy. It reflects the distinctness and identity of the tribe.

MUSIC: The tribal people of the district, including men and women are friendly and welcoming and they are aware of their cultural heritage. They enjoy singing songs praising and praying to rivers, lakes, mountains and waterfalls. They enjoy music and play various instruments. Instruments like "Duitara, Sarong (both are stringed instrument), Ka Nakra, Padiah, Katasa (different types of drums), Marynken (Harp), Chuwiang, Tangmuri (Flutes). Ka Duitara is considered as the queen of Music"

BEHDEINKHLAM FESTIVAL: The Behdeinkhlam (driving away the demons) Festival is a Jaintia Tribe religious festival held by non-Christians who follow the Niamtre faith. It is observed yearly in the month of July by praying to God for a successful harvest during the season of planting. Women do not take part in the festival's dancing, but they play a vital role at home by providing sacrificial food to the spirits of their forefathers. The Doloi execute religious ceremonies and rituals for three days, an on the last day in the afternoon, people congregate in a spot called Aitnar, where both young and old males dance to the rhythms of the drums and flutes. The contest for a giant undressed beam by two opposing groups of people is the centrepiece of the event. In the event at Wah Aitnar, participants smear mud on one another.

It is good if it rains during the festival because the dancers become emotional whn rall adorned structures called Rots and posts of wood called Khnongs are bought into the pool of Wah Aitnar. The tribes' women donned in the most exquisite traditional attire completed with gold ornaments. Dad-Iawakor, a game akin to football, is played with a wooden ball between the northernmost and southernmost of the region. As a result, whichever team wins the game will have a strong harvest the next year. The iconic gesture of this celebration is when young men beat the roofs of each household with sticks of bamboo to drive out the bad spirit, plagues and sickness.

POTTERY: Tyrshang village in the Thadlaskein Block of the West Jaintia Hills District, around 18kms from the district capital Jowai, Practice black clay pottery. They produce a wide range of earthern cookware, storing containers and decorative products. Their most unique cookware is the biodegradable Kchu Lyrnai (a black earthenware pot used for making popular specialities such as Tpusein-steam rice bread). One crucial aspect of this pottery production is that the clay is purchased from Sung Village, a place that belonged to the Doloi of the Nartiang Elaka. The clay cannot be sold in any manner because the Doloi gifts it once a year to be part of the Durga Puja, hence it is only utilized with the Doloi consent.

6.2 Natural Attractions

1) **Thlu Muwi:** It is 16kms from Jowai on the Jowai to Dawki Road. Because of the rise of the Muwistream during the monsoon a bridge was built for the king's journey from Jaintiapur to Nartiang. The bridge consists of huge granite stone slabs perched on equally similar pillars. Besides the stone bridge is also the Muwi waterfalls.
2) **Nartiang Monoliths:** Located about 24kms from Jowai. The village being the summer capital of erstwhile Jaintiapur King and the biggest collection of monoliths stones in one single area is found north of the Nartiang market. It is the only village in the State where people are following Hindu religion and Durga Puja is celebrated every year. These consists of Mehirs (Upright Stones) Moo Shynrang and Dolmens (Flat stones in the horizontal position) locally known as Moo Kynthai.
3) **Ialong Park:** The Park is located bout 8kms from Jowai. It is known for its sacred grove and for its setting which overlooks the beautiful Pynthorwah Valley (Paddy Fields) on the banks of River Myntdu.
4) **Phe Phe Falls:** It is located in Amlarem Sub Division and is about 18kms from Jowai. Descending a steep hike, through the shrub bamboo plants, the sight of the waterfall in the backdrop of the blue sky and the range of hills layered with light and shadow on the other side entered the realm of paradise. The water is as blue as the sky, clean transparent and above all cold.
5) **Thadlaskein Lake:** It is located about 8kms from Jowai. It is known locally as Pung Sajar Nangli, it is a man-mad natural lake. Traditions shares it that thelake was dug by the tip of the toes of bows by the followers of Jaintia Chieftain U Sing Nangli
6) **Jarain Pitcher Plant:** It is located in Jarain Village of Amlarem Civeil Sub-Division, the Lake is about 19 kms from Jarain. It is centered around Pitcher Palnt theme to attract attention of visitors to the Pitcher Plant which is not only unique to the area but also unique for its carnivorous qualities.
7) **Krangsuri Falls:** One of the most popular, beautiful and captivating waterfalls. It is located about 30kms from Jowai West Jaintia Hills District. Krangsuri waterfall is cradle in the valley between

the hills, majestic sight which can be glimpsed from viewpoints. Legends has it that after a battle between warring clans, the cave behind the waterfall also served as an armoury.

8) **Dawki:** Located about 54kms from Jowai and 91 kms from Shillong, it is famous for the Umngot River which receives most tourist attractions in from across the world and the heritage Dawki Bridge built by the British which straddles between East Khasi Hills and West Jaintia Hills District. Dawki is also located on the International Border with Bangladesh. It is a tourist destination for those interested in adventure tourism and provides for zip lining, scuba diving, outdoor camping, snorkelling, rock climbing and trekking. One of the unique features of Umngot River which flows through Shnongpdeng and Dawki is that it assumes aquamarine hues, in which the river bed is clearly visible from the surface.
9) **Amkoi:** It is located about 36kms towards South from District headquarters Jowai. It is well-known for having rock structures that caught the eye of many tourists and explores. The rocks there do not have the usual stony look and have been naturally smoothened. They look like they have been painted with different faded colours.
10) **Kudengrim Village:** Located about 52 kms from Jowai, 23 kms from Amlarem is another village in the Sub-Division famous for gushing waterfalls and unique living root bridge which is 20 minutes' hike away from the main village along the rushing streams and views of green valley. The following are the main natural attraction of the village:
    - **Amlamar Double Decker Living Root Bridge:** Locals name is LeU Chrai Amlamar. The Amlamar river is crossed by this 200-year-old double-decker living root bridge. It is Meghalaya's sole living root bridge with an arched entrance.
    - **Amkshar Living Root Bridge:** It is Meghalaya's only single highest living root bridge, standing 72.1 feet over the Amkshar river. During the period of the Jaintiapur Kings reign, locals of nearby village like Padu, Jarain and Syntung used this 400-year-old living root bridge for crossing to Dawki and Shylet for commercial purposes.
    - **Rbangpohpamlong Waterfall:** The Amlamar river distributes the Rbangpohpamlong waterfall, leading to have a high height, splattering and roars throughout its rocky route, then gradually rushes down. The waterfakk erupts in fill force.
    - **Mukre Multi-Tier Waterfall:** The glittering water flowing down in a succession of connected waterfalls, fed by the Amlamar river, makes Mukre Multi-Tier cascade an amazing sight to explore.
    - **Sacred Groves:** There are five Sacred Groves located within the village, they are Basan Thok Tekur, Basan Amlamar, Sasan Amkhlow, Basan Talang and Basan Pohshwet. The scared groves have been safeguarded by the locals since ages. They adhere to ancient religious beliefs that regards forests as sacred matters. The cutting of trees, picking and plucking of leaves, twigs, flowers, fruits, spitting and throwing anything in the forest are all prohibited. These sacred forests are home to a diverse array of flora (medicinal plants and trees, wild orchids and pitcher plants) and rare fauna.
    - **Fish Sanctuary:** The fish sanctuary is teeming with indigenous fish locally known as Kha Saw. Fishing is prohibited in this area of protection. The local inhabitants believed that the fish from this sanctuary aid in disease recovery. Furthermore, only the sick, elderly and pregnant women are permitted to consume the fish upon approval from the local headman.
11) **Trangblang/Rangkali Falls:** Trangblang is a village in Amlarem Sub-Division in West Jaintia Hills District. It is located at around 50 kms from Jowai. The village is covered with lush green

vegetation, forest and waterfalls. The Rangkali waterfall has the potential to be the highest waterfall in the whole Jaintia Hills Region.

12) **Rupasor Bathing Ghat (Royal Bathing Pool):** Located at around 60 kms of the Jowai-Amlarem-Muktapur Road, Rupasor Bathing Ghat also known as "Ka Mahadei swimming Pool" is a beautiful rectangular swimming pool. Equidistant from both the corners in the north-end of the pool there is a replica of an elephant carved out of the same rocks to serve as a platform for the bathing royals to sit upon. There are two huge rocks bearing the design of the Sun and the Moon. The findings of the figures together in a place expresses the symbol of fertility cult, as the Moon symbolises the Male and the Sun symbolises the Female.
13) **Krem Syndai:** It is about 47 kms away from Jowai, headquarters of the West Jaintia Hills District. It is the third longest cave in India. Krem Syndai lie deep inside the heart of the dense forests of Jaintia Hills. It is a major tourist attraction consisting of 11 independent caves, together form the Syndai Cave Systems.
14) **Mulieh Village:** It is 30kms away from the district headquarter Jowai. The highest-quality turmeric in the world is found in India, inside the untamed landscapes of West Jaintia Hills District, Mulieh village is famous for the production of Lakadong Turmeric. Tourist who visits the village will get learn about the production process of Lakadong turmeric, relish the local delicacies while admiring the picturesque landscape. It is well-known for its great quality and wide range of health benefits. Its high curcumin concentration of 7-12% has been beneficial for any health issues, it acts as an anti-inflammatory, antioxidants capabilities, and the treatment of many ailments. When compared to conventional turmeric, Lakadong Turmeric has a somewhat peppery and plain flavour.

7. CHALLENGES OF TOURISM FACED BY WEST JAINTIA HILLS DISTRICT

1) **Lack of Infrastructure:** Few years ago, there were no hotels/resorts or homestays in West Jaintia Hills District. Tourists had to stay in Shillong the capital of Meghalaya, to make their visit to different tourism destinations. Now-a-days, there are few homestays and hotels /resorts in West Jaintia Hills District and all of them are located far from the tourist destinations which makes the accessibility expensive and till date there is not a single budget hotel or homestay. The road connectivity to various tourist spot towards West Jaintia Hills District is very poor and rugged, which makes tourist discouraged to visit such destinations.
2) **Physical Seclusion:** The tourism in West Jaintia Hills District is still in its primitive stage. The entire tourism destinations are scattered all around with no proper shops or commercial areas, in most destinations there are no shops at all or no information centre to provide information to tourists.
3) **Inflexible Climate:** During the monsoon season, there is no tourism activities happening in West Jaintia Hills District because most of the activities are outdoor, tourists must spend their entire day in the hotel, resort or homestay itself. There are fun activities conducted by the hotel or resort to enhance their experience at the destination.
4) **Deficient Transportation:** In order to travel and explore the tourism in West Jaintia Hills District the transportation facility is very poor and minimal because the only mode of accessibility is by roadway. For visiting the tourist destinations, one has to hire a cab from Shillong and if stayed in any resorts or homestay, the tourist has to inform prior in advance for hiring taxis to the resort authority so that such transport facility is arranged to the tourist. There is no public transportation

and very few tourist taxi drivers who are will to take tourist to any of the tourist destinations in West Jaintia Hills District by charging a higher rate for reserving.

5) **Insufficient Support from the Government:** Despite being bestowed with unspoiled and natural beauty with rich cultural heritage. The local people inhabited in West Jaintia Hills District are well known for protecting the diversity of life, traditional beliefs and customs from ancient time but, the tourism industry here is deprived of getting proper support from government through various Schemes for the development of tourism.
6) **Lack of Tourism Marketing Strategy:** Improper tourism marketing strategy is one of the biggest challenges faced by West Jaintia Hills District tourism industry. In this age of online tourism, where pictures, videos, reviews and marketing acts as the selling point of any destination it is very important for any tourist destinations to have a well-designed website which contains all the necessary information about the region. In the case of West Jaintia Hills District tourism industry, it has been found that the tourism website lacks necessary information about the tourist attraction places, feedback and lack of creativity in the web advertisement.
7) **Amenities in the roadside:** because the journey to the tourist spots in West Jaintia Hills District is by roadway, it is critical to provide clean, comfortable, sanitise, and safe amenities and facilities along the major routes. However, services such as clean drinking water, sanitised restrooms, ample parking and medical support such as pharmacies are insufficiently available to cater to tourists travelling to key attractions.
8) **Signage:** proper signage provides clear, understandable communication as well as information such as dos and don'ts and a 24-hour emergency number. However, the bulk of tourist spots in West Jaintia Hills District lack sufficient and obvious signage to guide tourists throughout their visit to the region. As a result, the tourist is discouraged from learning about the significance of the destination and return will no knowledge to share with friends and relatives.

8. FINDINGS

West Jaintia Hills District (WJHD) has a total area of 1693 square kilometres and share borders with Assam and Bangladesh. The climate here is pleasing, warm and not too hot during summers or too cold during winter, with rains beginning in mid-May and lasting until the end of September. On non-rainy days, the sky is sunny and clear. WJHD's physical geographical location has resulted in a variety of natural attractions and treasures of nature such as sacred forest and minerals being discovered here. The regions inhabitants have been conserving their cultural legacy and ethnological uniqueness by retaining and adhering to the matrilineal social hierarchy.

1) Due to its geographical seclusion, tourism in West Jaintia Hills District has received less attention than other tourist destinations in Meghalaya.
2) With fewer details about the tourist destinations in WJHD on the internet and tourism marketing as a promotional approach, it has been discovered to be quite insufficient.
3) WJHD has a lot of potential for tourism because it has a lot of unspoiled natural attractions and an array of fascinating cultural history, but the tourist's infrastructure is still weak and neglected compared to other destinations for tourism in the State.

4) Another key hurdle or reason for less tourist growth in WJHD is a lack of effective government support and central sector schemes.
5) Some of the destinations are dispersed throughout WJHD, which is significant for any tourist attraction to establish souvenirs shops, restaurants and homestay in a certain region and in a particular spot, but this is not the case with WJHD.
6) WJHD has few and insufficient transportation options. To visit, one must hire a vehicle from Shillong, otherwise, it is difficult for tourist to depart from the area, where tourist drivers charge exorbitant fares.
7) There is no awareness about the importance of tourism as a source of income into the livelihood of the local people in WJHD since there are no workshops or seminars conducted to help train the benefits of tourism.
8) The media should portray the right image of North East India, since most tourist vision North East a region of militancy unsafe and dense forest which is practically not true because the North East India lies on the Eastern Himalayan Region which have the potentiality of untapped natural beauty.
9) WJHD has the potential for caving, one of the longest caves in India is found here, and many more independent caves. But no special promotion is done to attract adventure enthusiast to explore these caves.
10) Most of the tourist destinations in WJHD are managed and preserved by the local communities along with the village headman. Very few tourist properties are managed by the government.

9. SUGGESTIONS

1) It has been suggested that the government should give support and guidance for tourism development in WJHD because there is a wide scope for the tribal to showcase their practices in preserving their cultural heritage and various tourism development through central schemes assistance.
2) It has been suggested that a proper planning and an overview of the infrastructure gaps and loopholes must be undertaken for the success of the tourism industry in WJHD.
3) It has also been suggested that appropriate steps and procedures are taken to focus tourism in WJHD and bring it in the lime light, so that the tourism industry of the region will flourish and develop along with community participation.
4) Most of the destinations in WJHD are secluded and local people will provide the information and assistance to various tourists. In addition to this, the concern of security for tourist needs to be managed well.
5) Infrastructure should be developed. It is not only the sole responsibility of the government to develop tourism in WJHD but also the local people should give their major concerns and co-operation for this because tourism is the prime source of revenue for them. More homestays, souvenirs shops, restaurants etc. should be open.
6) Effective implementation and promotional marketing strategy will help change the perception of the tourists about North-East as well as attract more tourist to explore the tribes living in WJHD in the future.
7) Thes strategy development and execution that will be executed for the growth of the tourism industry in WJHD must follow comprehensive procedures and guidelines that take into account all aspects of tourism related challenges of the future.

10. CONCLUSION

The tourism destinations in WJHD Meghalaya, in the foothills of the Eastern Himalayas Region is gifted with wild natural beauty and its unique cultural heritage, but most of the tourist are not aware about this. There is a wide scope for tourism to flourish and grow in WJHD in the future. The tribal people of WJHD and their representative head (Headman, Chieftain) should be involved in fostering the growth of tourism. This participation will play a significant role in preparing the regions tourism potential by maintaining the regions age old cultural history, traditions, beliefs, art and craft. Furthermore, all of this will serve as an attraction for various tourists who are going to discover about the tribes past by witnessing their participation in various social events and day-to-day lifestyles that have been historically preserved from time immemorial and continue to be practiced till today. So, it is the prime responsibility of the local people and the government of Meghalaya to work together to take initiatives and come up with proper plans and strategies for the promotional and development of tourism industry in WJHD which will help to preserve the tribe's culture and traditions and improve the economy of the State as well as the country.

REFERENCES

Barman, A., Singh, R., & Yedla, V. R. (2010). Empowering Tribes Through Cultural Tourism in India- A dream Project on ICT Integration. SSRN *Electronic Journal*. DOI:. 1718126 doi:10.2139/ssrn

Bera G (2015). A Brief Account of Tourism Potentiality in Shillong and its Surroundings. *IOSR Journal, 20*(9).

Bhutia S. (2020). Sustainable Tourism for Regional Development in North-East States of India: Trends, Problems and Prospects. *Global Journal of Human Social Science.*

BillaChowdhary, S., & Tiwari, N. (2022). *Indian Tourism Diaspora perspectives*. Emerald Publishing Limited.

Kumar, S. (2014). *Alternate tourism in tribal destinations A study of Himachal Pradesh*. Shodhganga.

Laloo, D., Cajee, L., & Burman, D. (2007). *Potential for Cultural and eco-tourism in North-East India: a community-based approach*. Research Gate.

Lyngdoh, B. (2022). An Exploration of Tourism Development of Meghalaya, India: What is the way forward? *Atna-Journal of Tourism Studies, 17*(1), 143–166. doi:10.12727/ajts.27.6

Meena, K. (2021). *Role of Media in Development A case Study of Tribal Communities of Sirohi District.*

Nayak, P., & Mishra, K. S. (2013). *Problems and Prospects of Promoting Tourism in Meghalaya*. North Eastern Hill University.

Patel, A. (2021). *An Analytical Study of a Gujarat as a Tourism Destination with a Special reference to Tribal Areas of Gujarat*. Shodhganga.

Peinlang, B. L. (2019). The Scope of Tourism in Meghalaya. *J Tourism Hospit, 8*(1), 397. doi:10.35248/2167-0269.19.8.397

Raju, S. S. (2019). *Future Scope of Tourism Courses: India Perspective*. Research Gate.

Saha, G. K. (2015). Tourism Sector in the North Eastern States- Challenges and Opportunities [IJDR]. *International Journal of Developmental Research*, *5*(11), 6077–6081.

Saha, K. S., Babai, C. T., & Saha, S. (2018). *Adventure Tourism in Meghalaya*. Research Gate.

Saha, K. S., Suiam, S., & Sarangi, A. (2022). *Rural Tourism and its Impact on the Economy: A Study of Ialong Village*. doi:10.4018/978-1-7998-9664-7.ch015

Samsar, H. (2017). *Impact of Social Media on Tourism Industry*. Research Gate.

Sohtun M L (2022). *A Study of Adventure Tourism in The East Khasi Hills of Meghalaya*. Research Gate.

Verma, S. (2021). *An Analytical Study of Tribal Tourism in Southern Rajasthan*. Research Gate.

Vyas, N. (2016). *Tourism Growth Development and its impact – A study of Tribal Districts of Himachal Pradesh*. Research Gate.

Yogesh, K. (2019). Challenges and Solutions to the Development of the Tourism and Hospitality Industry in India. *African Journal of Hospitality Tourism and Leisure*.

Zou, J. M. M. L., & Prasain, G. P. (2017). North-Eastern States Tourism: Exploring the Unexplored on Earth. *IOSR Journal of Humanities and Social Science (IOSR-JHSS), 22*(8), 35-40.

Compilation of References

Abdollahzadeh, G., & Sharifzadeh, A. (2014). Rural Residents' Perceptions Toward Tourism Development: A Study from Iran. *International Journal of Tourism Research, 16*(2), 126–136. doi:10.1002/jtr.1906

Acharjee, M. L., & Barat, S. (2010). Impact of fishing methods on conservation of ichthyofauna of river Relli in Darjeeling Himalaya of West Bengal. *Journal of Environmental Biology, 31*(4), 431–434. PMID:21186715

Ajitha, S. (2004). Tourism - The Other Side. *The Journal of Business Studies, 2*(1), 87.

Amed, A., Rawat, J. S., & Rai, S. C. (1990). 'An analysis of the Himalayan environment and Guidelines for its Management and Ecologically Sustainable Development. *The Environmentalist, 10*(4), 281–298. doi:10.1007/BF02239722

An, W., & Alarcón, S. (2020, September 19). How Can Rural Tourism Be Sustainable? *A Systematic Review.* https://scite.ai/reports/10.3390/su12187758

Anand, A. (2017). Tourism Development in Munnar: A Socio-Cultural Perspective. In S. Das & R. N. Sahoo (Eds.), *Tourism Marketing* (pp. 301–311). Springer.

Anandan, (2004). Marketing of tourism products. *The Journal of Business Studies,* 2(1), 94.

Andereck, K. L., & Nyaupane, G. P. (2011). Exploring the nature of tourism and quality of life perceptions among residents. *Journal of Travel Research, 50*(3), 248–260. doi:10.1177/0047287510362918

Andrews, A. (2007). *Introduction to Tourism & Hospitality Industry.* Tata McGraw Hill, West Patel Nagar.

Apollo, M. (2016). The clash–social, environmental and economical changes in tourism destination areas caused by tourism the case of Himalayan villages (India and Nepal). *Current Issues in Tourism Research, 5*, 6–19.

Apollo, M., & Andreychouk, V. (2022). *Mountaineering Adventure Tourism and Local Communities: Social, Environmental and Economics Interactions.* Edward Elgar Publishing. doi:10.4337/9781802209389

Apollo, M., Andreychouk, V., Mostowska, J., Jin, Z., Maciuk, K., Rettinger, R., & Jones, T. E. (2021). Himalayan nature-based tourism. Potential, regional diversity, nature conservation and touristic load. *International Journal of Conservation Science, 12*(4).

Apostolopoulos, N., Liargovas, P., Stavroyiannis, S., Makris, I. A., Apostolopoulos, S., Petropoulos, D. P., & Anastasopoulou, E. (2020, September 17). *Sustaining Rural Areas, Rural Tourism Enterprises and EU Development Policies: A Multi-Layer Conceptualisation of the Obstacles in Greece.* ScitePress. https://scite.ai/reports/10.3390/su12187687

Arora, V. (2006). The forest of symbols embodied in the Tholung sacred landscape of North Sikkim, India. *Conservation & Society*, 55–83.

Arunmozhi, T., & Panneerselvam, A. (2013). Types of tourism in India. *International Journal of Current Research and Academic Review*, *1*(1), 84–88.

Badola, R., Hussain, S. A., Dobriyal, P., Manral, U., Barthwal, S., Rastogi, A., & Gill, A. K. (2018). Institutional arrangements for managing tourism in the Indian Himalayan protected areas. *Tourism Management*, *66*, 1–12. doi:10.1016/j.tourman.2017.10.020

Bagri, S. C., & Kala, D. (2016). Residents' Attitudes toward Tourism Development and Impacts in Koti -Kanasar, Indroli, Pattyur Tourism Circuit of Uttarakhand State, India. *Pasos (El Sauzal)*, *14*(1), 23–29. doi:10.25145/j.pasos.2016.14.002

Bahrami, R., & Noori, K. (2013). Analysis Of The Role Of Tourism And Its Impact On Rural Development (Case Study Of The Central Part Of Marivan). *Technical Journal of Engineering and Applied Sciences*, *3*(12), 1074–1080.

Baishya, K. (2017). Adventure Tourism in India. *EPRA International Journal of Economic and Business Review.*, *5*(10), 50–55.

Banerjee, S. K., Pal, D. K., & Banerjee, S. P. (1984). Soil characteristic and Floristic composition of some sub-alpine forests of Darjeeling (West Bengal). *Proc. Indian. Nat. Sci. Acad.*, *50A*(3), 269–277.

Bao, J., & Sun, J. (2003). The community participation in tourism planning. *Planners*, *19*(7), 32–38.

Barber, R. (1993). *Pilgrimages*. The Boydell Press.

Barman, A., Singh, R., & Yedla, V. R. (2010). Empowering Tribes Through Cultural Tourism in India- A dream Project on ICT Integration. SSRN *Electronic Journal*. DOI:. 1718126 doi:10.2139/ssrn

Barooah, S. (2018). Touristy feet in muddled Northeast India. Paper presented in *National Conference on Livelihoods through Rural Tourism in Northeast India: The Way Forward*. Research Gate.

Baruca, P. Z., & Civre, Z. (2012). *How do guests choose a hotel*. Academica Turistica- Tourism and Innovation Journal.

Bashir, S., & Muthueleckuvan, R. (2018). Effect of mountaineering training on selected physiological variable among moderate altitude inhabitants. *International Journal of Physiology. Nutrition and Physical Education*, *3*(2), 720–722.

Basu, S., & Starkel, L. (2000). Rains, Landslides and Floods in the Darjeeling Himalaya. Indian National Science Academy, New Delhi.

Basu, S., Singh, J., Shrivastava, S.K., Paliwal, A., & Chakravartty, A. (2016). *Remembering 2013 Uttarakhand floods.*

Batta, R. N. (2000). *Tourism & the Environment A Quest for Sustainability*. Indus Publishing Company.

Bennike, R. B., & Nielsen, M. R. (2023). Frontier tourism development and inequality in the Nepal Himalaya. *Journal of Sustainable Tourism*, *0*(0), 1–22. doi:10.1080/09669582.2023.2174129

Bera G (2015*)*. A Brief Account of Tourism Potentiality in Shillong and its Surroundings. *IOSR Journal, 20*(9).

Berle, P. A. A. (1990). Two faces of eco-tourism. *Audubon*, *92*(2), 6.

Bhalla, R., & Chowdhary, N. (2022). Green workers of Himalayas: Evidence of transformation induced regeneration. *Journal of Tourism Futures*, *8*(3), 380–392. doi:10.1108/JTF-12-2021-0273

Bhattacharya, D., & Mitra, A. (2013).. . *Paripex-Indian Journal of Research*, *2*(9), 222–224.

Bhat, Z. A. (2013). Tourism industry and pilgrimage tourism in Jammu and Kashmir: Prospects and challenges. *International Journal of Research in Management & Technology*, *2*, 105–113.

Bhutia S. (2020). Sustainable Tourism for Regional Development in North-East States of India: Trends, Problems and Prospects. *Global Journal of Human Social Science.*

Bhutia, K. D. (2018). Foxes, Yetis, and Bulls as Lamas: Human-Animal Interactions as a Resource for Exploring Buddhist Ethics in Sikkim. *Journal of Buddhist Ethics, 25*, 45–70.

Bhutia, S. (2014). Economic development and environmental issues in Darjeeling Himalaya of West Bengal, India: A theoretical perspective. *International Journal of Humanities and Social Science Invention, 3*(7), 42–47.

Bhutia, S. (2014, April). Human Response on Socio-Economic Development in Darjeeling Himalaya of West Bengal. *Asian Academic Research Journal of Social Sciences and Humanities, 1*(22), 166–173.

Bhuyan, R., & Deka, S. M. (2019). Homestay for whom-lessons of ASEAN countries for North East India. *International Journal of Tourism and Travel, 12*(1&2), 21–26.

BillaChowdhary, S., & Tiwari, N. (2022). *Indian Tourism Diaspora perspectives.* Emerald Publishing Limited.

Birkett, B., & McGrady, C. (Eds.). (2000). *Classic Treks: The 30 Most Spectacular Hikes in the World.* Bulfinch.

Biswas, S. K. (1990). Strategy of Development in the Himalaya: A Profile of Socio-Economic Change. Institute of Social Research and Applied Anthropology, Calcutta.

Blackstock, K. (2005). A critical look at community-based tourism. *Community Development Journal: An International Forum, 40*(1), 39–49. doi:10.1093/cdj/bsi005

Blesic, I., Pivac, T., Dordevic, J., Stamenkovic, I., & Janicevic, S. (2014). Cultural Events As Part Of Cultural Tourism Development. Case Study: Sombor And Apatin (Serbia). Geografski Zbornik. *Acta Geographica Slovenica, 54*(2), 381–390. doi:10.3986/AGS54406

Boo, E. (1990). *Ecotourism: the potentials and pitfalls* (Vols. 1–2). World Wildlife Fund.

Borst, B. 1990, Ecotourism: environment friendly international travel. *International Herald Tribune.*

Briassoulis, H. (2002). Sustainable tourism and the question of the commons. *Annals of Tourism Research, 29*(4), 1065–1085. doi:10.1016/S0160-7383(02)00021-X

Brida, J. G., Disegna, M., & Osti, L. (2014). Residents' perceptions of tourism impacts and attitudes towards tourism policies. *Tourismos:An International MultiDisplinary Journal of Tourism, 9*(1), 37–71.

Buckley, R., & Pandey, P. (2017). Trekking tourism in the Indian Himalayas: A case study of environmental impacts and management strategies. *Journal of Outdoor Recreation and Tourism, 19*, 40–50.

Burman, P. D., Cajee, L., & Laloo, D. D. (2007). Potential for cultural and eco-tourism in Northeast India: A community-based approach. *WIT Transactions on Ecology and the Environment*, 102.

Callan, Roger & Bowman, Lisa. (2000). Selecting a hotel and determining salient quality attributes: A preliminary study of mature British travelers. *International Journal of Tourism Research.*

Canter, W. L. (1996). *'Environmental Impact Assessment* (2nd ed.). McGraw Hill Inc.

Chakraborty, S. (1976, September). The Role of Transportation in the Economy of the Hill Areas of the Darjeeling District. *Indian Journal of Hill Economy, Darjeeling, 1*, 15–34.

Chakraborty, S. (1980, September). A Geographical Sketch of Darjeeling District. *Indian Journal of Hill Economy, Darjeeling, 5*, 1–7.

Chakravarty, S., Suresh, C. P., Puri, A., & Shukla, G. (2012). North-East India, the geographical gateway of India's phytodiversity. *Indian Forester, 138*(8), 702.

Chand, R., & Leimgruber, W. (2016). *Globalization and marginalization in Mountain Regions: Assets and Challenges in Marginal Regions*. Springer. doi:10.1007/978-3-319-32649-8

Chandra, P., & Kumar, J. (2021). Strategies for developing sustainable tourism business in the Indian Himalayan Region: Insights from Uttarakhand, the Northern Himalayan State of India. *Journal of Destination Marketing and Management, 19*, 100546. doi:10.1016/j.jdmm.2020.100546

Chang, L., & Beise-Zee, R. (2013). Consumer perception of healthfulness and appraisal of health-promoting tourist destinations. *Tourism Review, 68*(1), 34–47. doi:10.1108/16605371311310066

Chan, R. C. K., & Bhatta, K. D. (2023). Trans-Himalayan connectivity and sustainable tourism development in Nepal: A study of community perceptions of tourism impacts along the Nepal–China Friendship Highway. *Asian Geographer, 40*(1), 59–80. doi:10.1080/10225706.2021.1943470

Chaudhary, M., & Lama, R. (2014). Community Based Tourism Development in Sikkim of India- A study of Darap and Pastanga Villages. *Transnational Corporations Review, 6*(3), 228–237. doi:10.5148/tncr.2014.6302

Chauhan, N., Upadhyay, S. K., & Singh, R. (2021). The Himalayan natural resources: Challenges and conservation for sustainable development. *Article in Journal of Pharmacognosy and Phytochemistry, 10*(1), 1643–1648. www.phytojournal.com

Chettri, L., & Sharma, B. K. (2021). *Some Traditional Folk Songs and Dances of Sikkim Himalayas*.

Chettri, P. (2022). *Exploring the Potentiality of Cultural Tourism in Sikkim: A Study on Lepcha*.

Chirenje, L. I., Giliba, R. A., & Musamba, E. B. (2013). Local communities' participation in decision-making processes through planning and budgeting in African countries. *Zhongguo Renkou Ziyuan Yu Huanjing, 11*(1), 10–16. doi:10.1080/10042857.2013.777198

Chistyakova, O. V. (2020). Ethnic tourism as knowing other. *RUDN Journal of Philosophy, 24*(4), 720–729. doi:10.22363/2313-2302-2020-24-4-720-729

Chroneos-Krasavac, B., Radosavljević, K., & Bradić-Martinović, A. (2018, January 1). *SWOT analysis of the rural tourism as a channel of marketing for agricultural products in Serbia*. ScitePress. https://scite.ai/reports/10.5937/ekopolj1804573k

Chu, R., & Choi, T. (2000). *An importance-performance analysis of hotel selection factors in the Hong Kong hotel industry: A comparison of business and leisure travellers*. Tourism Management - Elsevier BV.

Clawson, D. L., & Johnson, M. L. (2004). World Regional Geography: A Development Approach, 8th ed. 36–37.

Clow, K. E., Garretson, J. A., & Kurtz, D. L. (1994). An exploratory study into the purchase decision process used by leisure travellers in hotel selection. *Journal of Hospitality & Leisure Marketing*.

Coelho, R. A., & George, B. P. (2018). Spiritual tourism in Kerala, India: An analysis of foreign tourist perspectives. *Journal of Heritage Tourism, 13*(2), 107–121.

Collins-Kreiner, N. (2018). Pilgrimage-Tourism: Common Themes in Different Religions. *International Journal of Religious Tourism and Pilgrimage, 6*, 3.

Critical Ecosystem Partnership Fund. (2005). *Ecosystem profile- Eastern Himalayan Region*. Critical Ecosystem Partnership Fund.

Critical Ecosystem Partnership. Fund. (2007). *Eastern Himalayas region.* Critical Ecosystem Partnership Fund.

D'souza, T. B., & Puttaraju, S. (2000). Examining the influence of trekking in nature of adaptive response to stress in individuals. *International Journal of Indian Psychology*, *8*(1), 913–923.

Dam, S. (2013). Issues of sustainable ecotourism development in Sikkim: An analysis. *South Asian Journal of Tourism and Heritage*, *6*(2), 32–47.

Dash, M. C., & Dash, S. P. (2010). Fundamentals of Ecology. McGrew-Hill Companies, Tata McGrew-Hill Education Private Limited, New Delhi.

Das, M. B., Kappor-Mehta, S., Tas, E. O., & Zumbyte, L. (2015). *Scaling the Heights: Social Inclusion and Sustainable Development in Himachal Pradesh. The World Bank.* International Bank for Reconstruction and Development / The World Bank.

Datta, P., & Behera, B. (2022). Assessment of adaptive capacity and adaptation to climate change in the farming households of Eastern Himalayan foothills of West Bengal, India. *Environmental Challenges*, *7*, 100462. doi:10.1016/j.envc.2022.100462

Dayananda, K. C., & Leelavathi, D. S. (2016). Tourism Development, Economic and Employment Growth in India. *IOSR Journal of Humanities and Social Science (IOSR-JHSS), 21*, 35-40.

de la Torre, M.-V., Genoveva, L. P.-N., & Martínez-Cárdenas, R. (2016). Factores que Determinan el Crecimiento del Turismo en Destinos Religiosos. [RCS]. *Revista de Ciencias Sociales*, *22*, 85–97.

Deyshappriya, N. P. R., & Nawarathna, A. M. D. B. (1970, January 1). Tourism and SME Development: Performance of tourism smes in coastal tourist destinations in southern Sri Lanka. *EconStor*. https://www.econstor.eu/handle/10419/238521

Dey, T., Pala, N. A., Shukla, G., Pal, P. K., & Chakravarty, S. (2017). Perception on impact of climate change on forest ecosystem in protected area of West Bengal, India. *Journal of Forest and Environmental Science*, *33*(1), 1–7. doi:10.7747/JFES.2017.33.1.1

Digance, J. (2006). Religious and Secular Pilgrimage: Journeys Redolent with Meaning. In *Tourism, Religion and Spiritual Journeys* (pp. 52–64). Routledge.

District of Lahaul and Spiti. (n.d.). *About District.* District of Lahaul and Spiti. https://hplahaulspiti.nic.in/geographical-conditions/

Dorsey, E. R., Steeves, H. L., & Porras, L. E. (2004). Advertising ecotourism on the internet: Commodifying environment and culture. *New Media & Society*, *6*(6), 753–779. doi:10.1177/146144804044328

Duran-Sanchen. (2018). *Religions*, *9*(9), 249. doi:10.3390/rel9090249

Dutta, M., & Dutta, P. K. (2019). Community-based tourism model for conservation: A case study of Arunachal Pradesh, part of Himalaya Biodiversity Hotspot. SSRN *Electronic Journal*. doi:10.2139/ssrn.3376193

Dutta, P. K. (2015). *Guidelines for the promotion of homestays in Arunachal Pradesh.* WWF, GOI-UNDP Project.

El Moslem Badr, D. M. (2022). Challenges and future of the development of sustainable ecotourism. *International Journal of Modern Agriculture and Environment*, *2*(2), 54–72. doi:10.21608/ijmae.2023.214937.1006

Erdoğan, S., Gedikli, A., Cevik, E. I., & Erdoğan, F. (2022). Eco-friendly technologies, international tourism and carbon emissions: Evidence from the most visited countries. *Technological Forecasting and Social Change*, *180*, 121705. doi:10.1016/j.techfore.2022.121705

Ewert, A. (1985). Why people climb: The relationship of participant motives and experience level to mountaineering. *Journal of Leisure Research*, *17*(3), 241–250. doi:10.1080/00222216.1985.11969634

Farrukh, B., Younis, I., & Longsheng, C. (2023). The impact of natural resource management, innovation, and tourism development on environmental sustainability in low-income countries. *Resources Policy*, *86*, 104088. doi:10.1016/j.resourpol.2023.104088

Fennel, D. (2008) *Ecotourism*. Rutledge.

Fernandes, J. P., & Guiomar, N. (2018). Nature-based solutions: The need to increase the knowledge on their potentialities and limits. *Land Degradation & Development*, *29*(6), 1925–1939. doi:10.1002/ldr.2935

Finnicum, P., & Zeiger, J. C. (1996). Tourism and wellness: A natural alliance in a natural state. *Parks & Recreation*, *31*(9), 84–90.

Gangotia, A. (2013). Home stay scheme in Himachal Pradesh: A successful story of community based tourism initiatives (CBTIS). *Global Research Analysis*, *2*(2), 206–207.

Gansser, A. (1974). Himalaya. *Special Publication - Geological Society of London*, *4*(1), 267–278. doi:10.1144/GSL.SP.2005.004.01.15

Gansser, A. (1981). The geodynamic history of the Himalaya. *Geodynamic Series*, *3*, 111–121. doi:10.1029/GD003p0111

Gaurav, R. & Kumar, S. (2016). *Prospects and Problems of Tourism Industry in Uttarakhand* (Vol. 4, pp. 277–284). Prashant Gautam, Sandeep Walia.

Gebreyesus, A., Tesfay, G., Tesfay, M., & Meheretu, Y. (2020). *Opportunities and challenges of community-based tourism development in Southern Tigray.*

George, M., & Prakash, G. (2021). Rural Tourism in Munnar: Challenges and Opportunities. *Journal of Sustainable Tourism*, *29*(8), 1090–1106.

Ghosh, M., & Ghosal, S. (2020). Determinants of household livelihood vulnerabilities to climate change in the Himalayan foothills of West Bengal, India. *International Journal of Disaster Risk Reduction*, *50*, 101706. doi:10.1016/j.ijdrr.2020.101706

Ghosh, M., & Ghosal, S. (2021). Climate change vulnerability of rural households in flood-prone areas of Himalayan foothills, West Bengal, India. *Environment, Development and Sustainability*, *23*(2), 2570–2595. doi:10.1007/s10668-020-00687-0

Gladstone, D. L. (2013). *From pilgrimage to package tour: Travel and tourism in the third world.* Routledge. doi:10.4324/9780203689363

Golzardi, F., & Sarvaramini, S., sadatasilan, K., & Sarvaramini, M. (2012). Residents Attitudes towards Tourism Development: A Case Study of Niasar, Iran. *Research Journal of Applied Sciences, Engineering and Technology*, •••, 863–868.

Gössling, S., Scott, D., & Hall, C. M. (2020). *Pandemics, tourism and global change: a rapid assessment of COVID-19.* doi:10.1080/09669582.2020.1758708

Gössling, S., Scott, D., & Hall, C. M. (2013). Tourism and water: Interactions, impacts, and challenges. *Annual Review of Environment and Resources*, *38*, 1–28.

Government of Sikkim. (2018). *Sikkim Tourism Policy 2018-2023*. Retrieved from. [URL]

Gunasekaran, N., & Anandkumar, V. (2012). Factors of Influence in Choosing Alternative Accommodation: A Study with Reference to Pondicherry, A Coastal Heritage Town. *Procedia: Social and Behavioral Sciences*, *62*, 1127–1132. doi:10.1016/j.sbspro.2012.09.193

Gupta, A., & Nair, S. (2017). Kerala's Western Ghats: A Biodiversity Hotspot. *Journal of Ecological Studies*, *10*(2), 78–91.

Hackman, J. R., & Wagerman, R. (2005). Tourism and Nature. *Academy of Management Review*, *30*(2), 269–287. doi:10.5465/amr.2005.16387885

Hall, C. M., & Weiler, B. (1992). Introduction: Tourism in the mountain environments. In Tourism in the Mountain South: A Double-Edged Sword (pp. 1-15). University of Tennessee Press.

Hall, C., & Weiler, B. (Eds.). (1992) Special Interest Tourism. Belhaven, London. Himalayan Mountaineering Institute. About us from https://hmidarjeeling.com/origin-and-growth/

Hall, C. M. (1992). Adventure, sport and health tourism. In *Special interest Tourism*. Belhaven Press.

Hall, C. M. (2016). Wellness tourism: A new paradigm for growth in the tourism and hospitality industry. *Journal of Hospitality and Tourism Management*, *27*, 67–73.

Hamzah, A., & Khalifah, Z. (2009). *Handbook on community based tourism 'How to develop and sustain CBT*. APEC Secretariat.

Haq, F., & Jackson, J. (2009). Spiritual journey to hajj: Australian and Pakistani experience and expectations. *Journal of Management, Spirituality & Religion*, *6*(2), 141–156. doi:10.1080/14766080902815155

Hardoy, J. E., & Satterthwaite, D. (1991). Environmental problems of Third World cities: A global issue ignored? *Public Administration and Development*, *11*(4), 341–361. doi:10.1002/pad.4230110405

Harrigan, P., Evers, U., Miles, M., & Daly, T. (2017). Customer engagement with tourism social media brands. *Tourism Management*, *59*, 597–609. doi:10.1016/j.tourman.2016.09.015

Harron, S., & Weiler, B. (1992). Review: Ethnic tourism. In B. Weiler & C. M. Hall (Eds.), *Special interest tourism* (pp. 83–92). Belhaven.

Hartline, M. D., Wooldridge, B. R., & Jones, K. C. (2003). Guest perceptions of hotel quality: Determining which employee groups count most. *The Cornell Hotel and Restaurant Administration Quarterly*, *44*(1), 43–52. doi:10.1016/S0010-8804(03)90045-0

Health and Wellness Tourism. (n.d.). Research Gate. https://www.researchgate.net/publication/302140629_Health_and_Wellness_Tourim

Health Promotion. (n.d.). WHO. https://www.who.int/teams/health-promotion/enhanced-wellbeing/first-global-conference

Health Tourism: An Overview By Kim Ross (HSMAI Marketing Review). (n.d.). Hospitality.net. https://www.hospitalitynet.org/news/4010521.html

Heritage, B. (n.d.). *Traditional Crafts*. Bharat Heritage. https://www.bharatheritage.in/himachal-pradesh/traditional_crafts.html

Hernandez, J. M., & Casimiro, L. A. (2011). Simulation model for a joint mass / rural tourism system. *Tourism and Hospitality Research*, *12*(1), 5–14. doi:10.1177/1467358411429525

Himachal Pradesh Tourism Development Corporation. (n.d.). *Himachal Handicrafts and Handloom*. Himachal Pradesh Tourism Development Coroporation. http://hptdc.in/index.php/himachal-handicrafts-and-handloom/

Himalaya, S. (n.d.). *Himachal Pradesh.* Secure Himalaya. https://securehimalaya.org/himachal/

Hussin, R., & Kunjuraman, V. (2014). Sustainable community based tourism (CBT) through homestay programme in Sabah, East Malaysia. *Proceeding of the social sciences research ICSSR*. Research Gate.

Hyde, K. F., & Harman, S. (2011). Motives for a Secular Pilgrimage to the Gallipoli Battlefields. *Tourism Management, 32*(6), 1343–1351. doi:10.1016/j.tourman.2011.01.008

India Craft. (n.d.). *Crafts of Himachal Pradesh.* India Craft. http://www.india-crafts.com/state_specials/himachal_handicrafts/

Ingram, C. D., & Durst, P. B. (1987). *Marketing nature-oriented tourism for rural development and wildlands management in developing countries: a bibliography, USDA Forest Service general technical report SE44*. US Department of Agriculture, Forest Service. doi:10.2737/SE-GTR-44

Iorio, M., & Corsale, A. (2010). Rural Tourism and livelihood strategies in Romania. *Journal of Rural Studies, 26*(2), 152–162. doi:10.1016/j.jrurstud.2009.10.006

Iseh, G. N., Yarhere, E. O., Ovharhe, O. J., & Apugo, O. E. (2021). Cultural and environmental factors influence on tourists' choice of destination in Delta State, Nigeria. *European Journal of Hospitality and Tourism Research, 9*(2), 8–15.

Ives, J. D. (2004). *Himalayan perceptions: Environmental change and the well-being of mountain peoples.* Routlege: Pxon.

Jafari, J., & Scott, N. (2014). Muslim world and its tourisms. *Annals of Tourism Research, 44*, 1–19. doi:10.1016/j.annals.2013.08.011

Jalilian, M. A., Danehkar, A., & Ali Fami, H. S. (2012). Determination of Indicators and Standards for Tourism Impacts in Protected Karaj River, Iran. *Tourism Management, 33*(1), 61–63. doi:10.1016/j.tourman.2011.01.024

Jamal, M. T., Sujood, M. A., & Arwab, M. (2019). Influence of Social Media on Travel and Tourism Industry: A Case Study of Pauri Garhwal District of Uttarakhand. *Amity Global Business Review.* https://www.researchgate.net/profile/Brighton-Nyagadza/publication/348297189_Embracing_new_banking_innovations_through_ICT_Firms_pursuing_growth_in_emerging_economies/links/5ff6d13245851553a026dad0/Embracing-new-banking-innovations-through-ICT-Firms-pursu

James, G. (2015). Rural Tourism and Biodiversity:A win-win Relationship. *Kurukshetra-A Journal on Rural Development*, 17-21.

Jamin, A., & Rahmafitria, F. (2022). Visitor Management Concept Through Carrying Capacity Analysis In Forest Recreation. *Journal of Indonesian Tourism. Hospitality and Recreation, 5*(1), 1–12. doi:10.17509/jithor.v5i1.46821

Jodha, N. S. (2001). *Life on the edge: Sustaining agriculture and community resources in fragile environments.* Oxford University Press.

Joshi, Pursotam & Saxena, S. (2019) Religious tourism in Uttarakhand. *JETIR1907C81 Journal of Emerging Technologies and Innovative Research (JETIR).* www.jetir.org

Joshi, P. (2020). Safety and risk management in adventure tourism: A case study of mountaineering expeditions in Himachal Pradesh. *International Journal of Adventure and Sports Tourism, 3*(1), 45–62.

Joshi, R., & Dhyani, P. P. (2009). Environmental sustainability and tourism - Implications of trend synergies of tourism in Sikkim Himalaya. *Current Science, 97*(1), 33–41.

Joveriya & Mariya. (2019) Problems and prospects of tourism industry in Uttarakhand. *International Journal of Geography, Geology and Environment, 1*(1), 10-16, 2706-7483

Joveriya & Mariya. (2019). Problems and prospects of tourism industry in Uttarakhand. International Journal of Geography. *Geology and Environment*, *1*(1), 10–16.

K C, A., Ghimire, S., & Dhakal, A.KC. (2021). Ecotourism and its impact on indigenous people and their local environment: Case of Ghalegaun and Golaghat of Nepal. *GeoJournal*, *86*(6), 2747–2765. doi:10.1007/s10708-020-10222-3

Kala, C. P., & Maikhuri, R. K. (2011). Mitigating people-park conflicts on resource use through ecotourism: A case of the Nanda Devi Biosphere Reserve, Indian Himalaya. *Journal of Mountain Science*, *8*(1), 87–95. doi:10.1007/s11629-011-1010-5

Kamble, U. P. (2019). Overview of adventure sports tourism in India. International Journal of Physicology. *Nutrition and Physical Education*, *4*(2), 155–158.

Kang, S. K., Hsu, C. H., & Wolfe, K. (2003). Family traveller segmentation by vacation decision-making patterns. *Journal of Hospitality & Tourism Research (Washington, D.C.)*, *27*(4), 448–469. doi:10.1177/10963480030274005

Kapoor, S., & Wangdus, J. (2022). Impact of overtourism on sustainable development and local community wellbeing in the Himalayan region. *International Journal of Leisure and Tourism Marketing*, *7*(3), 199–214. doi:10.1504/IJLTM.2022.122454

Karpagam, M. (1991). *Environmental Economics: A text Book.* Sterling Publishers Private Ltd.

Katoch, A. (2014). *The impact of rural tourism on local community development:A study of Himachal Pradesh, India.* [PhD Synopsis, Central university of Himachal Pradesh, Dharamshala].

Katoch, A. S. (2021). Adventure Tourism- A Case Study of Himachal Pradesh. [IJCRT]. *International Journal of Creative Research Thoughts*, *9*(6).

Katoch, A., & Gangotia, A. (2014). Rural Tourism: An instrument for local development in Hamlets around Dharamshala, Himachal Pradesh, India. AVAHAN-. *Journal of Hospitality and Tourism*, *2*(1), 12–24.

Kayat, K. (2014). Community-based rural tourism: A proposed sustainability framework. *SHS Web of Conferences*, *12*. IEEE. 10.1051/shsconf/20141201010

Kebete, Y., & Wondirad, A. (2019). Visitor management and sustainable destination management nexus in Zegie Peninsula, Northern Ethiopia. *Journal of Destination Marketing & Management*, *13*(March), 83–98. doi:10.1016/j.jdmm.2019.03.006

Khakpour, A. (2013). Methodology of comparative studies in education. *Contemporary Educational Researches Journal*, *2*(1), 20–26.

Khalid, Shahid, M., Natasha, Bibi, I., Sarwar, T., Shah, A., & Niazi, N. (2018). 2018, A Review of Environmental Contamination and Health Risk Assessment of Wastewater Use for Crop Irrigation with a Focus on Low and High-Income Countries. *International Journal of Environmental Research and Public Health*, *15*(5), 895. doi:10.3390/ijerph15050895 PMID:29724015

Khartishvili, L., Muhar, A., Dax, T., & Khelashvili, I. (2019, January 15). *Rural Tourism in Georgia in Transition: Challenges for Regional Sustainability.* ScitePress.https://scite.ai/reports/10.3390/su1102041

Kim, A., Kim, K. P., & Nguyen, T. H. D. (2021). The green accommodation management practices: The role of environmentally responsible tourist markets in understanding tourists' pro-environmental behaviour. Sustainability, 13(4), 2326.

Kiper, T. (2013). *Role of ecotourism in Sustainable Development.* Advances in Landscape Architecture. doi:10.5772/55749

Komppula, R. (2014). The role of individual entrepreneurs in the development of competitiveness for a rural tourism destination – A case study. *Tourism Management*, *40*, 361–371. doi:10.1016/j.tourman.2013.07.007

Konwar, J. (2014). Ethnic tourism- problems and prospects for community development in North East India. Mosaic: A collection of research articles (ed.) by Dibru College. Dibrugarh, Assam, India. 09–22.

Krishnan, R., & Rajan, S. (2018). Hydrological Significance of Kerala's Western Ghats Rivers. *Water Resources Research, 20*(3), 134–147.

Kumar, S. (2014). *Alternate tourism in tribal destinations A study of Himachal Pradesh.* Shodhganga.

Kutay, K. (1989). The new ethic in adventure travel. *Buzzworm: The Environmental Journal, 1*(4), 31–36.

Lall, J. S. (1981). *The Himalaya aspects of change.* Oxford University Press.

Laloo, D., Cajee, L., & Burman, D. (2007). *Potential for Cultural and eco-tourism in North-East India: a community-based approach.* Research Gate.

Lama, R. (2014). *Community based tourism development (a case study of Sikkim)* [Doctoral Dissertation, Kurukshetra University].

Lama, R., & Tyagi, N. (2019). Community-Based Tourism: A Tool for Sustainable Development (Case Study of Yuksom, Sikkim). *Emerging Trends in Indian Tourism and Hospitality: Transformation and Innovation*, 185.

Lanczkowski, G. (1982). *Die heilige Reise-Auf den Wegen von Gottern und Menschen.* Herder Verlag.

Lapeyre, R. (2010). Community-based tourism as a sustainable solution to maximise impacts locally? The Tsiseb Conservancy case, Namibia. *Development Southern Africa, 27*(5), 757–772. doi:10.1080/0376835X.2010.522837

Lee, C.-W., & Li, C. (2019). The Process of Constructing a Health Tourism Destination Index. *International Journal of Environmental Research and Public Health, 16*(22), 4579. Advance online publication. doi:10.3390/ijerph16224579 PMID:31752340

Lee, T. H. (2013). Influence analysis of community resident support for sustainable tourism development. *Tourism Management, 34*, 37–46. doi:10.1016/j.tourman.2012.03.007

Lee, T. J., Han, J. S., & Ko, T. G. (2020). Health-oriented tourists and sustainable domestic tourism. *Sustainability (Basel), 12*(12), 20–22. doi:10.3390/su12124988

Leksakundilok, A. (2004). *Ecotourism and community based tourism in the Mekong region,* (working paper series no. 10). Australia Mekong Resource Center.

Lepcha, O. (2022). *Youth Perception on Indigenous Knowledge and Ethnomedicinal Plants: A Study of Dzongu.*

Li, H., Nijkamp, P., Xie, X., & Liu, J. (2020, April 14). *A New Livelihood Sustainability Index for Rural Revitalization Assessment—A Modelling Study on Smart Tourism Specialization in China.* ScitePress. https://scite.ai/reports/10.3390/su12083148

Li, P., Ryan, C., & Cave, J. (2016). Chinese rural tourism development: Transition in the case of Qiyunshan, Anhui.-2008-2015. *Tourism Management, 55*, 240–260. doi:10.1016/j.tourman.2016.02.007

Liu, A. (2006). Tourism in rural areas: Kedah, Malaysia. *Tourism Management, 27*(5), 878–889. doi:10.1016/j.tourman.2005.05.007

Lois-González, R., & Santos, X. (2015). Tourists and Pilgrims on their Way to Santiago. Motives, Caminos and Final Destinations. *Journal of Tourism and Cultural Change, 13*(2), 149–164. doi:10.1080/14766825.2014.918985

Lorenz, D. F., & Dittmer, C. (2021). Disasters in the 'abode of gods'—Vulnerabilities and tourism in the Indian Himalaya. *International Journal of Disaster Risk Reduction, 55*, 102054. doi:10.1016/j.ijdrr.2021.102054

Lucas, P. H. C. 1984, How protected areas can help meet society's evolving needs. In J.A. McNeely, and K.R. Miller, (eds.), National parks, conservation, and development. Smithsonian Institution Press, Washington D.C.

Lyngdoh, B. (2022). An Exploration of Tourism Development of Meghalaya, India: What is the way forward? *Atna-Journal of Tourism Studies*, *17*(1), 143–166. doi:10.12727/ajts.27.6

MacKenzie, B. (2004). Ecotourism: It's good for your health. *Sustainable Tourism*, *9*, 161–170.

Macleod, D. (2006). Cultural commodification and tourism: A very special relationship. *Tourism, Culture & Communication*, *6*(2), 71–84. doi:10.3727/109830406777410580

Magar, D. A. (2021). Economic contribution of homestay tourism: A case study of Magar homestay of Naruwal, Lamjung. *Marsyangdi Journal*, 129–139. doi:10.3126/mj.v2i1.39972

Maharana Uprety. (2009). *Iyatta & Sharma, Ghanashyam*. Cultural Attributes, Economic Valuation and Community Conservation in Holy Khecheopalri Lake of Sikkim in the Eastern Himalaya.

Mandal, J., & Sengupta, P. (2015). The impact of tourism on livelihood and environment in West Sikkim: A case study of Pelling. In ILEE (Vol. 38, Issue 1).

Mangoch, S., & Jain, D. (2022). Adventure tourism in northern states of India. *Academy of Marketing Studies Journal*, *26*(1), 1–19.

Manning, R. E. (2007). *Parks and carrying capacity: commons without tragedy*. Island Press.

Mann, M. (2000). *The community tourism guide. Tourism concern*. Earthscan Publications.

Margry, P. (2008). *Shrines and Pilgrimage in the Modern World: New Itineraries into the Sacred*. University of Amsterdam Press. doi:10.5117/9789089640116

Martín Martín, J. M., Salinas Fernández, J. A., Rodríguez Martín, J. A., & Ostos Rey, M. (2020). Analysis of Tourism Seasonality as a Factor Limiting the Sustainable Development of Rural Areas. *Journal of Hospitality & Tourism Research (Washington, D.C.)*, *44*(1), 45–75. doi:10.1177/1096348019876688

Meena, K. (2021). *Role of Media in Development A case Study of Tribal Communities of Sirohi District*.

Mehta, P., Sekar, K. C., Bhatt, D., Tewari, A., Bisht, K., Upadhyay, S., Negi, V. S., & Soragi, B. (2020). Conservation and prioritization of threatened plants in Indian Himalayan Region. *Biodiversity and Conservation*, *29*(6), 1723–1745. doi:10.1007/s10531-020-01959-x

Mei, R., & Han, L. (2022, September 1). *Evaluation System of Rural Sustainable Tourism Land Based on Ecosystem Service Value*. ScitePress. https://scite.ai/reports/10.2478/eces-2022-0025

Mendoza, M. M. R. (2017). *How they do it: Community based tourism; case study of the community participation of Kikil, Yucatan, Mexico*.

Menon, P., & John, K. (2017). Impact of Cash Crop Cultivation on Biodiversity in Kerala. *Environmental Science & Policy*, *12*(5), 260–273.

Mieczkowski, Z. (1995). The tourism climatology: The state of the art. *Tourism Recreation Research*, *20*(2), 71–75.

Mitra, S., Roy, A. K., & Tamang, L. (2020). Assessing the status of changing channel regimes of Balason and Mahananda River in the Sub-Himalayan West Bengal, India. *Earth Systems and Environment*, *4*(2), 409–425. doi:10.1007/s41748-020-00160-y

Molnar, P. (1984). Structure and tectonics of the Himalaya: Constraints and implications of geophysical data. *Annual Review of Earth and Planetary Sciences*, *12*(1), 489–516. doi:10.1146/annurev.ea.12.050184.002421

MousimeXalxoa (2020). A Study on Tourist Behavior Towards Adventure Tourism. *Perspectives on Business Management & Economics*, *1*.

Mowforth, M., & Munt, I. (2015). *Tourism and Sustainability: New Tourism in the Third World*. Routledge. doi:10.4324/9781315795348

Muganda, M., Sahli, M., & A Smith, K. (2010). Tourism's contribution to poverty alleviation: A community perspective from Tanzania. *Development Southern Africa*, *27*(5), 629–646. doi:10.1080/0376835X.2010.522826

Mukherjee, D. (2012). Rural Tourism Path to Economic & Regional Development in India. Kurukshetra, A Journal on Rural Development, 62 (10), 3-6.

Myers, N. (1988). Threatened biotas: "hot spots" in tropical forests. *The Environmentalist*, *8*(3), 187–208. doi:10.1007/BF02240252 PMID:12322582

National Strategy For Adventure Tourism. (2022). Ministry of Tourism vide O.M. No. NT-1401/2/2021. https://tourism.gov.in

Nayak, P., & Mishra, K. S. (2013). *Problems and Prospects of Promoting Tourism in Meghalaya*. North Eastern Hill University.

Nepal, S. K. (2000). Tourism in protected areas. *Annals of Tourism Research*, *27*(3), 661–681. doi:10.1016/S0160-7383(99)00105-X

Newpaney, R., & Lee, S. (2016). Mountain Tourism Development and Impact of Tourism: A case study of Himalayan State Sikkim. *호텔경영학연구*, *25*(7), 329-349.

North East India People, history and culture. (2017). National Council of Educational Research and Training.

Nyaupane, G. P., & Timothy, D. J. (2022). *Tourism and development in the Himalaya: Social, environmental, and economic forces. Routledge cultural heritage and tourism services.*

Nyaupane, G. P., & Chhetri, N. (2009). Vulnerability to climate change of nature-based tourism in the Nepalese Himalayas. *Tourism Geographies*, *11*(1), 95–119. doi:10.1080/14616680802643359

Nyaupane, G. P., & Thapa, B. (2004). Evaluation of Ecotourism: A Comparative Assessment in the Annapurna Conservation Area Project. *Journal of Ecotourism*, *6*(1), 50–58. doi:10.1080/14724040408668148

Olson, D., Dinerstein, E., Wikramanayake, E. D., Burgess, N. D., Powell, G. V. N., Underwood, E. C., D'amico, J. A., Itoua, I., Strand, H. E., Morrison, J. C., Loucks, C. J., Allnutt, T. F., Ricketts, T. H., Kura, Y., Lamoreux, J. F., Wettengel, W. W., Hedao, P., & Kassem, K. R. (2001). Terrestrial eco regions of the world: A new map of life on earth. *Bioscience*, *51*(11), 933–938. doi:10.1641/0006-3568(2001)051[0933:TEOTWA]2.0.CO;2

Onuoha, C. A. (2022). ENVIRONMENTAL CHALLENGES AWARENESS IN NIGERIA. *RE:view*.

Page, S. J., & Hall, C. M. (2014). *The geography of tourism and recreation: Environment, place and space*. Routledge.

Parashar, A. (2012). *Feature of Health Tourism in Uttarakhand With Special Reference to Rishikesh- A Critical Study*. Dev Sanskriti Vishwavidhayalaya.

Park, E., & Kim, S. (2015). The potential of cittaslow for sustainable tourism development: Enhancing Local Community's empowerment. Tourism Planning & amp. *Tourism Planning & Development, 13*(3), 351–369. doi:10.1080/21568316.2015.1114015

Patel, A. (2021). *An Analytical Study of a Gujarat as a Tourism Destination with a Special reference to Tribal Areas of Gujarat*. Shodhganga.

Peinlang, B. L. (2019). The Scope of Tourism in Meghalaya. *J Tourism Hospit, 8*(1), 397. doi:10.35248/2167-0269.19.8.397

Pertiwi, P. & Sulistyawati, A. (2020). Factors of influence in choosing accommodation: a study with reference to Mas village, An Artistic Heritage Village In Ubud - Bali. *Journal of Business on Hospitality and Tourism.*

Phil, T. M. (2022). *Ecotourism: A Sustainable Development Connect to Nature and A Strategy for Balancing Economic Growth*. Socio-Cultural Development and Conservation.

Pradhan, P. (2016). *Culture and Tourism: A Study of Sikkim* [Doctoral dissertation, Sikkim University].

Price, M. F., & Price, L. W. (2000). *The geography of tourism and recreation: Environment, place, and space*. Routledge.

Priest, S. (1990). *'The semantics of adventure education', Adventure education.* Venture Publishing.

Quaranta, G., Citro, E., & Salvia, R. (2016). Economic and social sustainable synergies to promote innovations in rural tourism and Local Development. *Sustainability (Basel), 8*(7), 668. doi:10.3390/su8070668

Rajasekar, R., & Prasanna, A. (2019). Munnar: An Emerging Rural Tourism Destination. *Tourism Planning & Development, 16*(1), 53–68.

Raj, R., Griffin, K., & Blackwell, R. (2015). Motivations for Religious Tourism, Pilgrimage, Festivals and Events. In *Religious Tourism and Pilgrimage Management: An International Perspective* (pp. 103–117). Cabi. doi:10.1079/9781780645230.0103

Raju, S. S. (2019). *Future Scope of Tourism Courses: India Perspective*. Research Gate.

Rao, R. R. (1994). *Biodiversity in India: Floristics Aspects*. Bishen Singh Mahendra Pal Singh.

Rao, R. R., & Sharma, B. D. (1990). A Manual for Herbarium Collections, BSI. *Brabourne Road, Kolkata, 1*, 1990.

Rawat, A., Joshi, S., & Rai, S. K. (2023). Evaluating the issue of sustainable tourism with a system dynamic approach: Evidence from Uttarakhand, India. *Environment, Development and Sustainability, 3*(0123456789). doi:10.1007/s10668-023-03711-1

Rawat, D. S., Dash, S. S., Sinha, B. K., Kumar, V., Banerjee, A., & Singh, P. (2018). Community structure and regeneration status of tree species in Eastern Himalaya: A case study from Neora Valley National Park, West Bengal, India. *Taiwania, 63*(1).

Richards, G., & Hall, D. (2000). *Tourism and Sustainable Community Development*. Routledge.

Rinschede, G. (1992). Forms of Religious Tourism. *Annals of Tourism Research, 19*(1), 51–67. doi:10.1016/0160-7383(92)90106-Y

Rizal, P., & Asokan, R. (2016, September 16). *Measuring the impact of tourism industry on regional economy of Sikkim State, India*. SSRN. https://ssrn.com/abstract=2839694

Rodgers, W. A., & Panwar, H. S. (1988). *Planning a Wildlife Protected Area Network in India* (Vol. I and II). Wildlife Communications Trust of India.

Romario De Fretes, D., Caroline, C., Amin, A.-., Sirait, S., & Rizani, A. (2023). Exploring the economic impacts of sustainable tourism on locals: a comparative analysis between bali and north sumatera. *International journal of humanities, social sciences and business (INJOSS), 2*(3), 387–417. https://doi.org/. doi:10.54443/injoss.v2i3.88

Rongali, A. (2017). *Issue 11* [Year - 2017]. Vol. 8). Visiting Purposes of Tourists in Uttarakhand: A Comparative Study of Kumaun & Garhwal Regions, CASIRJ. ISSN.

Roy, S. (2022). *Rural homestay as a driver of community development and ecological sustainability, a case study of Buxa tiger reserve forest*, West Bengal. *International Journal of Mechanical Engineering, 7*(7).

Saha, K. S., Babai, C. T., & Saha, S. (2018). *Adventure Tourism in Meghalaya.* Research Gate.

Saha, G. K. (2015). Tourism Sector in the North Eastern States- Challenges and Opportunities [IJDR]. *International Journal of Developmental Research, 5*(11), 6077–6081.

Saha, G. K. (2015). Tourism sector in the north eastern states-challenges and opportunities. *International Journal of Developmental Research, 5*(11), 6077–6081.

Saha, K. S., Suiam, S., & Sarangi, A. (2022). *Rural Tourism and its Impact on the Economy: A Study of Ialong Village.* doi:10.4018/978-1-7998-9664-7.ch015

Sahani, N. (2021). Application of hybrid SWOT-AHP-FuzzyAHP model for formulation and prioritization of ecotourism strategies in Western Himalaya, India. *International Journal of Geoheritage and Parks, 9*(3), 349–362. doi:10.1016/j.ijgeop.2021.08.001

Sajwan, R. (2019), Rural tourism in Uttarakhand, Importance, Scope and Challenges. *JETIR, 6*(3), 2349-5162.

Samal, P. K., Palni, L. M. S., & Agrawal, D. K. (2003). Ecology, ecological poverty and sustainable development in Central Himalayan region of India. *International Journal of Sustainable Development and World Ecology, 10*(2), 157–168. doi:10.1080/13504500309469794

Samant, S. S., Vidyarthi, S., Pant, S., Sharma, P., Marpa, S., & Sharma, P. (2011). Diversity, distribution, indigenous uses and conservation of the medicinal plants of Indian Himalayan region used in cancer. *Journal of Biodiversity, 2*(2), 117–125. doi:10.1080/09766901.2011.11884732

Samsar, H. (2017). *Impact of Social Media on Tourism Industry*. Research Gate.

Sarkar, S. (2022). Home stay tourism and its sustainable approach in Rural Development: An appraisal from eastern Himalaya. *International Research Journal of Modernization in Engineering Technology and Science.*

Sattar, S. (2020). Culture, Creativity and Tourism: A Study of Sikkim. In The Cultural Heritage of Sikkim (pp. 289-302). Routledge.

Sattar, S. (2013). Potential of Tourism in Poverty Alleviation and Development of Sikkim. *IASSI-Quarterly, 32*(3), 74–87.

Schelling, D. (1992). The tectonostratigraphy and structure of the eastern Nepal Himalaya. *Tectonics, 11*(5), 925–943. doi:10.1029/92TC00213

Scheyvens, R. (2003). Local involvement in managing tourism. In *Tourism in destination communities* (pp. 229–252). CABI Publishing. doi:10.1079/9780851996110.0229

Scheyvens, R., & Momsen, J. H. (2008). Tourism and poverty reduction: Issues for Small Island states. *Tourism Geographies, 10*(1), 22–41. doi:10.1080/14616680701825115

Sharma, N., & Kalotra, A. (2016). Hospitality Industry in India: A Big Contributor to India's Growth. *International Journal of Emerging Research in Management &Technology*, *5*(6), 202–210.

Sharma, P. S. R. (2008). *Swasthay Raksha – Prakarti ke Anusaran Se Sambhav*. Yug Nirman Yojna, Gayatri Tapobhumi.

Sharpley, R. (2009). Tourism, Religion and Spirituality. In T. Jamal & M. Robinson (Eds.), *The SAGE Handbook of Tourism Studies* (pp. 237–253). SAGE Publications Ltd. doi:10.4135/9780857021076.n14

Sharpley, R. (2014). Host perceptions of tourism: A review of the research. *Tourism Management*, *42*, 37–49. doi:10.1016/j.tourman.2013.10.007

Shekhar, M. S., Chand, H., Kumar, S., Srinivasan, K., & Ganju, A. (2010). Climate-change studies in the western Himalaya. *Annals of Glaciology*, *51*(54), 105–112. doi:10.3189/172756410791386508

Shimla. (2021). Tribal crafts fair becomes hit among tourists in Kinnaur. *The Statesman*. https://www.thestatesman.com/cities/shimla/tribal-crafts-fair-becomes-hit-among-tourists-kinnaur-1503026016.html

Shin, H., Kim, H. N., & Son, J. (2017, April 18). *Measuring the Economic Impact of Rural Tourism Membership on Local Economy: A Korean Case Study*. ScitePress. https://scite.ai/reports/10.3390/su9040639

Shinde, K. (2018). Governance and Management of Religious Tourism in India. *International Journal of Religious Tourism and Pilgrimage*, *6*(1), 7.

Shrestha, B. G. (2015). *The Newars of Sikkim: reinventing language, culture and identity in the diaspora*. Vajra Books.

Shroder, J. F. Jr, & Bishop, M. P. (1998). Mass movement in the Himalaya: New insights and research directions. *Geomorphology*, *26*(1-3), 13–35. doi:10.1016/S0169-555X(98)00049-X

Shubhyatra. (n.d.). *Arts and Crafts*. Shubhyatra. http://www.shubhyatra.com/himachal-pradesh/arts-crafts.html

Singh, J. S., & Singh, S. P. (1987). Forest vegetation of the Himalaya. *Botanical Review*, *53*(1), 80–192. doi:10.1007/BF02858183

Singh, R. B., Mal, S., & Kala, C. P. (2009). Community responses to mountain tourism: A case in Bhyundar Valley, Indian Himalaya. *Journal of Mountain Science*, *6*(4), 394–404. doi:10.1007/s11629-009-1054-y

Singh, R. P. B. (2006). Pilgrimage in Hinduism: historical context and modern perspectives. In D. J. Timothy & D. H. Olsen (Eds.), *Tourism and Religious Journeys* (pp. 220–236). Routledge.

Smith, S. L. J., & Gupta, M. (2016). Community-based ecotourism and sustainable development in Ladakh, India. *Journal of Ecotourism*, *15*(3), 204–220.

Sohtun M L (2022). *A Study of Adventure Tourism in The East Khasi Hills of Meghalaya*. Research Gate.

Sood, J., Lynch, P., & Anastasiadou, C. (2017). Community non-participation in homestays in Kullu, Himachal Pradesh, India. *Tourism Management*, *60*, 332–347. doi:10.1016/j.tourman.2016.12.007

Stankey, G. H., & Cole, D. N. (1998). Historical development of limits of acceptable change: conceptual clarifications and possible extensions. In *Limits of acceptable change and related planning processes: progress and future directions. Ogden* (pp. 5–9). EUA.

Stoddard & Morinis. (1997). *The Geographic Contribution to Studies of Pilgrimage: Introduction to Sacred Places, Sacred Spaces*. University of Nebraska – Lincoln.

Sufi, T. (2015). Indian Hotel Industry: Past, Present and Future. Journal of Hotel management. ResearchGate.

Sujith, T. S., & Jisha, M. K. (2017). Tourism in India: Opportunities & Challenges . *Imperical Journal of Interdisciplinary Research, 3*, 201–2033.

Sung, H. H., Morrison, A. M., & O'Leary, J. T. (1996). Definition of adventure travel: Conceptual framework for empirical application from the providers' perspective. *Asia Pacific Journal of Tourism Research, 1*(2), 47–67. doi:10.1080/10941669708721975

Suresh, M., & Menon, A. (2019). Protected Areas in Kerala: Biodiversity Conservation and Challenges. *The Journal of Wildlife Management, 45*(4), 321–335.

Swarbrooke, J., & Horner, S. (1999). *Consumer Behavior in Tourism.* Butterworth-Heinemann.

Takahashi, L. Y., & Cegana, A. C. (2005). Como monitorar o impacto dos visitants utilizando o sistema LAC-Limite Aceitável de Câmbio. *Revista Turismo Em Análise, 16*(2), 206–222. doi:10.11606/issn.1984-4867.v16i2p206-222

Tambe, S., Arrawatia, M. L., & Ganeriwala, A. K. (2012). Managing rural development in the mountain state of Sikkim, India. *Mountain Research and Development, 32*(2), 242–252. doi:10.1659/MRD-JOURNAL-D-12-00005.1

Terzidou, M., Scarles, C., & Saunders, M. (2017). Religiousness as Tourist Performances: A Case Study of Greek Orthodox Pilgrimage. *Annals of Tourism Research, 66*, 116–129. doi:10.1016/j.annals.2017.06.009

Thakur, J., & Monga, N. (2022). Role of homestay in reviving tourism in Shimla. *PUSA Journal of Hospitality and Applied Sciences, 8*(2), 49–59. doi:10.48165/pjhas.2022.8.2.7

The 10 Biggest Benefits of Nature Travel. (n.d.). Smarter Travel. https://www.smartertravel.com/nature-travel-benefits/

The Travel & Tourism Competitiveness Report. (2015). World Economic Forum.

Thomas, A., & Joseph, B. (2020). Kerala's Involvement in Western Ghats Conservation Initiatives. *Environmental Management, 25*(1), 87–98.

Timlasana, P. (2012). Homestay tourism boosts Ghale Gaon"s Economy". http://www.gorkhapatra.org.np/risin Retrieved 10/6/2016

Tisdell, C. 1984, Tourism, the environment, international trade and public economics. Asian-Australia Joint Research Project, Australian National University, Canberra.

Tiwari, A., & Verma, S. (2018). Marketing strategies for promoting adventure tourism in the Indian Himalayan region. *Journal of Travel & Tourism Marketing, 35*(6), 726–740.

Tourism Council of Bhutan. (2022). *Final Tourism Rules and Regulations.* Tourism Council of Bhutan. https://www.tourism.gov.bt/uploads/attachment_files/tcb_7uivTQtN_23.09.22%20Final%20Tourism%20Rules%20and%20Regulations%202022.pdf

Tuan, L. (2019). *Study the factors affecting the choice of accommodation for new students: case in Da Nang City.* International Research Journal of Advanced Engineering and Science.

Turner, V. (1973). The Center out There: Pilgrim's Goal. *History of Religions, 12*(3), 191–230. doi:10.1086/462677

Turner, V., & Turner, E. (1978). *Image and Pilgrimage in Christian Culture: Anthropological Perspective.* Basil Blackwell.

U. N. D. P. (n.d.). *Biodiversity Conservation and Impacts of Nature-Based Tourism in Annapurna Conservation Area.* United Nations Development Programme. https://www.undp.org/nepal/biodiversity-conservation-and-impacts-nature-based-tourism-annapurna-conservation-area

Uddin, H., Ahammed, S., Rana, M. M., & Majumder, S. C. (2023). Investigating the relationship between environmental quality and tourism industry in Thailand. *Environment, Development and Sustainability*. Advance online publication. doi:10.1007/s10668-023-03801-0

Ueda, A. (2003). *Culture and modernisation: From the perspectives of young people in Bhutan.*

UNESCO. (2021). *Khangchendzonga National Park*. UNESCO. https://whc.unesco.org/en/list/1513

UNWTO. (2004). *A guidebook on indicators of sustainable development for tourism destinations*. World Tourism Organization.

UNWTO. (2014). *Global Report on Adventure Tourism*. UN. https://www.e-unwto.org/doi/pdf/10.18111/9789284416622

UNWTO. (2018). *UNWTO/UNESCO World Conference on Tourism and Culture: Building a New Partnership*. UNWTO & UNESCO.

Upreti, B. N. (1999). An overview of the stratigraphy and tectonics of the Nepal Himalaya. *Journal of Asian Earth Sciences*, *17*(5-6), 577–606. doi:10.1016/S1367-9120(99)00047-4

Valdiya, K. S. (1980). The two intracrustal boundary thrusts of the Himalaya. *Tectonophysics*, *66*(4), 323–348. doi:10.1016/0040-1951(80)90248-6

Van den Berghe, P. L. (1992). Tourism and the ethnic division of labor. *Annals of Tourism Research*, *19*(2), 234–249. doi:10.1016/0160-7383(92)90079-5

Van Den Bosch, M., & Sang, O. (2017). Urban natural environments as nature-based solutions for improved public health-A systematic review of reviews A R T I C L E I N F O. *Environmental Research*, *158*, 373–384. doi:10.1016/j.envres.2017.05.040 PMID:28686952

Verma, S. (2021). *An Analytical Study of Tribal Tourism in Southern Rajasthan*. Research Gate.

Vidović, B. (2018, January 1). *Ecotourism management for development countries*. ScitePress. https://scite.ai/reports/10.5937/tehnika1802293v

Vyas, N. (2016). *Tourism Growth Development and its impact – A study of Tribal Districts of Himachal Pradesh*. Research Gate.

Wearing, S., & Neil, J. (2009). *Ecotourism: Impacts, potentials, and possibilities?* Routledge. doi:10.4324/9780080940182

Weber, K. (2001). Outdoor adventure tourism: A review of research approaches. *Annals of Tourism Research*, *28*(2), 360–377. doi:10.1016/S0160-7383(00)00051-7

WHO/Europe. (n.d.). | *Urban health - Protecting nature protects health – lessons for the future from COVID-19*. WHOEuro. https://www.euro.who.int/en/health-topics/environment-and-health/urban-health/news/news/2020/6/protecting-nature-protects-health-lessons-for-the-future-from-covid-19

Wilkins, H. C. (2005). *A structural model of satisfaction and brand attitude in hotels. Griffith*. Griffith Business School, Griffith University.

Wilson, J. A. J., & Grant, J. (2013). Islamic marketing – a challenger to the classical marketing canon? *Journal of Islamic Marketing*, *4*(1), 7–21. doi:10.1108/17590831311306327

Wuest, B. E. S., Tas, R. F., & Emenheiser, D. A. (1996). *What do mature travellers perceive as important hotel/motel customer service?* Hospitality Research Journal.

Yang, L. (2007). *Planning for ethnic tourism: Case studies from Xishuangbanna, Yunnan, China*. [Doctoral Dissertation, University of Waterloo].

Yavas, U., & Babakus, E. (2005). Dimensions of hotel choice criteria: Congruence between business and leisure travelers. *International Journal of Hospitality Management*, *24*(3), 359–367. doi:10.1016/j.ijhm.2004.09.003

Yogesh, K. (2019). Challenges and Solutions to the Development of the Tourism and Hospitality Industry in India. *African Journal of Hospitality Tourism and Leisure*.

Young, A. M. (1986). Eco-enterprises: Eco-tourism and farming of exotics in the tropics. *Ambio*, *15*, 361–363.

Ziegler, A. D., Wasson, R. J., Sundriyal, Y., Srivastava, P., Sasges, G., Ramchunder, S. J., Ong, C. E., Nepal, S. K., McAdoo, B. G., Gillen, J., Bishwokarma, D., Bhardwaj, A., & Apollo, M. (2023). A call for reducing tourism risk to environmental hazards in the Himalaya. *Environmental Hazards*, *22*(1), 1–28. doi:10.1080/17477891.2021.1984196

Zou, J. M. M. L., & Prasain, G. P. (2017). North-Eastern States Tourism: Exploring the Unexplored on Earth. *IOSR Journal of Humanities and Social Science (IOSR-JHSS)*, *22*(8), 35-40.

Related References

To continue our tradition of advancing academic research, we have compiled a list of recommended IGI Global readings. These references will provide additional information and guidance to further enrich your knowledge and assist you with your own research and future publications.

Ajiboye, O. E., & Yusuff, O. S. (2017). Foreign Land Acquisition: Food Security and Food Chains – The Nigerian Experience. In I. Management Association (Ed.), Natural Resources Management: Concepts, Methodologies, Tools, and Applications (pp. 1524-1545). Hershey, PA: IGI Global. https://doi.org/ doi:10.4018/978-1-5225-0803-8.ch072

Alapiki, H. E., & Amadi, L. A. (2018). Sustainable Food Consumption in the Neoliberal Order: Challenges and Policy Implications. In A. Obayelu (Ed.), *Food Systems Sustainability and Environmental Policies in Modern Economies* (pp. 90–123). Hershey, PA: IGI Global. doi:10.4018/978-1-5225-3631-4.ch005

Altaş, A. (2018). Geographical Information System Applications Utilized in Museums in Turkey Within the Scope of the Cultural Heritage Tourism: A Case Study of Mobile Application of Müze Asist. In S. Chaudhuri & N. Ray (Eds.), *GIS Applications in the Tourism and Hospitality Industry* (pp. 42–60). Hershey, PA: IGI Global. doi:10.4018/978-1-5225-5088-4.ch002

Andreea, I. R. (2018). Beyond Macroeconomics of Food and Nutrition Security. *International Journal of Sustainable Economies Management*, *7*(1), 13–22. doi:10.4018/IJSEM.2018010102

Anwar, J. (2017). Reproductive and Mental Health during Natural Disaster: Implications and Issues for Women in Developing Nations – A Case Example. In I. Management Association (Ed.), Gaming and Technology Addiction: Breakthroughs in Research and Practice (pp. 446-472). Hershey, PA: IGI Global. https://doi.org/ doi:10.4018/978-1-5225-0778-9.ch021

Awadh, H., Aksissou, M., Benhardouze, W., Darasi, F., & Snaiki, J. (2018). Socioeconomic Status of Artisanal Fishers in the West Part of Moroccan Mediterranean. *International Journal of Social Ecology and Sustainable Development*, *9*(1), 40–52. doi:10.4018/IJSESD.2018010104

Aye, G. C., & Haruna, R. F. (2018). Effect of Climate Change on Crop Productivity and Prices in Benue State, Nigeria: Implications for Food Security. In V. Erokhin (Ed.), *Establishing Food Security and Alternatives to International Trade in Emerging Economies* (pp. 244–268). Hershey, PA: IGI Global. doi:10.4018/978-1-5225-2733-6.ch012

Azizan, S. A., & Suki, N. M. (2017). Consumers' Intentions to Purchase Organic Food Products. In T. Esakki (Ed.), *Green Marketing and Environmental Responsibility in Modern Corporations* (pp. 68–82). Hershey, PA: IGI Global. doi:10.4018/978-1-5225-2331-4.ch005

Barakabitze, A. A., Fue, K. G., Kitindi, E. J., & Sanga, C. A. (2017). Developing a Framework for Next Generation Integrated Agro Food-Advisory Systems in Developing Countries. In I. Management Association (Ed.), Agri-Food Supply Chain Management: Breakthroughs in Research and Practice (pp. 47-67). Hershey, PA: IGI Global. doi:10.4018/978-1-5225-1629-3.ch004

Beachcroft-Shaw, H., & Ellis, D. (2018). Using Successful Cases to Promote Environmental Sustainability: A Social Marketing Approach. In I. Management Association (Ed.), Sustainable Development: Concepts, Methodologies, Tools, and Applications (pp. 936-953). Hershey, PA: IGI Global. doi:10.4018/978-1-5225-3817-2.ch042

Behnassi, M., Kahime, K., Boussaa, S., Boumezzough, A., & Messouli, M. (2017). Infectious Diseases and Climate Vulnerability in Morocco: Governance and Adaptation Options. In I. Management Association (Ed.), Public Health and Welfare: Concepts, Methodologies, Tools, and Applications (pp. 91-109). Hershey, PA: IGI Global. https://doi.org/ doi:10.4018/978-1-5225-1674-3.ch005

Bekele, F., & Bekele, I. (2017). Social and Environmental Impacts on Agricultural Development. In W. Ganpat, R. Dyer, & W. Isaac (Eds.), *Agricultural Development and Food Security in Developing Nations* (pp. 21–56). Hershey, PA: IGI Global. doi:10.4018/978-1-5225-0942-4.ch002

Benaouda, A., & García-Peñalvo, F. J. (2018). Towards an Intelligent System for the Territorial Planning: Agricultural Case. In F. García-Peñalvo (Ed.), *Global Implications of Emerging Technology Trends* (pp. 158–178). Hershey, PA: IGI Global. doi:10.4018/978-1-5225-4944-4.ch010

Bhaskar, A., Rao, G. B., & Vencatesan, J. (2017). Characterization and Management Concerns of Water Resources around Pallikaranai Marsh, South Chennai. In P. Rao & Y. Patil (Eds.), *Reconsidering the Impact of Climate Change on Global Water Supply, Use, and Management* (pp. 102–121). Hershey, PA: IGI Global. doi:10.4018/978-1-5225-1046-8.ch007

Bhyan, P., Shrivastava, B., & Kumar, N. (2022). Requisite Sustainable Development Contemplating Buildings: Economic and Environmental Sustainability. In A. Hussain, K. Tiwari, & A. Gupta (Eds.), *Addressing Environmental Challenges Through Spatial Planning* (pp. 269–288). IGI Global. https://doi.org/10.4018/978-1-7998-8331-9.ch014

Bogataj, D., & Drobne, D. (2017). Control of Perishable Goods in Cold Logistic Chains by Bionanosensors. In S. Joo (Ed.), *Applying Nanotechnology for Environmental Sustainability* (pp. 376–402). Hershey, PA: IGI Global. doi:10.4018/978-1-5225-0585-3.ch016

Bogataj, D., & Drobne, D. (2017). Control of Perishable Goods in Cold Logistic Chains by Bionanosensors. In I. Management Association (Ed.), Materials Science and Engineering: Concepts, Methodologies, Tools, and Applications (pp. 471-497). Hershey, PA: IGI Global. https://doi.org/ doi:10.4018/978-1-5225-1798-6.ch019

Bogueva, D., & Marinova, D. (2018). What Is More Important: Perception of Masculinity or Personal Health and the Environment? In D. Bogueva, D. Marinova, & T. Raphaely (Eds.), *Handbook of Research on Social Marketing and Its Influence on Animal Origin Food Product Consumption* (pp. 148–162). Hershey, PA: IGI Global. doi:10.4018/978-1-5225-4757-0.ch010

Bouzid, M. (2017). Waterborne Diseases and Climate Change: Impact and Implications. In M. Bouzid (Ed.), *Examining the Role of Environmental Change on Emerging Infectious Diseases and Pandemics* (pp. 89–108). Hershey, PA: IGI Global. doi:10.4018/978-1-5225-0553-2.ch004

Bowles, D. C. (2017). Climate Change-Associated Conflict and Infectious Disease. In M. Bouzid (Ed.), *Examining the Role of Environmental Change on Emerging Infectious Diseases and Pandemics* (pp. 68–88). Hershey, PA: IGI Global. doi:10.4018/978-1-5225-0553-2.ch003

Buck, J. J., & Lowry, R. K. (2017). Oceanographic Data Management: Quills and Free Text to the Digital Age and "Big Data". In P. Diviacco, A. Leadbetter, & H. Glaves (Eds.), *Oceanographic and Marine Cross-Domain Data Management for Sustainable Development* (pp. 1–22). Hershey, PA: IGI Global. doi:10.4018/978-1-5225-0700-0.ch001

Buse, C. G. (2017). Are Climate Change Adaptation Policies a Game Changer?: A Case Study of Perspectives from Public Health Officials in Ontario, Canada. In M. Bouzid (Ed.), *Examining the Role of Environmental Change on Emerging Infectious Diseases and Pandemics* (pp. 230–257). Hershey, PA: IGI Global. doi:10.4018/978-1-5225-0553-2.ch010

Calderon, F. A., Giolo, E. G., Frau, C. D., Rengel, M. G., Rodriguez, H., Tornello, M., ... Gallucci, R. (2018). Seismic Microzonation and Site Effects Detection Through Microtremors Measures: A Review. In N. Ceryan (Ed.), *Handbook of Research on Trends and Digital Advances in Engineering Geology* (pp. 326–349). Hershey, PA: IGI Global. doi:10.4018/978-1-5225-2709-1.ch009

Carfì, D., Donato, A., & Panuccio, D. (2018). A Game Theory Coopetitive Perspective for Sustainability of Global Feeding: Agreements Among Vegan and Non-Vegan Food Firms. In I. Management Association (Ed.), Game Theory: Breakthroughs in Research and Practice (pp. 71-104). Hershey, PA: IGI Global. https://doi.org/ doi:10.4018/978-1-5225-2594-3.ch004

Castagnolo, V. (2018). Analyzing, Classifying, Safeguarding: Drawing for the Borgo Murattiano Neighbourhood of Bari. In G. Carlone, N. Martinelli, & F. Rotondo (Eds.), *Designing Grid Cities for Optimized Urban Development and Planning* (pp. 93–108). Hershey, PA: IGI Global. doi:10.4018/978-1-5225-3613-0.ch006

Chekima, B. (2018). The Dilemma of Purchase Intention: A Conceptual Framework for Understanding Actual Consumption of Organic Food. *International Journal of Sustainable Economies Management*, *7*(2), 1–13. doi:10.4018/IJSEM.2018040101

Chen, Y. (2017). Sustainable Supply Chains and International Soft Landings: A Case of Wetland Entrepreneurship. In B. Christiansen & F. Kasarcı (Eds.), *Corporate Espionage, Geopolitics, and Diplomacy Issues in International Business* (pp. 232–247). Hershey, PA: IGI Global. doi:10.4018/978-1-5225-1031-4.ch013

Çıtak, L., Akel, V., & Ersoy, E. (2018). Investors' Reactions to the Announcement of New Constituents of BIST Sustainability Index: An Analysis by Event Study and Mean-Median Tests. In M. Risso & S. Testarmata (Eds.), *Value Sharing for Sustainable and Inclusive Development* (pp. 270–289). Hershey, PA: IGI Global. doi:10.4018/978-1-5225-3147-0.ch012

D'Aleo, V., D'Aleo, F., & Bonanno, R. (2018). New Food Industries Toward a New Level of Sustainable Supply: Success Stories, Business Models, and Strategies. In V. Erokhin (Ed.), *Establishing Food Security and Alternatives to International Trade in Emerging Economies* (pp. 74–97). Hershey, PA: IGI Global. doi:10.4018/978-1-5225-2733-6.ch004

Dagevos, H., & Reinders, M. J. (2018). Flexitarianism and Social Marketing: Reflections on Eating Meat in Moderation. In D. Bogueva, D. Marinova, & T. Raphaely (Eds.), *Handbook of Research on Social Marketing and Its Influence on Animal Origin Food Product Consumption* (pp. 105–120). Hershey, PA: IGI Global. doi:10.4018/978-1-5225-4757-0.ch007

Danisman, G. O. (2022). What Drives Eco-Design Innovations in European SMEs? In U. Akkucuk (Ed.), *Disruptive Technologies and Eco-Innovation for Sustainable Development* (pp. 191–206). IGI Global. https://doi.org/10.4018/978-1-7998-8900-7.ch011

Deenapanray, P. N., & Ramma, I. (2017). Adaptations to Climate Change and Climate Variability in the Agriculture Sector in Mauritius: Lessons from a Technical Needs Assessment. In I. Management Association (Ed.), Natural Resources Management: Concepts, Methodologies, Tools, and Applications (pp. 655-680). Hershey, PA: IGI Global. https://doi.org/ doi:10.4018/978-1-5225-0803-8.ch030

Deenapanray, P. N., & Ramma, I. (2017). Adaptations to Climate Change and Climate Variability in the Agriculture Sector in Mauritius: Lessons from a Technical Needs Assessment. In I. Management Association (Ed.), Natural Resources Management: Concepts, Methodologies, Tools, and Applications (pp. 655-680). Hershey, PA: IGI Global. https://doi.org/ doi:10.4018/978-1-5225-0803-8.ch030

Deshpande, S., Basu, S. K., Li, X., & Chen, X. (2017). Smart, Innovative and Intelligent Technologies Used in Drug Designing. In I. Management Association (Ed.), Pharmaceutical Sciences: Breakthroughs in Research and Practice (pp. 1175-1191). Hershey, PA: IGI Global. https://doi.org/ doi:10.4018/978-1-5225-1762-7.ch045

Dlamini, P. N. (2017). Use of Information and Communication Technologies Tools to Capture, Store, and Disseminate Indigenous Knowledge: A Literature Review. In P. Ngulube (Ed.), *Handbook of Research on Theoretical Perspectives on Indigenous Knowledge Systems in Developing Countries* (pp. 225–247). Hershey, PA: IGI Global. doi:10.4018/978-1-5225-0833-5.ch010

Dolejsova, M., & Kera, D. (2017). The Fermentation GutHub Project and the Internet of Microbes. In S. Konomi & G. Roussos (Eds.), *Enriching Urban Spaces with Ambient Computing, the Internet of Things, and Smart City Design* (pp. 25–46). Hershey, PA: IGI Global. doi:10.4018/978-1-5225-0827-4.ch002

Dolunay, O. (2018). A Paradigm Shift: Empowering Farmers to Eliminate the Waste in the Form of Fresh Water and Energy Through the Implementation of 4R+T. In I. Management Association (Ed.), Sustainable Development: Concepts, Methodologies, Tools, and Applications (pp. 882-892). Hershey, PA: IGI Global. https://doi.org/ doi:10.4018/978-1-5225-3817-2.ch039

Dube, P., Heijman, W. J., Ihle, R., & Ochieng, J. (2018). The Potential of Traditional Leafy Vegetables for Improving Food Security in Africa. In V. Erokhin (Ed.), *Establishing Food Security and Alternatives to International Trade in Emerging Economies* (pp. 220–243). Hershey, PA: IGI Global. doi:10.4018/978-1-5225-2733-6.ch011

Duruji, M. M., & Urenma, D. F. (2017). The Environmentalism and Politics of Climate Change: A Study of the Process of Global Convergence through UNFCCC Conferences. In I. Management Association (Ed.), *Natural Resources Management: Concepts, Methodologies, Tools, and Applications* (pp. 77-108). Hershey, PA: IGI Global. 10.4018/978-1-5225-0803-8.ch004

Dutta, U. (2017). Agro-Geoinformatics, Potato Cultivation, and Climate Change. In S. Londhe (Ed.), *Sustainable Potato Production and the Impact of Climate Change* (pp. 247–271). Hershey, PA: IGI Global. doi:10.4018/978-1-5225-1715-3.ch012

Edirisinghe, R., Stranieri, A., & Wickramasinghe, N. (2017). A Taxonomy for mHealth. In N. Wickramasinghe (Ed.), *Handbook of Research on Healthcare Administration and Management* (pp. 596–615). Hershey, PA: IGI Global. doi:10.4018/978-1-5225-0920-2.ch036

Ekpeni, N. M., & Ayeni, A. O. (2018). Global Natural Hazard and Disaster Vulnerability Management. In A. Eneanya (Ed.), *Handbook of Research on Environmental Policies for Emergency Management and Public Safety* (pp. 83–104). Hershey, PA: IGI Global. doi:10.4018/978-1-5225-3194-4.ch005

Ene, C., Voica, M. C., & Panait, M. (2017). Green Investments and Food Security: Opportunities and Future Directions in the Context of Sustainable Development. In M. Mieila (Ed.), *Measuring Sustainable Development and Green Investments in Contemporary Economies* (pp. 163–200). Hershey, PA: IGI Global. doi:10.4018/978-1-5225-2081-8.ch007

Escamilla, I., Ruíz, M. T., Ibarra, M. M., Soto, V. L., Quintero, R., & Guzmán, G. (2018). Geocoding Tweets Based on Semantic Web and Ontologies. In M. Lytras, N. Aljohani, E. Damiani, & K. Chui (Eds.), *Innovations, Developments, and Applications of Semantic Web and Information Systems* (pp. 372–392). Hershey, PA: IGI Global. doi:10.4018/978-1-5225-5042-6.ch014

Escribano, A. J. (2018). Marketing Strategies for Trendy Animal Products: Sustainability as a Core. In F. Quoquab, R. Thurasamy, & J. Mohammad (Eds.), *Driving Green Consumerism Through Strategic Sustainability Marketing* (pp. 169–203). Hershey, PA: IGI Global. doi:10.4018/978-1-5225-2912-5.ch010

Eudoxie, G., & Roopnarine, R. (2017). Climate Change Adaptation and Disaster Risk Management in the Caribbean. In W. Ganpat & W. Isaac (Eds.), *Environmental Sustainability and Climate Change Adaptation Strategies* (pp. 97–125). Hershey, PA: IGI Global. doi:10.4018/978-1-5225-1607-1.ch004

Farmer, L. S. (2017). Data Analytics for Strategic Management: Getting the Right Data. In V. Wang (Ed.), *Encyclopedia of Strategic Leadership and Management* (pp. 810–822). Hershey, PA: IGI Global. doi:10.4018/978-1-5225-1049-9.ch056

Fattal, L. R. (2017). Catastrophe: An Uncanny Catalyst for Creativity. In R. Shin (Ed.), *Convergence of Contemporary Art, Visual Culture, and Global Civic Engagement* (pp. 244–262). Hershey, PA: IGI Global. doi:10.4018/978-1-5225-1665-1.ch014

Forti, I. (2017). A Cross Reading of Landscape through Digital Landscape Models: The Case of Southern Garda. In A. Ippolito (Ed.), *Handbook of Research on Emerging Technologies for Architectural and Archaeological Heritage* (pp. 532–561). Hershey, PA: IGI Global. doi:10.4018/978-1-5225-0675-1.ch018

Gharbi, A., De Runz, C., & Akdag, H. (2017). Urban Development Modelling: A Survey. In S. Faiz & K. Mahmoudi (Eds.), *Handbook of Research on Geographic Information Systems Applications and Advancements* (pp. 96–124). Hershey, PA: IGI Global. doi:10.4018/978-1-5225-0937-0.ch004

Ghosh, I., & Ghoshal, I. (2018). Implications of Trade Liberalization for Food Security Under the ASEAN-India Strategic Partnership: A Gravity Model Approach. In V. Erokhin (Ed.), *Establishing Food Security and Alternatives to International Trade in Emerging Economies* (pp. 98–118). Hershey, PA: IGI Global. doi:10.4018/978-1-5225-2733-6.ch005

Glaves, H. M. (2017). Developing a Common Global Framework for Marine Data Management. In P. Diviacco, A. Leadbetter, & H. Glaves (Eds.), *Oceanographic and Marine Cross-Domain Data Management for Sustainable Development* (pp. 47–68). Hershey, PA: IGI Global. doi:10.4018/978-1-5225-0700-0.ch003

Godulla, A., & Wolf, C. (2018). Future of Food: Transmedia Strategies of National Geographic. In R. Gambarato & G. Alzamora (Eds.), *Exploring Transmedia Journalism in the Digital Age* (pp. 162–182). Hershey, PA: IGI Global. doi:10.4018/978-1-5225-3781-6.ch010

Gomes, P. P. (2018). Food and Environment: A Review on the Sustainability of Six Different Dietary Patterns. In A. Obayelu (Ed.), *Food Systems Sustainability and Environmental Policies in Modern Economies* (pp. 15–31). Hershey, PA: IGI Global. doi:10.4018/978-1-5225-3631-4.ch002

Gonzalez-Feliu, J. (2018). Sustainability Evaluation of Green Urban Logistics Systems: Literature Overview and Proposed Framework. In A. Paul, D. Bhattacharyya, & S. Anand (Eds.), *Green Initiatives for Business Sustainability and Value Creation* (pp. 103–134). Hershey, PA: IGI Global. doi:10.4018/978-1-5225-2662-9.ch005

Goodland, R. (2017). A Fresh Look at Livestock Greenhouse Gas Emissions and Mitigation. In I. Management Association (Ed.), Natural Resources Management: Concepts, Methodologies, Tools, and Applications (pp. 124-139). Hershey, PA: IGI Global. doi:10.4018/978-1-5225-0803-8.ch006

Goundar, S., & Appana, S. (2018). Mainstreaming Development Policies for Climate Change in Fiji: A Policy Gap Analysis and the Role of ICTs. In I. Management Association (Ed.), Sustainable Development: Concepts, Methodologies, Tools, and Applications (pp. 402-432). Hershey, PA: IGI Global. doi:10.4018/978-1-5225-3817-2.ch020

Granell-Canut, C., & Aguilar-Moreno, E. (2018). Geospatial Influence in Science Mapping. In M. Khosrow-Pour, D.B.A. (Ed.), Encyclopedia of Information Science and Technology, Fourth Edition (pp. 3473-3483). Hershey, PA: IGI Global. doi:10.4018/978-1-5225-2255-3.ch302

Grigelis, A., Blažauskas, N., Gelumbauskaitė, L. Ž., Gulbinskas, S., Suzdalev, S., & Ferrarin, C. (2017). Marine Environment Data Management Related to the Human Activity in the South-Eastern Baltic Sea (The Lithuanian Sector). In P. Diviacco, A. Leadbetter, & H. Glaves (Eds.), *Oceanographic and Marine Cross-Domain Data Management for Sustainable Development* (pp. 282–302). Hershey, PA: IGI Global. doi:10.4018/978-1-5225-0700-0.ch012

Guma, I. P., Rwashana, A. S., & Oyo, B. (2018). Food Security Policy Analysis Using System Dynamics: The Case of Uganda. *International Journal of Information Technologies and Systems Approach*, *11*(1), 72–90. doi:10.4018/IJITSA.2018010104

Guma, I. P., Rwashana, A. S., & Oyo, B. (2018). Food Security Indicators for Subsistence Farmers Sustainability: A System Dynamics Approach. *International Journal of System Dynamics Applications*, *7*(1), 45–64. doi:10.4018/IJSDA.2018010103

Gupta, P., & Goyal, S. (2017). Wildlife Habitat Evaluation. In A. Santra & S. Mitra (Eds.), *Remote Sensing Techniques and GIS Applications in Earth and Environmental Studies* (pp. 258–264). Hershey, PA: IGI Global. doi:10.4018/978-1-5225-1814-3.ch013

Hanson, T., & Hildebrand, E. (2018). GPS Travel Diaries in Rural Transportation Research: A Focus on Older Drivers. In I. Management Association (Ed.), Intelligent Transportation and Planning: Breakthroughs in Research and Practice (pp. 609-625). Hershey, PA: IGI Global. doi:10.4018/978-1-5225-5210-9.ch027

Hartman, M. B. (2017). Research-Based Climate Change Public Education Programs. In I. Management Association (Ed.), Natural Resources Management: Concepts, Methodologies, Tools, and Applications (pp. 992-1003). Hershey, PA: IGI Global. doi:10.4018/978-1-5225-0803-8.ch046

Hashim, N. (2017). Zanzibari Seaweed: Global Climate Change and the Promise of Adaptation. In I. Management Association (Ed.), Natural Resources Management: Concepts, Methodologies, Tools, and Applications (pp. 365-391). Hershey, PA: IGI Global. https://doi.org/ doi:10.4018/978-1-5225-0803-8.ch019

Herrera, J. E., Argüello, L. V., Gonzalez-Feliu, J., & Jaimes, W. A. (2017). Decision Support System Design Requirements, Information Management, and Urban Logistics Efficiency: Case Study of Bogotá, Colombia. In G. Jamil, A. Soares, & C. Pessoa (Eds.), *Handbook of Research on Information Management for Effective Logistics and Supply Chains* (pp. 223–238). Hershey, PA: IGI Global. doi:10.4018/978-1-5225-0973-8.ch012

Huizinga, T., Ayanso, A., Smoor, M., & Wronski, T. (2017). Exploring Insurance and Natural Disaster Tweets Using Text Analytics. *International Journal of Business Analytics*, *4*(1), 1–17. doi:10.4018/IJBAN.2017010101

Hung, K., Kalantari, M., & Rajabifard, A. (2017). An Integrated Method for Assessing the Text Content Quality of Volunteered Geographic Information in Disaster Management. *International Journal of Information Systems for Crisis Response and Management*, *9*(2), 1–17. doi:10.4018/IJISCRAM.2017040101

Husnain, A., & Avdic, A. (2018). Identifying the Contemporary Status of E-Service Sustainability Research. In I. Management Association (Ed.), Sustainable Development: Concepts, Methodologies, Tools, and Applications (pp. 467-485). Hershey, PA: IGI Global. https://doi.org/ doi:10.4018/978-1-5225-3817-2.ch022

Iarossi, M. P., & Ferro, L. (2017). “The Past is Never Dead. It’s Not Even Past”: Virtual Archaeological Promenade. In A. Ippolito & M. Cigola (Eds.), *Handbook of Research on Emerging Technologies for Digital Preservation and Information Modeling* (pp. 228–255). Hershey, PA: IGI Global. doi:10.4018/978-1-5225-0680-5.ch010

Ignjatijević, S., & Cvijanović, D. (2018). Analysis of Serbian Honey Production and Exports. In *Exploring the Global Competitiveness of Agri-Food Sectors and Serbia’s Dominant Presence: Emerging Research and Opportunities* (pp. 109–139). Hershey, PA: IGI Global. doi:10.4018/978-1-5225-2762-6.ch005

Jana, S. K., & Karmakar, A. K. (2017). Globalization, Governance, and Food Security: The Case of BRICS. In I. Management Association (Ed.), Natural Resources Management: Concepts, Methodologies, Tools, and Applications (pp. 692-712). Hershey, PA: IGI Global. https://doi.org/ doi:10.4018/978-1-5225-0803-8.ch032

Jana, S. K., & Karmakar, A. K. (2017). Food Security in Asia: Is There Convergence? In I. Management Association (Ed.), Natural Resources Management: Concepts, Methodologies, Tools, and Applications (pp. 109-123). Hershey, PA: IGI Global. https://doi.org/ doi:10.4018/978-1-5225-0803-8.ch005

John, J., & Kumar, S. (2018). A Locational Decision Making Framework for Shipbreaking Under Multiple Criteria. In I. Management Association (Ed.), Operations and Service Management: Concepts, Methodologies, Tools, and Applications (pp. 504-527). Hershey, PA: IGI Global. doi:10.4018/978-1-5225-3909-4.ch024

John, J., & Srivastava, R. K. (2018). Decision Insights for Shipbreaking using Environmental Impact Assessment: Review and Perspectives. *International Journal of Strategic Decision Sciences*, *9*(1), 45–62. doi:10.4018/IJSDS.2018010104

Joshi, Y., & Rahman, Z. (2018). Determinants of Sustainable Consumption Behaviour: Review and Conceptual Framework. In A. Paul, D. Bhattacharyya, & S. Anand (Eds.), *Green Initiatives for Business Sustainability and Value Creation* (pp. 239–262). Hershey, PA: IGI Global. doi:10.4018/978-1-5225-2662-9.ch011

Juma, D. W., Reuben, M., Wang, H., & Li, F. (2018). Adaptive Coevolution: Realigning the Water Governance Regime to the Changing Climate. In I. Management Association (Ed.), Hydrology and Water Resource Management: Breakthroughs in Research and Practice (pp. 346-357). Hershey, PA: IGI Global. https://doi.org/ doi:10.4018/978-1-5225-3427-3.ch014

K., S., & Tripathy, B. K. (2018). Neighborhood Rough-Sets-Based Spatial Data Analytics. In M. Khosrow-Pour, D.B.A. (Ed.), *Encyclopedia of Information Science and Technology, Fourth Edition* (pp. 1835-1844). Hershey, PA: IGI Global. https://doi.org/ doi:10.4018/978-1-5225-2255-3.ch160

Kabir, F. (2018). Towards a More Gender-Inclusive Climate Change Policy. In N. Mahtab, T. Haque, I. Khan, M. Islam, & I. Wahid (Eds.), *Handbook of Research on Women’s Issues and Rights in the Developing World* (pp. 354–369). Hershey, PA: IGI Global. doi:10.4018/978-1-5225-3018-3.ch022

Kabir, F. (2018). Towards a More Gender-Inclusive Climate Change Policy. In I. Management Association (Ed.), Climate Change and Environmental Concerns: Breakthroughs in Research and Practice (pp. 525-540). Hershey, PA: IGI Global. https://doi.org/ doi:10.4018/978-1-5225-5487-5.ch027

Kanyamuka, J. S., Jumbe, C. B., & Ricker-Gilbert, J. (2018). Making Agricultural Input Subsidies More Effective and Profitable in Africa: The Role of Complementary Interventions. In A. Obayelu (Ed.), *Food Systems Sustainability and Environmental Policies in Modern Economies* (pp. 172–187). Hershey, PA: IGI Global. doi:10.4018/978-1-5225-3631-4.ch008

Karimi, H., & Gholamrezafahimi, F. (2017). Study of Integrated Coastal Zone Management and Its Environmental Effects: A Case of Iran. In R. Singh, A. Singh, & V. Srivastava (Eds.), *Environmental Issues Surrounding Human Overpopulation* (pp. 64–88). Hershey, PA: IGI Global. doi:10.4018/978-1-5225-1683-5.ch004

Kaya, I. R., Hutabarat, J., & Bambang, A. N. (2018). "Sasi": A New Path to Sustain Seaweed Farming From Up-Stream to Down-Stream in Kotania Bay, Molucass. *International Journal of Social Ecology and Sustainable Development*, *9*(2), 28–36. doi:10.4018/IJSESD.2018040103

Khader, V. (2018). Technologies for Food, Health, Livelihood, and Nutrition Security. In I. Management Association (Ed.), Food Science and Nutrition: Breakthroughs in Research and Practice (pp. 94-112). Hershey, PA: IGI Global. https://doi.org/ doi:10.4018/978-1-5225-5207-9.ch005

Kocadağlı, A. Y. (2017). The Temporal and Spatial Development of Organic Agriculture in Turkey. In W. Ganpat, R. Dyer, & W. Isaac (Eds.), *Agricultural Development and Food Security in Developing Nations* (pp. 130–156). Hershey, PA: IGI Global. doi:10.4018/978-1-5225-0942-4.ch006

Koundouri, P., Giannouli, A., & Souliotis, I. (2017). An Integrated Approach for Sustainable Environmental and Socio-Economic Development Using Offshore Infrastructure. In I. Management Association (Ed.), Renewable and Alternative Energy: Concepts, Methodologies, Tools, and Applications (pp. 1581-1601). Hershey, PA: IGI Global. doi:10.4018/978-1-5225-1671-2.ch056

Kumar, A., & Dash, M. K. (2017). Sustainability and Future Generation Infrastructure on Digital Platform: A Study of Generation Y. In N. Ray (Ed.), *Business Infrastructure for Sustainability in Developing Economies* (pp. 124–142). Hershey, PA: IGI Global. doi:10.4018/978-1-5225-2041-2.ch007

Kumar, A., Mukherjee, A. B., & Krishna, A. P. (2017). Application of Conventional Data Mining Techniques and Web Mining to Aid Disaster Management. In A. Kumar (Ed.), *Web Usage Mining Techniques and Applications Across Industries* (pp. 138–167). Hershey, PA: IGI Global. doi:10.4018/978-1-5225-0613-3.ch006

Kumar, C. P. (2017). Impact of Climate Change on Groundwater Resources. In I. Management Association (Ed.), Natural Resources Management: Concepts, Methodologies, Tools, and Applications (pp. 1094-1120). Hershey, PA: IGI Global. doi:10.4018/978-1-5225-0803-8.ch052

Kumari, S., & Patil, Y. (2017). Achieving Climate Smart Agriculture with a Sustainable Use of Water: A Conceptual Framework for Sustaining the Use of Water for Agriculture in the Era of Climate Change. In P. Rao & Y. Patil (Eds.), *Reconsidering the Impact of Climate Change on Global Water Supply, Use, and Management* (pp. 122–143). Hershey, PA: IGI Global. doi:10.4018/978-1-5225-1046-8.ch008

Kumari, S., & Patil, Y. (2018). Achieving Climate Smart Agriculture With a Sustainable Use of Water: A Conceptual Framework for Sustaining the Use of Water for Agriculture in the Era of Climate Change. In I. Management Association (Ed.), Climate Change and Environmental Concerns: Breakthroughs in Research and Practice (pp. 111-133). Hershey, PA: IGI Global. doi:10.4018/978-1-5225-5487-5.ch006

Kursah, M. B. (2017). Least-Cost Pipeline using Geographic Information System: The Limit to Technicalities. *International Journal of Applied Geospatial Research*, *8*(3), 1–15. doi:10.4018/ijagr.2017070101

Lahiri, S., Ghosh, D., & Bhakta, J. N. (2017). Role of Microbes in Eco-Remediation of Perturbed Aquatic Ecosystem. In J. Bhakta (Ed.), *Handbook of Research on Inventive Bioremediation Techniques* (pp. 70–107). Hershey, PA: IGI Global. doi:10.4018/978-1-5225-2325-3.ch004

Lallo, C. H., Smalling, S., Facey, A., & Hughes, M. (2018). The Impact of Climate Change on Small Ruminant Performance in Caribbean Communities. In I. Management Association (Ed.), Climate Change and Environmental Concerns: Breakthroughs in Research and Practice (pp. 193-218). Hershey, PA: IGI Global. doi:10.4018/978-1-5225-5487-5.ch010

Laurini, R. (2017). Nature of Geographic Knowledge Bases. In S. Faiz & K. Mahmoudi (Eds.), *Handbook of Research on Geographic Information Systems Applications and Advancements* (pp. 29–60). Hershey, PA: IGI Global. doi:10.4018/978-1-5225-0937-0.ch002

Lawrence, J., Simpson, L., & Piggott, A. (2017). Protected Agriculture: A Climate Change Adaptation for Food and Nutrition Security. In I. Management Association (Ed.), Natural Resources Management: Concepts, Methodologies, Tools, and Applications (pp. 140-158). Hershey, PA: IGI Global. doi:10.4018/978-1-5225-0803-8.ch007

Leadbetter, A., Cheatham, M., Shepherd, A., & Thomas, R. (2017). Linked Ocean Data 2.0. In P. Diviacco, A. Leadbetter, & H. Glaves (Eds.), *Oceanographic and Marine Cross-Domain Data Management for Sustainable Development* (pp. 69–99). Hershey, PA: IGI Global. doi:10.4018/978-1-5225-0700-0.ch004

Lucas, M. R., Rego, C., Vieira, C., & Vieira, I. (2017). Proximity and Cooperation for Innovative Regional Development: The Case of the Science and Technology Park of Alentejo. In L. Carvalho (Ed.), *Handbook of Research on Entrepreneurial Development and Innovation Within Smart Cities* (pp. 199–228). Hershey, PA: IGI Global. doi:10.4018/978-1-5225-1978-2.ch010

Ma, X., Beaulieu, S. E., Fu, L., Fox, P., Di Stefano, M., & West, P. (2017). Documenting Provenance for Reproducible Marine Ecosystem Assessment in Open Science. In P. Diviacco, A. Leadbetter, & H. Glaves (Eds.), *Oceanographic and Marine Cross-Domain Data Management for Sustainable Development* (pp. 100–126). Hershey, PA: IGI Global. doi:10.4018/978-1-5225-0700-0.ch005

Mabe, L. K., & Oladele, O. I. (2017). Application of Information Communication Technologies for Agricultural Development through Extension Services: A Review. In T. Tossy (Ed.), *Information Technology Integration for Socio-Economic Development* (pp. 52–101). Hershey, PA: IGI Global. doi:10.4018/978-1-5225-0539-6.ch003

Malomo, B. I. (2018). A Review of Psychological Resilience as a Response to Natural Hazards in Nigeria. In A. Eneanya (Ed.), *Handbook of Research on Environmental Policies for Emergency Management and Public Safety* (pp. 147–165). Hershey, PA: IGI Global. doi:10.4018/978-1-5225-3194-4.ch008

Manchiraju, S. (2018). Predicting Behavioral Intentions Toward Sustainable Fashion Consumption: A Comparison of Attitude-Behavior and Value-Behavior Consistency Models. In I. Management Association (Ed.), Fashion and Textiles: Breakthroughs in Research and Practice (pp. 1-21). Hershey, PA: IGI Global. doi:10.4018/978-1-5225-3432-7.ch001

Manzella, G. M., Bartolini, R., Bustaffa, F., D'Angelo, P., De Mattei, M., Frontini, F., ... Spada, A. (2017). Semantic Search Engine for Data Management and Sustainable Development: Marine Planning Service Platform. In P. Diviacco, A. Leadbetter, & H. Glaves (Eds.), *Oceanographic and Marine Cross-Domain Data Management for Sustainable Development* (pp. 127–154). Hershey, PA: IGI Global. doi:10.4018/978-1-5225-0700-0.ch006

Mbonigaba, J. (2018). Comparing the Effects of Unsustainable Production and Consumption of Food on Health and Policy Across Developed and Less Developed Countries. In A. Obayelu (Ed.), *Food Systems Sustainability and Environmental Policies in Modern Economies* (pp. 124–158). Hershey, PA: IGI Global. doi:10.4018/978-1-5225-3631-4.ch006

McKeown, A. E. (2017). Nurses, Healthcare, and Environmental Pollution and Solutions: Breaking the Cycle of Harm. In I. Management Association (Ed.), Natural Resources Management: Concepts, Methodologies, Tools, and Applications (pp. 392-415). Hershey, PA: IGI Global. https://doi.org/ doi:10.4018/978-1-5225-0803-8.ch020

Mili, B., Barua, A., & Katyaini, S. (2017). Climate Change and Adaptation through the Lens of Capability Approach: A Case Study from Darjeeling, Eastern Himalaya. In I. Management Association (Ed.), Natural Resources Management: Concepts, Methodologies, Tools, and Applications (pp. 1351-1365). Hershey, PA: IGI Global. https://doi.org/ doi:10.4018/978-1-5225-0803-8.ch064

Mir, S. A., Shah, M. A., Mir, M. M., & Iqbal, U. (2017). New Horizons of Nanotechnology in Agriculture and Food Processing Industry. In B. Nayak, A. Nanda, & M. Bhat (Eds.), *Integrating Biologically-Inspired Nanotechnology into Medical Practice* (pp. 230–258). Hershey, PA: IGI Global. doi:10.4018/978-1-5225-0610-2.ch009

Moallem, M., Sterrett, W. L., Gordon, C. R., Sukhera, S. M., Mahmood, A., & Bashir, A. (2021). An Investigation of the Effects of Integrating Computing and Project- or Problem-Based Learning in the Context of Environmental Sciences: A Case of Pakistani STEM Teachers. In S. Schroth & J. Daniels (Eds.), *Building STEM Skills Through Environmental Education* (pp. 49–89). IGI Global. https://doi.org/10.4018/978-1-7998-2711-5.ch003

Mujere, N., & Moyce, W. (2017). Climate Change Impacts on Surface Water Quality. In W. Ganpat & W. Isaac (Eds.), *Environmental Sustainability and Climate Change Adaptation Strategies* (pp. 322–340). Hershey, PA: IGI Global. doi:10.4018/978-1-5225-1607-1.ch012

Mukherjee, A. B., Krishna, A. P., & Patel, N. (2018). Geospatial Technology for Urban Sciences. In *Geospatial Technologies in Urban System Development: Emerging Research and Opportunities* (pp. 99–120). Hershey, PA: IGI Global. doi:10.4018/978-1-5225-3683-3.ch005

Nagarajan, S. K., & Sangaiah, A. K. (2017). Vegetation Index: Ideas, Methods, Influences, and Trends. In N. Kumar, A. Sangaiah, M. Arun, & S. Anand (Eds.), *Advanced Image Processing Techniques and Applications* (pp. 347–386). Hershey, PA: IGI Global. doi:10.4018/978-1-5225-2053-5.ch016

Naraine, L., & Meehan, K. (2017). Strengthening Food Security with Sustainable Practices by Smallholder Farmers in Lesser Developed Economies. In W. Ganpat, R. Dyer, & W. Isaac (Eds.), *Agricultural Development and Food Security in Developing Nations* (pp. 57–81). Hershey, PA: IGI Global. doi:10.4018/978-1-5225-0942-4.ch003

Naraine, L., & Meehan, K. (2017). Strengthening Food Security with Sustainable Practices by Smallholder Farmers in Lesser Developed Economies. In W. Ganpat, R. Dyer, & W. Isaac (Eds.), *Agricultural Development and Food Security in Developing Nations* (pp. 57–81). Hershey, PA: IGI Global. doi:10.4018/978-1-5225-0942-4.ch003

Naraine, L., & Meehan, K. (2017). Strengthening Food Security with Sustainable Practices by Smallholder Farmers in Lesser Developed Economies. In W. Ganpat, R. Dyer, & W. Isaac (Eds.), *Agricultural Development and Food Security in Developing Nations* (pp. 57–81). Hershey, PA: IGI Global. doi:10.4018/978-1-5225-0942-4.ch003

Nikolaou, K., Tsakiridou, E., Anastasiadis, F., & Mattas, K. (2017). Exploring Alternative Distribution Channels of Agricultural Products. *International Journal of Food and Beverage Manufacturing and Business Models*, *2*(2), 36–66. doi:10.4018/IJFBMBM.2017070103

Nishat, K. J., & Rahman, M. S. (2018). Disaster, Vulnerability, and Violence Against Women: Global Findings and a Research Agenda for Bangladesh. In N. Mahtab, T. Haque, I. Khan, M. Islam, & I. Wahid (Eds.), *Handbook of Research on Women's Issues and Rights in the Developing World* (pp. 235–250). Hershey, PA: IGI Global. doi:10.4018/978-1-5225-3018-3.ch014

O'Hara, S., Jones, D., & Trobman, H. B. (2018). Building an Urban Food System Through UDC Food Hubs. In A. Burtin, J. Fleming, & P. Hampton-Garland (Eds.), *Changing Urban Landscapes Through Public Higher Education* (pp. 116–143). Hershey, PA: IGI Global. doi:10.4018/978-1-5225-3454-9.ch006

Obayelu, A. E. (2018). Integrating Environment, Food Systems, and Sustainability in Feeding the Growing Population in Developing Countries. In A. Obayelu (Ed.), *Food Systems Sustainability and Environmental Policies in Modern Economies* (pp. 1–14). Hershey, PA: IGI Global. doi:10.4018/978-1-5225-3631-4.ch001

Othman, R., Nath, N., & Laswad, F. (2018). Environmental Reporting and Accounting: Sustainability Hybridisation. In G. Azevedo, J. da Silva Oliveira, R. Marques, & A. Ferreira (Eds.), *Handbook of Research on Modernization and Accountability in Public Sector Management* (pp. 130–158). Hershey, PA: IGI Global. doi:10.4018/978-1-5225-3731-1.ch007

Ouadi, A., & Zitouni, A. (2021). Phasor Measurement Improvement Using Digital Filter in a Smart Grid. In A. Recioui & H. Bentarzi (Eds.), *Optimizing and Measuring Smart Grid Operation and Control* (pp. 100–117). IGI Global. https://doi.org/10.4018/978-1-7998-4027-5.ch005

Padigala, B. S. (2018). Traditional Water Management System for Climate Change Adaptation in Mountain Ecosystems. In I. Management Association (Ed.), Climate Change and Environmental Concerns: Breakthroughs in Research and Practice (pp. 630-655). Hershey, PA: IGI Global. doi:10.4018/978-1-5225-5487-5.ch033

Panda, C. K. (2018). Mobile Phone Usage in Agricultural Extension in India: The Current and Future Perspective. In F. Mtenzi, G. Oreku, D. Lupiana, & J. Yonazi (Eds.), *Mobile Technologies and Socio-Economic Development in Emerging Nations* (pp. 1–21). Hershey, PA: IGI Global. doi:10.4018/978-1-5225-4029-8.ch001

Pandian, S. L., Yarrakula, K., & Chaudhury, P. (2018). GIS-Based Decision Support System for Village Level: A Case Study in Andhra Pradesh. In S. Chaudhuri & N. Ray (Eds.), *GIS Applications in the Tourism and Hospitality Industry* (pp. 275–295). Hershey, PA: IGI Global. doi:10.4018/978-1-5225-5088-4.ch012

Quaranta, G., & Salvia, R. (2017). Social-Based Product Innovation and Governance in The Milk Sector: The Case of Carciocacio and Innonatura. In T. Tarnanidis, M. Vlachopoulou, & J. Papathanasiou (Eds.), *Driving Agribusiness With Technology Innovations* (pp. 293–310). Hershey, PA: IGI Global. doi:10.4018/978-1-5225-2107-5.ch015

Rahman, M. K., Schmidlin, T. W., Munro-Stasiuk, M. J., & Curtis, A. (2017). Geospatial Analysis of Land Loss, Land Cover Change, and Landuse Patterns of Kutubdia Island, Bangladesh. *International Journal of Applied Geospatial Research, 8*(2), 45–60. doi:10.4018/IJAGR.2017040104

Rajack-Talley, T. A. (2017). Agriculture, Trade Liberalization and Poverty in the ACP Countries. In W. Ganpat, R. Dyer, & W. Isaac (Eds.), *Agricultural Development and Food Security in Developing Nations* (pp. 1–20). Hershey, PA: IGI Global. doi:10.4018/978-1-5225-0942-4.ch001

Rajamanickam, S. (2018). Exploring Landscapes in Regional Convergence: Environment and Sustainable Development in South Asia. In I. Management Association (Ed.), Sustainable Development: Concepts, Methodologies, Tools, and Applications (pp. 1051-1087). Hershey, PA: IGI Global. https://doi.org/ doi:10.4018/978-1-5225-3817-2.ch047

Rizvi, S. M., & Dearden, A. (2018). KHETI: ICT Solution for Agriculture Extension and Its Replication in Open and Distance Learning. In U. Pandey & V. Indrakanti (Eds.), *Open and Distance Learning Initiatives for Sustainable Development* (pp. 163–174). Hershey, PA: IGI Global. doi:10.4018/978-1-5225-2621-6.ch008

Roşu, L., & Macarov, L. I. (2017). Management of Drought and Floods in the Dobrogea Region. In I. Management Association (Ed.), Agri-Food Supply Chain Management: Breakthroughs in Research and Practice (pp. 372-403). Hershey, PA: IGI Global. https://doi.org/ doi:10.4018/978-1-5225-1629-3.ch016

Rouzbehani, K., & Rouzbehani, S. (2018). Mapping Women's World: GIS and the Case of Breast Cancer in the US. *International Journal of Public Health Management and Ethics, 3*(1), 14–25. doi:10.4018/IJPHME.2018010102

Roy, D. (2018). Success Factors of Adoption of Mobile Applications in Rural India: Effect of Service Characteristics on Conceptual Model. In M. Khosrow-Pour, D.B.A. (Ed.), Green Computing Strategies for Competitive Advantage and Business Sustainability (pp. 211-238). Hershey, PA: IGI Global. https://doi.org/ doi:10.4018/978-1-5225-5017-4.ch010

Sajeva, M., Lemon, M., & Sahota, P. S. (2017). Governance for Food Security: A Framework for Social Learning and Scenario Building. *International Journal of Food and Beverage Manufacturing and Business Models, 2*(2), 67–84. doi:10.4018/IJFBMBM.2017070104

Sambhanthan, A., & Potdar, V. (2017). A Study of the Parameters Impacting Sustainability in Information Technology Organizations. *International Journal of Knowledge-Based Organizations, 7*(3), 27–39. doi:10.4018/IJKBO.2017070103

Sanga, C., Kalungwizi, V. J., & Msuya, C. P. (2017). Bridging Gender Gaps in Provision of Agricultural Extension Service Using ICT: Experiences from Sokoine University of Agriculture (SUA) Farmer Voice Radio (FVR) Project in Tanzania. In I. Management Association (Ed.), Discrimination and Diversity: Concepts, Methodologies, Tools, and Applications (pp. 682-697). Hershey, PA: IGI Global. https://doi.org/ doi:10.4018/978-1-5225-1933-1.ch031

Santra, A., & Mitra, D. (2017). Role of Remote Sensing in Potential Fishing Zone Forecast. In A. Santra & S. Mitra (Eds.), *Remote Sensing Techniques and GIS Applications in Earth and Environmental Studies* (pp. 243–257). Hershey, PA: IGI Global. doi:10.4018/978-1-5225-1814-3.ch012

Schaap, D. (2017). SeaDataNet: Towards a Pan-European Infrastructure for Marine and Ocean Data Management. In P. Diviacco, A. Leadbetter, & H. Glaves (Eds.), *Oceanographic and Marine Cross-Domain Data Management for Sustainable Development* (pp. 155–177). Hershey, PA: IGI Global. doi:10.4018/978-1-5225-0700-0.ch007

Seckin-Celik, T. (2017). Sustainability Reporting and Sustainability in the Turkish Business Context. In U. Akkucuk (Ed.), *Ethics and Sustainability in Global Supply Chain Management* (pp. 115–132). Hershey, PA: IGI Global. doi:10.4018/978-1-5225-2036-8.ch006

Segbefia, A. Y., Barnes, V. R., Akpalu, L. A., & Mensah, M. (2018). Environmental Location Assessment for Seaweed Cultivation in Ghana: A Spatial Multi-Criteria Approach. *International Journal of Applied Geospatial Research, 9*(1), 51–64. doi:10.4018/IJAGR.2018010104

Sen, Y. (2018). How to Manage Sustainability: A Framework for Corporate Sustainability Tools. In I. Management Association (Ed.), Sustainable Development: Concepts, Methodologies, Tools, and Applications (pp. 568-589). Hershey, PA: IGI Global. doi:10.4018/978-1-5225-3817-2.ch026

Shamshiry, E., Abdulai, A. M., Mokhtar, M. B., & Komoo, I. (2015). Regional Landfill Site Selection with GIS and Analytical Hierarchy Process Techniques: A Case Study of Langkawi Island, Malaysia. In P. Thomas, M. Srihari, & S. Kaur (Eds.), *Handbook of Research on Cultural and Economic Impacts of the Information Society* (pp. 248–282). Hershey, PA: IGI Global. doi:10.4018/978-1-4666-8598-7.ch011

Sharma, Y. K., Mangla, S. K., Patil, P. P., & Uniyal, S. (2018). Analyzing Sustainable Food Supply Chain Management Challenges in India. In M. Ram & J. Davim (Eds.), *Soft Computing Techniques and Applications in Mechanical Engineering* (pp. 162–180). Hershey, PA: IGI Global. doi:10.4018/978-1-5225-3035-0.ch008

Silvestrelli, P. (2018). The Impact of Events: To Which Extent Are They Sustainable for Tourist Destinations? Some Evidences From Expo Milano 2015. In M. Risso & S. Testarmata (Eds.), *Value Sharing for Sustainable and Inclusive Development* (pp. 185–204). Hershey, PA: IGI Global. doi:10.4018/978-1-5225-3147-0.ch008

Silvius, G. (2018). Sustainability Evaluation of IT/IS Projects. In I. Management Association (Ed.), Sustainable Development: Concepts, Methodologies, Tools, and Applications (pp. 26-40). Hershey, PA: IGI Global. https://doi.org/ doi:10.4018/978-1-5225-3817-2.ch002

Singh, R., Srivastava, P., Singh, P., Upadhyay, S., & Raghubanshi, A. S. (2017). Human Overpopulation and Food Security: Challenges for the Agriculture Sustainability. In R. Singh, A. Singh, & V. Srivastava (Eds.), *Environmental Issues Surrounding Human Overpopulation* (pp. 12–39). Hershey, PA: IGI Global. doi:10.4018/978-1-5225-1683-5.ch002

Singh, R., Srivastava, P., Singh, P., Upadhyay, S., & Raghubanshi, A. S. (2017). Human Overpopulation and Food Security: Challenges for the Agriculture Sustainability. In R. Singh, A. Singh, & V. Srivastava (Eds.), *Environmental Issues Surrounding Human Overpopulation* (pp. 12–39). Hershey, PA: IGI Global. doi:10.4018/978-1-5225-1683-5.ch002

Srivastava, N. (2017). Climate Change Mitigation: Collective Efforts and Responsibly. In I. Management Association (Ed.), Natural Resources Management: Concepts, Methodologies, Tools, and Applications (pp. 64-76). Hershey, PA: IGI Global. https://doi.org/ doi:10.4018/978-1-5225-0803-8.ch003

Stanganelli, M., & Gerundo, C. (2017). Understanding the Role of Urban Morphology and Green Areas Configuration During Heat Waves. *International Journal of Agricultural and Environmental Information Systems*, *8*(2), 50–64. doi:10.4018/IJAEIS.2017040104

Stewart, M. K., Hagood, D., & Ching, C. C. (2017). Virtual Games and Real-World Communities: Environments that Constrain and Enable Physical Activity in Games for Health. *International Journal of Game-Based Learning*, *7*(1), 1–19. doi:10.4018/IJGBL.2017010101

Stone, R. J. (2017). Modelling the Frequency of Tropical Cyclones in the Lower Caribbean Region. In W. Ganpat & W. Isaac (Eds.), *Environmental Sustainability and Climate Change Adaptation Strategies* (pp. 341–349). Hershey, PA: IGI Global. doi:10.4018/978-1-5225-1607-1.ch013

Syed, A., & Jabeen, U. A. (2018). Climate Change Impact on Agriculture and Food Security. In A. Eneanya (Ed.), *Handbook of Research on Environmental Policies for Emergency Management and Public Safety* (pp. 223–237). Hershey, PA: IGI Global. doi:10.4018/978-1-5225-3194-4.ch012

Tam, G. C. (2017). The Global View of Sustainability. In *Managerial Strategies and Green Solutions for Project Sustainability* (pp. 1–24). Hershey, PA: IGI Global. doi:10.4018/978-1-5225-2371-0.ch001

Tam, G. C. (2017). Understanding Project Sustainability. In *Managerial Strategies and Green Solutions for Project Sustainability* (pp. 110–139). Hershey, PA: IGI Global. doi:10.4018/978-1-5225-2371-0.ch005

Tam, G. C. (2017). Perspectives on Sustainability. In *Managerial Strategies and Green Solutions for Project Sustainability* (pp. 53–76). Hershey, PA: IGI Global. doi:10.4018/978-1-5225-2371-0.ch003

Tang, M., & Karunanithi, A. T. (2018). Visual Logic Maps (vLms). In *Advanced Concept Maps in STEM Education: Emerging Research and Opportunities* (pp. 108–149). Hershey, PA: IGI Global. doi:10.4018/978-1-5225-2184-6.ch005

Taşçıoğlu, M., & Yener, D. (2018). The Value and Scope of GIS in Marketing and Tourism Management. In S. Chaudhuri & N. Ray (Eds.), *GIS Applications in the Tourism and Hospitality Industry* (pp. 189–211). Hershey, PA: IGI Global. doi:10.4018/978-1-5225-5088-4.ch009

Thanh Tung, B. (2021). Pharmacology and Therapeutic Applications of Resveratrol. In A. Hussain & S. Behl (Eds.), *Treating Endocrine and Metabolic Disorders With Herbal Medicines* (pp. 321–333). IGI Global. https://doi.org/10.4018/978-1-7998-4808-0.ch014

Tianming, G. (2018). Food Security and Rural Development on Emerging Markets of Northeast Asia: Cases of Chinese North and Russian Far East. In V. Erokhin (Ed.), *Establishing Food Security and Alternatives to International Trade in Emerging Economies* (pp. 155–176). Hershey, PA: IGI Global. doi:10.4018/978-1-5225-2733-6.ch008

Tiftikçigil, B. Y., Yaşgül, Y. S., & Güriş, B. (2017). Sustainability of Foreign Trade Deficit in Energy: The Case of Turkey. In N. Ray (Ed.), *Business Infrastructure for Sustainability in Developing Economies* (pp. 94–109). Hershey, PA: IGI Global. doi:10.4018/978-1-5225-2041-2.ch005

Tiwari, S., Vaish, B., & Singh, P. (2018). Population and Global Food Security: Issues Related to Climate Change. In I. Management Association (Ed.), Climate Change and Environmental Concerns: Breakthroughs in Research and Practice (pp. 41-64). Hershey, PA: IGI Global. https://doi.org/ doi:10.4018/978-1-5225-5487-5.ch003

Toujani, A., & Achour, H. (2018). A Data Mining Framework for Forest Fire Mapping. In I. Management Association (Ed.), Information Retrieval and Management: Concepts, Methodologies, Tools, and Applications (pp. 771-794). Hershey, PA: IGI Global. doi:10.4018/978-1-5225-5191-1.ch033

Tripathy, B., & K., S. B. (2017). Rough Fuzzy Set Theory and Neighbourhood Approximation Based Modelling for Spatial Epidemiology. In I. Management Association (Ed.), *Public Health and Welfare: Concepts, Methodologies, Tools, and Applications* (pp. 1257-1268). Hershey, PA: IGI Global. https:// doi.org/ doi:10.4018/978-1-5225-1674-3.ch058

Trukhachev, A. (2017). New Approaches to Regional Branding through Green Production and Utilization of Existing Natural Advantages. In I. Management Association (Ed.), Advertising and Branding: Concepts, Methodologies, Tools, and Applications (pp. 1758-1778). Hershey, PA: IGI Global. doi:10.4018/978-1-5225-1793-1.ch081

Tsobanoglou, G. O., & Vlachopoulou, E. I. (2017). Social-Ecological Systems in Local Fisheries Communities. In G. Korres, E. Kourliouros, & M. Michailidis (Eds.), *Handbook of Research on Policies and Practices for Sustainable Economic Growth and Regional Development* (pp. 306–316). Hershey, PA: IGI Global. doi:10.4018/978-1-5225-2458-8.ch026

Tuydes-Yaman, H., & Karatas, P. (2018). Evaluation of Walkability and Pedestrian Level of Service. In I. Management Association (Ed.), Intelligent Transportation and Planning: Breakthroughs in Research and Practice (pp. 264-291). Hershey, PA: IGI Global. doi:10.4018/978-1-5225-5210-9.ch012

Uddin, S., Chakravorty, S., Ray, A., & Sherpa, K. S. (2018). Optimal Location of Sub-Station Using Q-GIS and Multi-Criteria Decision Making Approach. *International Journal of Decision Support System Technology*, *10*(2), 65–79. doi:10.4018/IJDSST.2018040104

Uddin, S., Chakravorty, S., Sherpa, K. S., & Ray, A. (2018). Power Distribution System Planning Using Q-GIS. *International Journal of Energy Optimization and Engineering*, *7*(2), 61–75. doi:10.4018/IJEOE.2018040103

Uzun, F. V. (2018). Natural Resources Management. In A. Eneanya (Ed.), *Handbook of Research on Environmental Policies for Emergency Management and Public Safety* (pp. 1–21). Hershey, PA: IGI Global. doi:10.4018/978-1-5225-3194-4.ch001

V, M., Agrawal, R., Sharma, V., & T.N., K. (2018). Supply Chain Social Sustainability and Manufacturing. In I. Management Association (Ed.), *Technology Adoption and Social Issues: Concepts, Methodologies, Tools, and Applications* (pp. 226-252). Hershey, PA: IGI Global. https://doi.org/ doi:10.4018/978-1-5225-5201-7.ch011

van der Vliet-Bakker, J. M. (2017). Environmentally Forced Migration and Human Rights. In C. Akrivopoulou (Ed.), *Defending Human Rights and Democracy in the Era of Globalization* (pp. 146–180). Hershey, PA: IGI Global. doi:10.4018/978-1-5225-0723-9.ch007

Vargas-Hernández, J. G. (2022). A Comprehensive Entrepreneurship Model for the Internationalization of Green Innovation Businesses. In U. Akkucuk (Ed.), *Disruptive Technologies and Eco-Innovation for Sustainable Development* (pp. 131–149). IGI Global. https://doi.org/10.4018/978-1-7998-8900-7.ch008

Vaskov, A. G., Lin, Z. Y., Tyagunov, M. G., Shestopalova, T. A., & Deryugina, G. V. (2018). Design of Renewable Sources GIS for ASEAN Countries. In V. Kharchenko & P. Vasant (Eds.), *Handbook of Research on Renewable Energy and Electric Resources for Sustainable Rural Development* (pp. 1–25). Hershey, PA: IGI Global. doi:10.4018/978-1-5225-3867-7.ch001

Vázquez, D. G., & Gil, M. T. (2017). Sustainability in Smart Cities: The Case of Vitoria-Gasteiz (Spain) – A Commitment to a New Urban Paradigm. In L. Carvalho (Ed.), *Handbook of Research on Entrepreneurial Development and Innovation Within Smart Cities* (pp. 248–268). Hershey, PA: IGI Global. doi:10.4018/978-1-5225-1978-2.ch012

Wahab, I. N., & Soonthodu, S. (2018). Geographical Information System in Eco-Tourism. In S. Chaudhuri & N. Ray (Eds.), *GIS Applications in the Tourism and Hospitality Industry* (pp. 61–75). Hershey, PA: IGI Global. doi:10.4018/978-1-5225-5088-4.ch003

Weiss-Randall, D. (2018). Cultivating Environmental Justice. In *Utilizing Innovative Technologies to Address the Public Health Impact of Climate Change: Emerging Research and Opportunities* (pp. 110–143). Hershey, PA: IGI Global. doi:10.4018/978-1-5225-3414-3.ch004

Weiss-Randall, D. (2018). Cultivating Resilience. In *Utilizing Innovative Technologies to Address the Public Health Impact of Climate Change: Emerging Research and Opportunities* (pp. 204–235). Hershey, PA: IGI Global. doi:10.4018/978-1-5225-3414-3.ch007

Weiss-Randall, D. (2018). Climate Change Solutions: Where Do We Go From Here? In *Utilizing Innovative Technologies to Address the Public Health Impact of Climate Change: Emerging Research and Opportunities* (pp. 236–268). Hershey, PA: IGI Global. doi:10.4018/978-1-5225-3414-3.ch008

Whyte, K. P., List, M., Stone, J. V., Grooms, D., Gasteyer, S., Thompson, P. B., . . . Bouri, H. (2018). Uberveillance, Standards, and Anticipation: A Case Study on Nanobiosensors in U.S. Cattle. In I. Management Association (Ed.), Biomedical Engineering: Concepts, Methodologies, Tools, and Applications (pp. 577-596). Hershey, PA: IGI Global. https://doi.org/ doi:10.4018/978-1-5225-3158-6.ch025

Wulff, E. (2017). Data and Operational Oceanography: A Review in Support of Responsible Fisheries and Aquaculture. In P. Diviacco, A. Leadbetter, & H. Glaves (Eds.), *Oceanographic and Marine Cross-Domain Data Management for Sustainable Development* (pp. 303–324). Hershey, PA: IGI Global. doi:10.4018/978-1-5225-0700-0.ch013

Yener, D. (2017). Geographic Information Systems and Its Applications in Marketing Literature. In S. Faiz & K. Mahmoudi (Eds.), *Handbook of Research on Geographic Information Systems Applications and Advancements* (pp. 158–172). Hershey, PA: IGI Global. doi:10.4018/978-1-5225-0937-0.ch006

Yu, K., Liu, Y., & Sharma, A. (2021). Analyze the Effectiveness of the Algorithm for Agricultural Product Delivery Vehicle Routing Problem Based on Mathematical Model. *International Journal of Agricultural and Environmental Information Systems*, *12*(3), 26–38. https://doi.org/10.4018/IJAEIS.2021070103

Zhou, X., Sharma, A., & Mohindru, V. (2021). Research on Linear Programming Algorithm for Mathematical Model of Agricultural Machinery Allocation. *International Journal of Agricultural and Environmental Information Systems*, *12*(3), 1–12. https://doi.org/10.4018/IJAEIS.2021070101

About the Contributors

Soumendra Nath Biswas is presently working as an Associate Professor and Head of the Department of Hospitality & Tourism Management, Assam University, Silchar (a Central University). Prior to that, he had served in Indian Institute of Tourism and Travel Management, Bhubaneswar (An Autonomous body under Ministry of Tourism, Govt. of India) as an Assistant Professor (senior scale) in last 14 years. He has been awarded PhD in the year 2010 in Business Administration (Tourism) by The University of Burdwan, W.B. He has obtained Master in Tourism Management and bachelor in Hotel management (BHM). He has also qualified UGC-NET and UGC-JRF. Dr. Biswas has more than 20 years of teaching experience in MBA (Tourism), BBA (Tourism), BHM and DHM, in various government and private institutions. Dr. Biswas has written numbers of text book in Tourism Studies by international and national publishers and contributed numbers of research papers in reputed international and national peer-reviewed journals and edited books. He has contributed in many research projects sponsored by UNDP, Government of India and Government of Odisha.

***

Prachi Agarwal is presently a research scholar in the pursuit of Tourism and travel Management at Dev Sanskriti Vishwavidyalaya, Haridwar. She is UGC- NET qualified in Tourism Administration. Her area of research is in Nature-based Health Tourism with a commitment of sustainable growth and future in the field of tourism.

Dillip Kumar Das is currently working as an Associate Professor in the Department of Tourism, Sikkim Central University. He has been awarded PhD in the year 2008 in Tourism from Utkal University, Bhubaneswar, Orissa on the topic "Economic, Socio Cultural and Environmental Impact of Tourism: A case Study on Puri, Konark and Bhubaneswar". He has Completed Masters in Tourism Management (MTM) from IGNOU, and also Two times Qualified UGC- NET (National Eligibility test for Lectureship) in December 2003and in 2004. He has Completed Master Degree in Economics, from Ravenshaw University, Cuttack, and also Completed Diploma in Tourism Management, from Indian Institute of Tourism and Travel Management, Bhubaneswar, Orissa. The author's areas of research interest are Eco-Tourism, Tourism Impact Studies and Travel Agency Management. He has altogether 15 years of Academic experience in tourism education. The author is now currently associated with different university and institute of repute in India including I.G.N.O.U, IITTM, Magadh University, Utkal University etc.

Jy harath Reddy G K is a tourism professional with four years of Teaching and Research experience. He have participated in 7 National and 10 International conferences, 4 Workshops during his research period. He has completed a certification programme on Sustainable Tourism – promoting environmental public health from University of Copenhagen. He is a UGC, NET/JRF holder in Tourism Administration and Management. He completed his Master of Business Administration in Tourism and Hospitality Management from National Institute of Tourism and Hospitality Management (NITHM), Hyderabad with Distinction. While studying his masters he has won second prize in 2015 Inter- Collegiate quiz competition, Government City College (Hyderabad).

Dankan Gowda V is currently working as an Assistant Professor in the Department of Electronics and Communication Engineering at BMS Institute of Technology and Management in Bangalore. Previously, he worked as a Research Fellow at ADA DRDO and as a Software Engineer at Robert Bosch. With a total experience of 14 years, including teaching and industry, Dr. Gowda has made significant contributions to both academia and research. He has published over 60 research papers in renowned international journals and conferences. In recognition of his innovative work, Dr. Gowda has been granted six patents, including four from Indian authorities and two international patents. His research interests primarily lie in the fields of IoT and Signal Processing, where he has conducted workshops and handled industry projects. Dr. Gowda is passionate about teaching and strives to create an engaging and effective learning environment for his students. He consistently seeks opportunities to enhance his knowledge and stay updated with the latest advancements in his field.

Ai Rahaman Jamader is working as an Assistant Professor in the department of Hospitality & Hotel Administration at Pailan College of Management & Technology, Kolkata, India, also he is the Corporate Advisory Board Member of the Smart Journal of Business Management Studies indexed by Emerging Sources Citation Index (ESCI) - Web of Science (Clarivate Analytics) with 5.748 Impact factor. His research interest is in Hospitality and innovative Technique implemented in Hospitality Industry. He is a scientist by having 23 numbers of International granted patents & 12 numbers of registered and published national & International patents also have a good number of authored Book/Book Chapters publications, including some SCOPUS/SCIE/ESCI/WOS publications. Recently he joined as an Ad Hoc reviewer of the International Journal of Business Intelligence Research (IJBIR), IGI Global publishing indexed by WOS and Scopus.

KN. Lokesh Kumar is currently working as assistant professor in the Department of Hospitality and Tourism Management, Assam University (A Central University), Silchar, Assam. He has about 10 years of teaching experience. He completed his MBA from Prestigious Indian institute of tourism and travel management, Gwalior (An autonomous institute under Ministry of Tourism, Govt. of India). He is a UGC, NET& JRF holder in Tourism Management, Gold Medallist & S.K. Mishra merit awardee in MBA (Tourism & Travel). His areas of interest include tour operations, Travel agency, Airline operations and Management, Destination Marketing and Human resource in tourism. He has to his credit 05 publications and also 04 Published chapters in books and 10 participated national and international conferences. He developed E-content for Entrepreneurship in Tourism, Research Methods in Tourism to Centre for Distance Education, Acharya Nagarjuna University, developed the E-Content for the Diploma in Travel and Tourism Management (TTM) course which is being offered by the MOOCs Platform ANU and also edited the self-learning material for Board of Open Schooling & Skill Education (BOSSE), Sikkim. In addition, he also completed Basic Mountaineering course from Atal Bihari Vajpayee Institute of Mountaineering and Allied Sports (ABVIMAS), Manali, Himachal Pradesh. With regards to tourism activities and other related assignments he had travelled extensively in States of India and abroad to countries like Thailand and Taiwan

Amit Kumar possesses more than 13 years of academic and 2 years of industry experience. Currently, he is working as an Assistant Professor in the Department of Tourism and Hotel Management at Central University of Haryana, Mahendragarh. He is a well-structured academic professional demonstrating proven success in fostering student learning outcomes through creative and innovative curriculum delivery methods. He adapts well in driving thought-provoking class debates to promote student engagement and learning. He has also participated and presented research papers in various international and National Seminars/Conferences/Workshops along with research publications in international and National reputed journals and books. He is also on the panel list of paper setting and evaluation for UG and PG courses in Tourism and Hotel Management at various Universities.

Jtendra Mohan Mishra is an accomplished academic with eighteen years of teaching experience. He currently serves as Dean of the School of Management Studies and is the Head of the Department of Tourism & Hospitality Management at Central Tribal University of Andhra Pradesh. His contributions include 8 completed and 5 ongoing Ph.D. studies, 60 paper presentations, and leadership in major research projects. Prof. Mishra's passion for outdoor activities, including trekking and yoga, complements his academic pursuits. He has authored 40 research articles and five books including the one, Tourism Principles and Practices published by Oxford University Press.

Shreeansh Mishra is working as an Assistant Professor of Tourism in Central University of Tamil Nadu, India. His research highlights the transformative role of self-help groups, microfinance, and community readiness in advancing community-based tourism and fostering sustainable development. His work examines factors that synergize marginalized communities and foster sustainable development. With a commitment to global collaboration and continuous efforts, Shreeansh Mishra continues his mission to create a more inclusive and equitable tourism landscape, where gender disparities are addressed and women's empowerment takes center stage, catalyzing a brighter future for all.

AihMondal is presently engaged in the pursuit of a Master of Science degree in Hospitality Administration, specializing in sales and marketing, at the prestigious National Council for Hotel Management and Catering Technology - Institute of Hotel Management (NCHM-IH), located in Noida. His academic journey prior to this has been marked by academic excellence, having successfully completed a Bachelor of Science degree in Hotel and Hospitality Administration at the esteemed Institute of Hotel Management (IHM) in Mumbai. His areas of expertise and passion encompass Market Research, Tourism Management, Sales and Marketing, Sustainable Tourism, Eco-tourism, Pilgrimage Tourism, Medical Tourism, and Smart Tourism. These areas of focus have not only enriched his academic pursuits but have also driven his career aspirations towards contributing meaningfully to the dynamic and ever-evolving tourism and hospitality industry.

Arunesh Parashar is an eminent figure in the realm of academia and a distinguished professional in the field of tourism and hospitality management. As the Associate Professor and Head of Department of Tourism and Travel Management at Dev Sanskriti Vishwavidyalaya, Haridwar, Dr. Parashar brings a wealth of knowledge, expertise, and leadership to the vibrant academic community. He has carved a niche for himself in the specialized domain of health tourism. His academic journey is adorned with more than 50 research papers published in both national and international journals. Through extensive contributions to various books and chapters, he has left an indelible mark on the academic landscape, showcasing a deep understanding of the complexities within tourism, travel and hospitality sector. Beyond the rigors of academia, Dr. Arunesh Parashar is a versatile writer with a penchant for creative expression. His prowess extends to e-books and self-learning books, reflecting a commitment to disseminating knowledge through diverse mediums. A stalwart in the academic arena, Dr. Parashar actively participates as a member of different academic bodies, industry forums, and professional organizations. This multifaceted engagement underscores his dedication to advancing the discourse within the tourism and hospitality industry by organizing more than 40 national and international conferences in the field of tourism and the hospitality sector. Notably, he has been honored with more than 30 awards and has served as a distinguished guest of honor on numerous occasions, celebrating his outstanding contributions to the tourism and hospitality sector. His contributions, both scholarly and practical, continue to shape the future of the industry while inspiring the next generation of leaders in this dynamic field.

Wahengbam Priyalakshmi is a Research Scholar in Department of Hospitality and Tourism Management.

Vrinder Singh Rana is a Department Chair of Hospitality and Tourism Management.

Pramita Sarkar is working as an Assistant Professor with the Department of Fash-ion Design, NIFT, Bengaluru. She is a PhD in the area of study of tribal costumes of Tripura and specializes in ethnographic research. She specializes in subjects like Sustainable Design, Traditional Indian Textiles, Styling for performance art & fictional character, Design Process & Design Research. She has several publications in the area of Tribal textiles & costume of Tripura in various national and international journals. She has presented her research work in IAFOR, Osaka Japan, 2013 & IFFTI, Manchester 2019. She was awarded for the category The Best Design for Innovation at FTC, IIT, New Delhi, 2020.

Shipra Sharma is an Associate Professor with the Department of Fashion Technology, NIFT, Panchkula. She is a PhD in Management and specializes in Apparel manufacturing operations and management development. With the background of industry and academics, the interest in imparting the knowledge through teaching brought her to NIFT. Apart from holding various portfolios and administrative responsibilities at Bangalore, Kangra and Panchkula centers of NIFT, she has an academic contribution of various publications in the area of management, apparel manufacturing operations, craft clusters in national and international journals.

Index

O

P

R

S

T

V

www.ingramcontent.com/pod-product-compliance
Lightning Source LLC
LaVergne TN
LVHW060305011026
849161LV00011B/2778